AF838431

POWER QUALITY ISSUES

Edited by **Ahmed Zobaa**

Power Quality Issues
http://dx.doi.org/10.5772/3373
Edited by Ahmed Zobaa

Contributors

Belkacem Mahdad, Bedri Kekezoglu, Celal Kocatepe, Recep Yumurtacı, Oktay Arıkan, Mustafa Baysal, Altuğ Bozkurt, Celal Fadıl Kumru, Moisés Vidal Ribeiro, Zbigniew Hanzelka, Soliman Abdel-Hady Soliman, Mariacristina Roscia, Cristian George Lazaroiu, Nicolae Golovanov, António Pina Martins, Gabriel Gasparesc, Jaroslaw Luszcz, Patricio Salmerón, Alejandro Pérez, Joao Luiz Afonso, José Gabriel Pinto, Henrique Gonçalves

Published by InTech

Janeza Trdine 9, 51000 Rijeka, Croatia

Edition 2014

Copyright © 2013 InTech

All chapters are distributed under the Creative Commons Attribution 3.0 license, which allows users to download, copy and build upon published articles even for commercial purposes, as long as the author and publisher are properly credited, which ensures maximum dissemination and a wider impact of our publications. However, users who aim to disseminate and distribute copies of this book as a whole must not seek monetary compensation for such service (excluded InTech representatives and agreed collaborations). After this work has been published by InTech, authors have the right to republish it, in whole or part, in any publication of which they are the author, and to make other personal use of the work. Any republication, referencing or personal use of the work must explicitly identify the original source.

Notice

Statements and opinions expressed in the chapters are these of the individual contributors and not necessarily those of the editors or publisher. No responsibility is accepted for the accuracy of information contained in the published chapters. The publisher assumes no responsibility for any damage or injury to persons or property arising out of the use of any materials, instructions, methods or ideas contained in the book.

Publishing Process Manager Sandra Bakic
Technical Editor InTech DTP team
Cover InTech Design team

Additional hard copies can be obtained from orders@intechopen.com

Power Quality Issues, Edited by Ahmed Zobaa
p. cm.
ISBN 978-953-51-1068-2

INTECH
open science | open minds

Contents

Chapter 1

Power Quality Data Compression

Gabriel Găşpăresc

Additional information is available at the end of the chapter

http://dx.doi.org/10.5772/53059

1. Introduction

Nowadays we assist to the increasing of devices and equipments connected to power systems (non-linear loads, industrial rectifiers and inverters, solid-state switching devices, computers, peripheral devices etc). Hence, the parameters of the power supply should be accurately estimated and monitorized. For this purpose have been proposed power quality monitoring systems that are abble to automatically detect and classify disturbances. They are using the most recent signal processing techniques for power quality analysis (Bollen et al., 2006), (Dungan et al., 2004), (Lin et al., 2009).

A power quality monitoring system provides huge volume of raw data from different locations, acquired during long periods of time and the amount of data is increasing daily. The hardware of a power quality monitoring systems should have a high sampling rate because the power quality events cover a broad frequency range, starting from a few Hz (flicker) to a few MHz (transient phenomenon). A high sampling rate leads to large volume of aquired data (for example, one recorded event could requires megabytes of storage space) which should be transferred and stored. Therefore, it is necessary data compression to save storing space and to reduce the communication time. Any compression methode is a compromise between the resulted volume of data and the remained information. The aim is to obtain the smalest size with the highest information level (Barrera Nunez et al., 2008), (Lorio et al., 2004), (Wang et al., 2005).

In order to compress data there are many approaches used in digital communications and image compression. These may be divided in two broad categories: lossless and lossy techniques. The first category keep the signal information intact. The second category remove redundant information from signals to achieve a higher compression ratio (Ribeiro et al., 2004).

INTECH
open science | open minds

© 2013 Găşpăresc; licensee InTech. This is a paper distributed under the terms of the Creative Commons Attribution License (http://creativecommons.org/licenses/by/3.0), which permits unrestricted use, distribution, and reproduction in any medium, provided the original work is properly cited.

In recent years, the results presented in scientific literature show that the most used compression methods in power quality are based on wavelet transform and Slantlet transform. This chapter will provide an overview of their applications for power quality signals.

2. Data compression using wavelet transform

2.1. Wavelet transform

The wavelet transform ensures a progressive resolution in time-frequency domain, suitable to track the nonstationary signals dynamics properly. It use a variable window size, wide for low frequencies and narrow for high frequencies, to achieve a good localization in time and frequency domains (Ribeiro et al., 2007), (Zhang et al., 2011).

The Continuous Wavelet Transform (CWT) of a signal $f(t) \in L^2[R]$ is defined as

$$CWT(\tau,\gamma) = \int_{-\infty}^{\infty} f(t)\overline{\Psi_{\tau,\gamma}(\frac{t-\tau}{\gamma})}dt \tag{1}$$

where τ is the scale factor, γ is the translation factor and $\Psi_{\tau,\gamma}$ is the mother wavelet.

The Fourier analysis decomposes a signal into a sum of harmonics and wavelet analysis into set of functions called wavelets. A wavelet is a waveform of limited duration, usually irregular and asymmetric. These functions are obtained by dilations and translations of a unique function called mother wavelet $\Psi_{\tau,\gamma}$ and the function set $(\Psi_{\tau,\gamma})$ is called the wavelet family

$$\Psi_{\tau,\gamma}(t) = \frac{1}{\sqrt{\gamma}}\Psi(\frac{t-\tau}{\gamma}), \gamma > 0, \tau \in R \tag{2}$$

The Inverse Continuous Wavelet Transform (ICWT) is given as

$$f(t) = \frac{1}{C_\Psi}\int_{-\infty}^{\infty}\int_{-\infty}^{\infty} CWT(\tau,\gamma)\Psi_{\tau,\gamma}(t)\frac{d\tau d\gamma}{\tau^2} \tag{3}$$

where C_Ψ is the normalized constant.

In power quality analysis we work with acquired signals. These are discrete-time signals. Moreover, the CWT provides a redundant signal reprezentation in continuous-time, because the initial signal is possible to be reconstructed by a discrete version of CWT. The CWT is evaluated at dyadic intervals: the factor τ and γ are discretezed as $\tau=2^k$, $\gamma=2^k n$ where $n,k \in Z$. The relation (2) becomes (Dash et al., 2007), (Qian, 2002)

$$\Psi_{k,n}(t) = 2^{k/2}\Psi(2^k t - n) \tag{4}$$

The wavelet transform is the most used multiresolution analysis (MRA) technique of signals. Multiresolutions signal decompositon is based on subbads decomposition using low-pass filtering and high-pass filtering.

In muliresolution analysis a continuous function *x(t)* is decomposed as follows

$$x(t) = A_{j_0}(t) + \sum_{j=0}^{j_0} D_j(t) \tag{5}$$

where

$$A_j(t) = \sum_k c_j(k)\varphi_{j,k}(t) \tag{6}$$

$$D_j(t) = \sum_k d_j(k)\psi_{j,k}(t) \tag{7}$$

$c_j(k)$ are the scaling function coefficients, $d_j(k)$ are the wavelet function coefficients, j_0 is the scale, $\varphi(t)$ is the scaling function, $A_j(t)$ is called approximation at level j and $D_j(t)$ is called the detail at level j (Azam et al., 2004), (Zhang et al., 2011).

For a given signal *x(t)* and a three levels wavelet decomposition the relation (5) become

$$\begin{aligned} x(t) &= A_1 + D_1 \\ &= A_2 + D_2 + D_1 \\ &= A_3 + D_3 + D_2 + D_1. \end{aligned} \tag{8}$$

The decomposition of signal *x(t)* in A_1 and D_1 is the first decomposition level. At each decomposition level the signal is decomposed into an approximation and a detail.

Each detail D_j reveals details of the signal and each approximation A_j shows corse information. If the analysed signal contains a high frequency event (for instance, a transient phenomenon), the magnitude of details D_j associated with the event are significant larger than the rest of the coefficients. This observation is useful for compressing power quality signals: only the details D_j associated with the events are retained and all other coefficients are discarded. Moreover, the approximation coefficient is also kept for signal reconstruction (Santoso et al., 1997).

2.2. Data compression with wavelet transform

A general data compression method based on wavelet decomposition and reconstruction is shown in Fig. 1. That is a lossy compression method which includes certain steps (Wu et al.,

2003), (Hamid et al., 2002), (Littler et al., 1999) : first, the signal is decomposed into several wavelet transform coefficients (WTCs) using the DWT, thresholding of WTCs (useful to extract information and remove redundancy) and finally the signal is reconstructed from the retained WTCs.

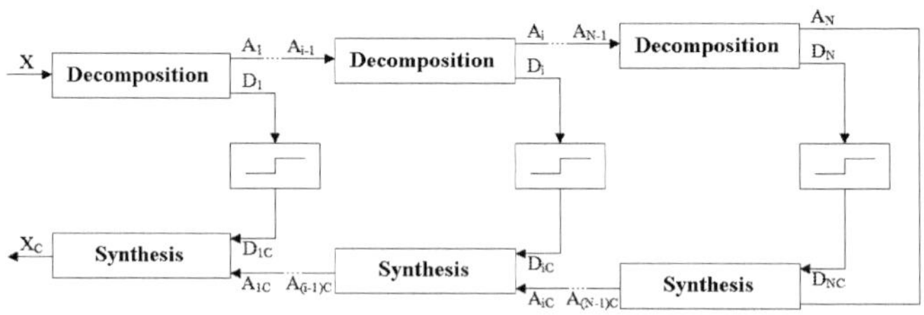

Figure 1. A general multi-scale wavelet compression method

One of the methods for the thresholding of WTCs is to set a threshold than only the coefficients above threshold are retained. Those below threshold are set to zero and are discarded (almost 90% of WTCs, some information will be lost). As a result, the amount of stored data is reduced.

The threshold is calculated based on absolute maximum value of the WTCs as

$$\eta_S = (1-u) \times \max\{|D_i(n)|\} \tag{9}$$

where u take values in the range $0 \le u \le 1$ and s is the associated scale.

Thresholding of WTCs is given by

$$D_{iS}(n) = \begin{cases} D_i(n), & |D_i(n)| \ge \eta_s \\ 0, & |D_i(n)| < \eta_s \end{cases} \tag{10}$$

and the retained WTCs are stored together along with their temporal positions.

To evaluate the performance of signal compression are used the compression ratio (CR) and the normalized mean-square error (NMSE).

The data compression ratio is defined by

$$CR = \frac{S_o}{S_c} \tag{11}$$

where S_o is the size of original file and S_c is the size of compressed file.

The quality of reconstructed signal is evaluated using the normalized mean-square error which is defined as

$$NMSE = \frac{\left\| X(n) - X_C(n) \right\|^2}{\left\| X(n) \right\|^2} \tag{12}$$

where $X(n)$ is the original signal and $X_c(n)$ is the compressed signal. A low value of NMSE corresponds to a small error between the original and reconstructed signal.

In the sections 2.2.1-2.2.2 is tested the performances of the general multi-scale wavelet compression method for transient phenomena and voltage swell. The influence of the order of Daubechies scaling function and the number of decomposition levels on data compression are analysed. The signals are simulated in Matlab environment. The details and the results are presented below.

2.2.1. Transient phenomena

Transient phenomena are sudden and short-duration change in the steady-state condition of the voltage, current or both. These are classified in two categories: impulsive and oscillatory transient (Fig. 2). The first category has exponential rise and falling fronts and it is characterized by magnitude, rise time (the time required for a signal to rise from 10% to 90% of final value), decay time (the time until a signal is greater than ½ from its magnitude) and its spectral content. The second category is characterized by magnitude, decay time and predominant frequence (Dungan et al., 2004), (Găşpăresc, 2011).

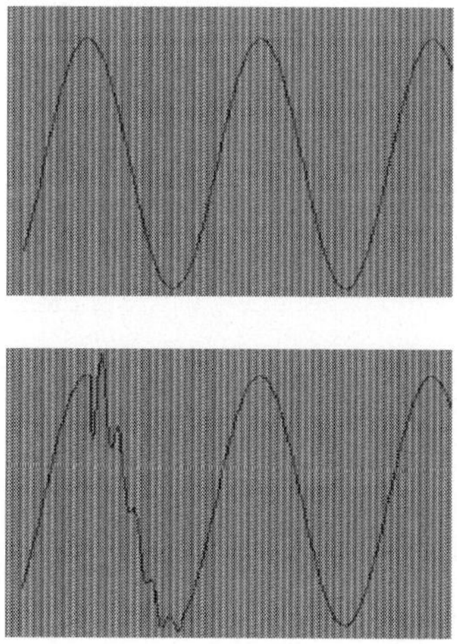

Figure 2. Transient phenomena

Fig. 3 shows the first test signal, an impulsive transient with magnitude of 1000 V superimposed on a sinusoidal signal with amplitude of 230 V and frequency of 50 Hz, corrupted with additive white noise. The sampling rate is 20 kHz. The signal is decomposed into three, four and five levels based on wavelet decomposition (Daubechies scaling function of order 3rd, 4th and 5th is used). Than the signal it is compressed using the threshold values 1, 5, 7 and 10. The results are presented in table 1.

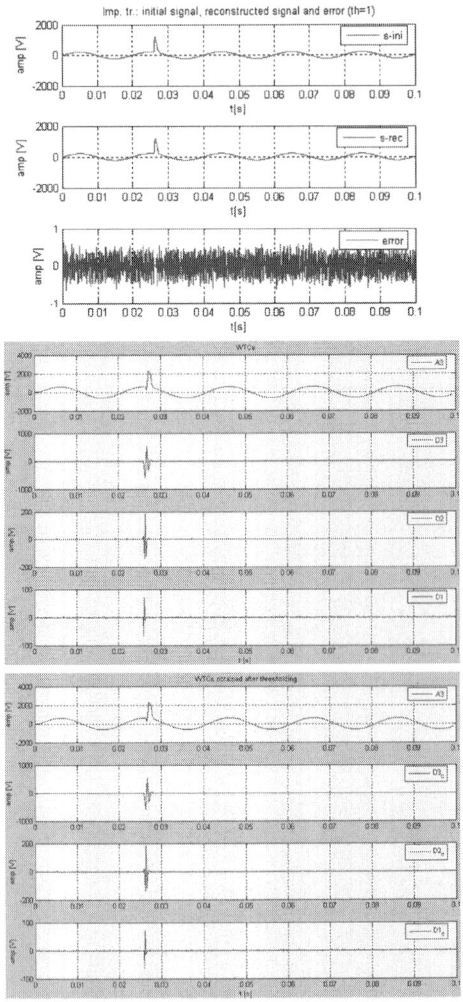

Figure 3. Impulsive transient compression with threshold value 1

Signal	Ψ(t)	Levels	η_s	NMSE [%]	CR
Impulsive transient	Db3	3	1	5.5154e-006	1.47
	Db3	3	5	2.7151e-005	6.54
	Db3	3	7	2.7827e-005	7.09
	Db3	3	10	2.6502e-005	7.14
	Db3	4	5	2.9240e-005	10.7
	Db3	4	7	3.1081e-005	11.17
	Db3	4	10	3.1515e-005	11.7
	Db3	5	5	3.6143e-005	9.66
	Db3	5	7	4.0148e-005	12.42
	Db3	5	10	5.6165e-005	15.04
	Db4	3	5	2.9386e-005	6.94
	Db4	3	7	2.9246e-005	7.14
	Db4	3	10	2.9481e-005	7.3
	Db4	4	5	3.2845e-005	10.47
	Db4	4	7	3.0742e-005	10.93
	Db4	4	10	3.1029e-005	11.05
	Db5	4	5	3.0875e-005	9.85
	Db5	4	7	2.8626e-005	10.36
	Db5	4	10	3.1029e-005	10.69

Table 1. Compression results for impulsive transient

The initial value of threshold is 1. The size of the WTCs obtained after thresholding using the relations (9) and (10) is reduced as follows (first line from table 1): the coefficient *D1* at scale 1 has 2*303=606 samples (303 samples nonzero and their temporal positions), the coefficient *D2* at scale 2 has 2*167=334 samples, the coefficient *D3* has 2*85=170 samples and the coefficient A3 has 250 samples. The compressed signal has 1360 samples and the initial test signal has 2000. The compression ratio is 2000/1360=1.47.

From table 1 can be observed that the highest compression rate is 15.04. This value is obtained for Daubechies scaling function of order 3 (*Db3*), 5 levels of decomposition and the threshold value of 10. The order of *NMSE* error is 10^{-6}.

If the threshold value is greater than or equal to 5 (Fig. 4), the signal distortions start to rise especially in the area of the overlapped impulsive transient. The enlargement of threshold leads to more and more information discarded and NMSE grow up.

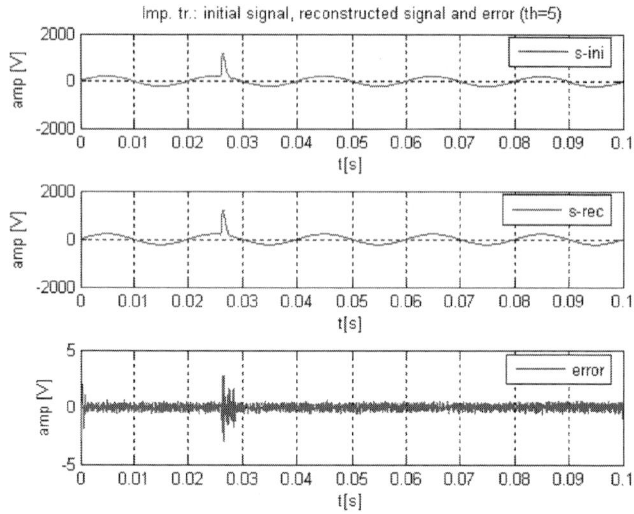

Figure 4. Impulsive transient compression with threshold value 5

Fig. 5 shows the second test signal, an oscillatory transient superimposed on a sinusoidal signal. The signal parameters and the decomposition parameters have the same values as the first test signal. The results are presented in table 2.

From table 2 the highest compression rate is 7.84. The value is lower than for the first test signal. This compression rate is obtained using the same settings: Daubechies scaling function of order 3 (*Db3*), 5 levels of decomposition and the threshold value of 10. The order of *NMSE* error is 10^{-5}.

Again, if the threshold value is greater than or equal to 5 (Fig. 6), the signal distortions start to rise especially in the area of the overlapped oscillatory transient and NMSE grow up too.

http://dx.doi.org/10.5772/53059

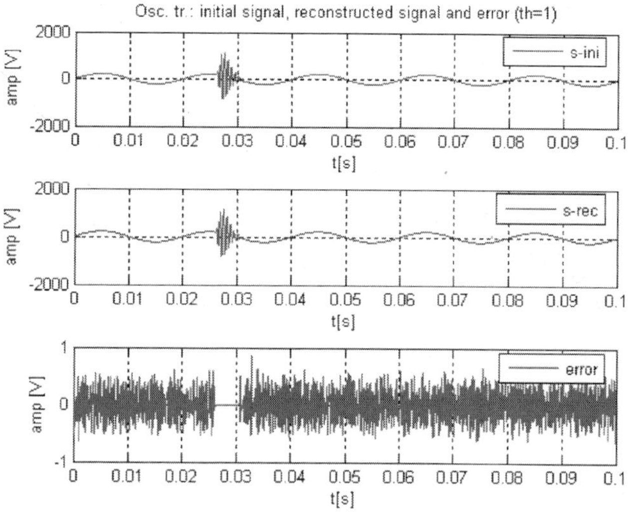

Figure 5. Oscillatory transient compression with threshold value 1

Signal	Ψ(t)	Levels	η_s	NMSE [%]	CR
Oscillatory transient	Db3	3	1	4.7257e-006	1.35
	Db3	3	5	2.9394e-005	4.85
	Db3	3	7	2.6282e-005	5.02
	Db3	3	10	3.7811e-005	5.21
	Db3	4	5	2.9285e-005	6.31
	Db3	4	7	3.3209e-005	6.6
	Db3	4	10	4.5275e-005	7.07
	Db3	5	5	3.3699e-005	6.08
	Db3	5	7	4.4197e-005	6.69
	Db3	5	10	6.7879e-005	8.3
	Db4	3	5	2.8067e-005	5.05
	Db4	3	7	3.0176e-005	5.18
	Db4	3	10	4.0081e-005	5.43
	Db4	4	5	2.9924e-005	6.47
	Db4	4	7	3,2733e-005	6.87
	Db4	4	10	4.2280e-005	7.38
	Db5	4	5	2,9223e-005	6.89
	Db5	4	7	3.6702e-005	7.22
	Db5	4	10	4.7778e-005	7.84

Table 2. Compression results for oscillatory transient

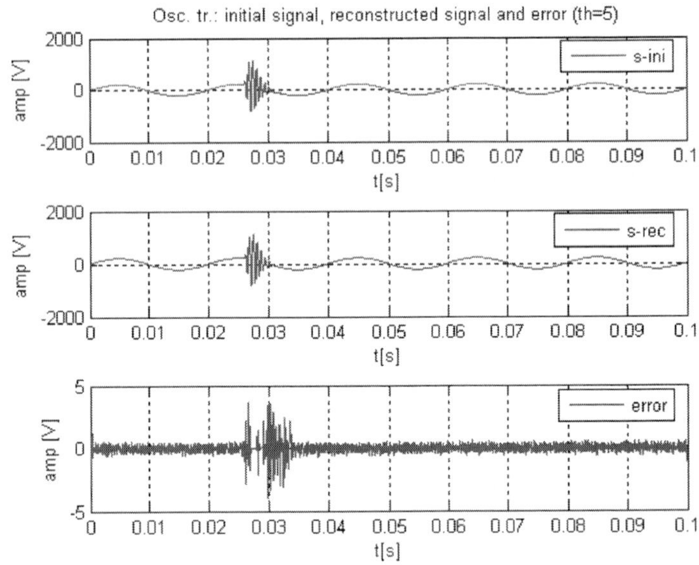

Figure 6. Oscillatory transient compression with threshold value 5

2.2.2. Voltage swell

Fig. 7 shows the third test signal, a swell with magnitude of 375 V superimposed on a sinusoidal signal. The rest of signal parameters and the decomposition parameters have the same values as the previous test signals. The results are presented in table 3.

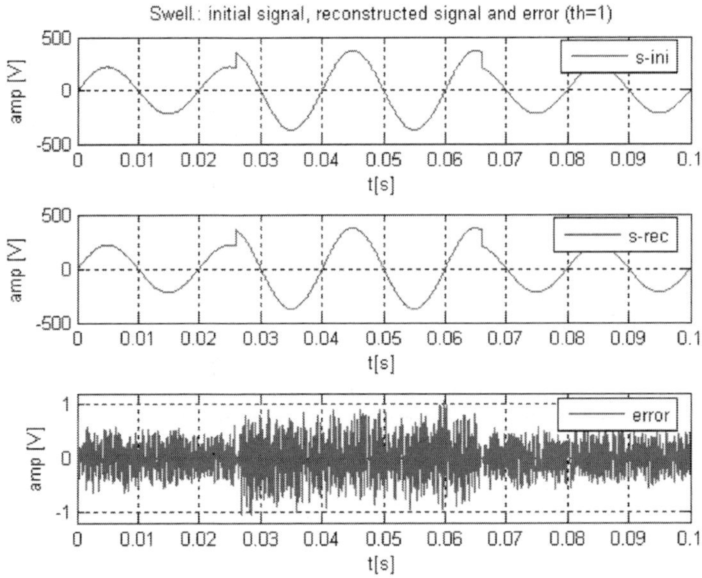

Figure 7. Voltage swell compression with threshold value 1

Signal	Ψ(t)	Levels	η_s	NMSE [%]	CR
Voltage sag	Db3	3	1	3.6484e-006	1.21
	Db3	3	5	3.3454e-005	6.9
	Db3	3	7	3.4048e-005	7.14
	Db3	3	10	3.5194e-005	7.19
	Db3	4	5	3.7718e-005	10,47
	Db3	4	7	4.0528e-005	11.17
	Db3	4	10	4.1284e-005	11.7
	Db3	5	5	3.7621e-005	9.05
	Db3	5	7	4.4344e-005	10.47
	Db3	5	10	5.8101e-005	13.42
	Db4	3	5	3.4595e-005	6.62
	Db4	3	7	3.2270e-005	6.99
	Db4	3	10	3.5533e-005	7.14
	Db4	4	5	3.8236e-006	9.3
	Db4	4	7	6.4098e-006	10.47
	Db4	4	10	3.9005e-005	10.81
	Db5	4	5	3.7183e-005	9.13
	Db5	4	7	3.9994e-0056	9.95
	Db5	4	10	4.1631e-005	10.36

Table 3. Compression results for swell

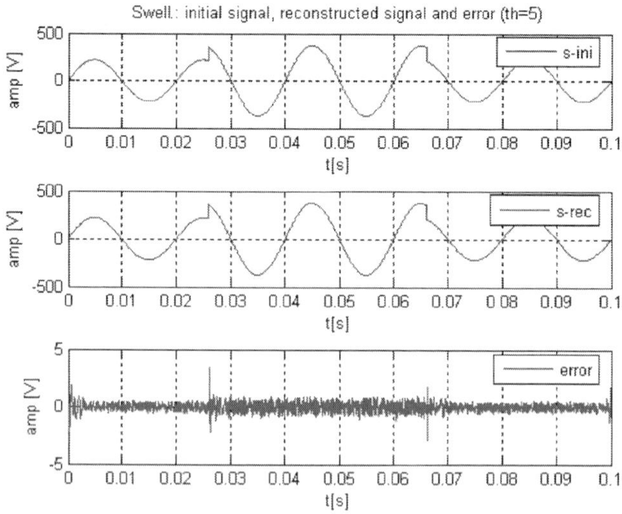

Figure 8. Volatge swell compression with threshold value 5

From table 3 the highest compression rate is 13.42. The value is higher than for the second test signal. This compression rate is obtained using the same settings as for previous test signals: Daubechies scaling function of order 3 (*Db3*), 5 levels of decomposition and the threshold value of 10. The order of *NMSE* error is 10^{-5}.

Again, if the threshold value is greater than or equal to 5 (Fig. 8), the signal distortions start to rise especially in the area of the overlapped disturbance and NMSE grow up too.

A few conclusions are described below:

- using 5 levels of decomposition, Daubechies scaling function of order 3 (*Db3*) and 5 (*Db5*) is obtained the highest compression rate;
- for a higher threshold value the compression rate will be higher, but NMSE and signal distorsions grow up;
- for different types of disturbances using the same settings the compression rate is different.

2.3. New wavelet-based data compression technique using decimation and spline interpolation

This chapter describes a new technique proposed for signal compression based on wavelet decomposition and spline interpolation method (Găşpăresc, 2010). It follows to obtain a higher compression ratio than the general data compression method used for the test signals analysed before, where for a given signal it is applied a signal decomposition and than thresholding of WTCs D_i, $i=1,...,N$. Using this method the coefficient A_N is not thresholded and it has the largest number of samples from all the coefficients of signal decomposition. In order to obtain a higher sample rate this coefficient is decimated with a decimation factor *Fd* and at signal reconstruction will be interpolated. The cost is the increase of *NMSE* error.

Given an interval *[a,b]* and a divizion Δ:$a=x_0<x_1<...<x_n=b$, a function $S : [a,b] \rightarrow R$ is called cubic spline interpolation function if this function meets the next conditions:

- S is a polynomial of degree at most 3 on any interval (x_k, x_{k+1}), $k=1,\ldots,N$ (relation 13);
- $S \epsilon C^2([a,b])$;
- $S(x_i)=f(x_i)$, $i \epsilon (o,1,..., n)$, where $f(x)$ is the interpolated function.

$$S(x) = a_i + b_i x + c_i x^2 + d_i x^3, \forall x \in [x_{i-1}, x_i] \tag{13}$$

The proposed technique is tested using an impulsive transient with magnitude of 700 V superimposed on a sinusoidal signal (Fig. 9-10). The sampling rate is 5 MHz in this case. The signal is decomposed using a Daubechies scaling function of order 4 and 5 and respectively 4 levels of decomposition. Than the signal it is compressed using a threshold (Table 4).

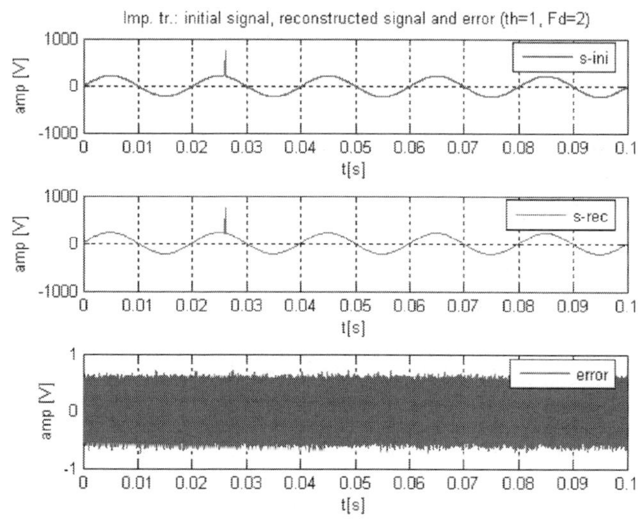

Figure 9. Impulsive transient compression with decimation factor *Fd=2*

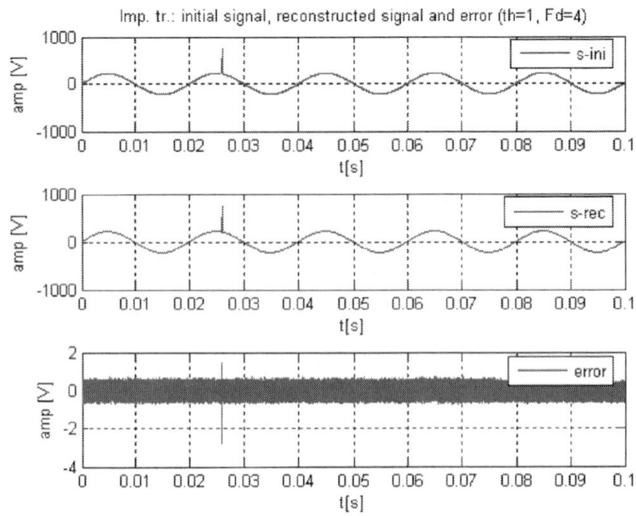

Figure 10. Impulsive transient compression with decimation factor *Fd=4*

Signal	Ψ(t)	η_s	Fd	NMSE [%]	CRa
Impulsive transient	Db4	3	2	2.2901e-004	32
	Db5	3	2	2.2901e-004	32
	Db4	3	4	9.8005e-004	63.99
	Db5	3	4	2.2830e-004	63.98

Table 4. Compression results for impulsive transient using the proposed technique

From table 4 can be observed that the highest compression ratio is 63.99. This value is obtained for Daubechies scaling function of order 4 (*Db4*), 4 levels of decomposition, threshold value of 3 and the decimation factor value of 4. The order of *NMSE* error is 10^{-4}. The resulted compresion ratio is 4 times higher than the values from the prevoius tables, but the NMSE error is higher also.

This proposed technique is efficient especially for signals acquired at high sample rates, when are acquired a sufficient number of samples of the disturbance overlapped on the power supply signal. If this number is small, after the decimation of coefficient A_N are losed disturbance details which cannot be reconstructed by interpolation and the reconstructed signal will contain distortions on the disturbance area.

3. Data compression using slantlet transform

3.1. Slantlet transform

The Slantlet transform (SLT) is a relatively new multiresolution technique base on DWT. In fact, it is an orthogonal DWT with two zero moments and compared to DWT provides better time localization (Selesnick, 1999), (Panda et al., 2002), (Duda, 2008).

In (Panda et al., 2002) is proposed a new approach for power quality data compression based on SLT. The technique is compared with the discrete cosine transform (DCT) and the discrete wavelet transform (DWT) using various types of power quality disturbances (impulse, sag, swell, harmonics, momentary interruption, oscillatory transient, voltage flicker).

In order to compare DWT and SLT is considered a two-scale iterated filterbank (Fig. 11) and a two-scale slantlet filterbank (Fig. 12). First three blocks from Fig. 12 are not products. The filters have shorter length and the difference grows with the number of stages. The time localization is improved but SLT filterbank is less frequency selective.

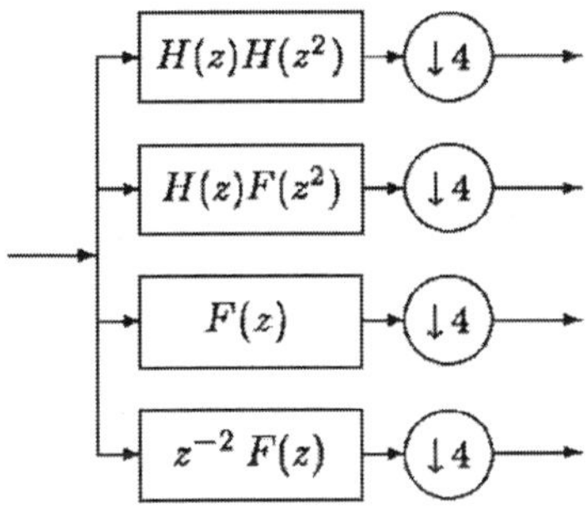

Figure 11. Two-scale iterated filterbank

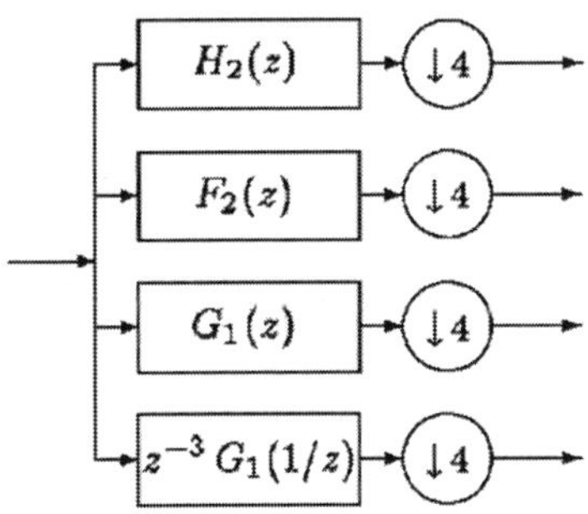

Figure 12. Two-scale slantlet filterbank

The SLT is based on the principle of designing different filters for different scales unlike iterated filterbank approaches for DWT. In (Selesnick, 1999) are described the basis for filterbank design, polynomial expresions to determine the filter coefficients and an algorithm to calculate the transform.

The filter coefficients are

$$
\begin{aligned}
&E_0(z)=G_1(z)=(-\frac{\sqrt{10}}{20}-\frac{\sqrt{2}}{4})+(\frac{3\sqrt{10}}{20}+\frac{\sqrt{2}}{4})z^{-1}+\left(-\frac{3\sqrt{10}}{20}+\frac{\sqrt{2}}{4}\right)z^{-2}\\
&+\left(\frac{\sqrt{10}}{20}-\frac{\sqrt{2}}{4}\right)z^{-3}\\
&E_1(z)=F_2(z)=(\frac{7\sqrt{5}}{80}-\frac{3\sqrt{55}}{80})+(-\frac{\sqrt{5}}{80}-\frac{\sqrt{55}}{80})z^{-1}+\left(-\frac{9\sqrt{5}}{80}+\frac{\sqrt{55}}{80}\right)z^{-2}\\
&+\left(-\frac{17\sqrt{5}}{80}+\frac{3\sqrt{55}}{80}\right)z^{-3}+\left(\frac{17\sqrt{5}}{80}+\frac{3\sqrt{55}}{80}\right)z^{-4}+\left(\frac{9\sqrt{5}}{80}+\frac{\sqrt{55}}{80}\right)z^{-5}\\
&+\left(\frac{\sqrt{5}}{80}-\frac{\sqrt{55}}{80}\right)z^{-6}+\left(-\frac{7\sqrt{5}}{80}-\frac{3\sqrt{55}}{80}\right)z^{-7}\\
&E_2(z)=H_2(z)=(\frac{1}{16}+\frac{\sqrt{11}}{16})+(\frac{3}{16}+\frac{\sqrt{11}}{16})z^{-1}+(\frac{5}{16}+\frac{\sqrt{11}}{16})z^{-2}\\
&+(\frac{7}{16}+\frac{\sqrt{11}}{16})z^{-3}+(\frac{7}{16}-\frac{\sqrt{11}}{16})z^{-4}+(\frac{5}{16}-\frac{\sqrt{11}}{16})z^{-5}\\
&+(\frac{3}{16}-\frac{\sqrt{11}}{16})z^{-6}+(\frac{1}{16}-\frac{\sqrt{11}}{16})z^{-7}\\
&E_3(z)=z^{-3}E_2(\frac{1}{z})
\end{aligned}
\tag{14}
$$

Table 5 displays the test results obtained using DCT, DWT and SLT (Panda et al., 2002). The compression performance is analysed based on percentage of energy retained (relation 15) and mean square error (MSE) in decibels (relation 16). The compression rate is 10. The results shows improved values for energy retained (near 4%) and MSEs.

$$
\left[\frac{\textit{Vector}\text{ norm of the retained SLT coefficients after thresholding}}{\text{Vector norm of the original SLT coefficients}}\right]\times 100 \tag{15}
$$

$$MSE[dB] = 10\left[\log_{10}\left(\frac{1}{N}\sum_{i=1}^{N}\left\| x(i) - \hat{x}(i) \right\|^2\right) \right] \tag{16}$$

Signal	Energy Retained [%]			MSE [dB]		
	DCT	DWT	SLT	DCT	DWT	SLT
Impulse	88.01	91.13	94.01	-10.67	-13.54	-16.98
Sag	87.81	90.01	93.20	-10.08	-13.04	-17.54
Swell	89.46	91.01	94.44	-11.88	-13.77	-17.95
Harmonics	87.69	90.89	93.14	-11.04	-13.31	-17.68
Momentary Interruption	90.44	91.10	94.11	-12.27	-15.89	-18.79
Oscillatory Transient	91.63	90.88	95.04	-12.98	-14.45	-19.07
Voltage Flicker	90.75	91.34	95.18	-10.76	-14.74	-19.78

Table 5. Test results obtained using DCT, DWT and SLT (CR=10)

4. Conclusions

The research results on data compression using DWT presented in this work show the optimal order of Daubechies scaling function recommended in order to achieve the best compression ratio for three types of power quality disturbances and the necessary number of decomposition levels. An compression algorithm base on spline interpolation method that allows higher compression rates is also presented.

The Slantlet transform is analysed as a new approach for power quality data compression. The compression performance using SLT was compared based on percentage of energy retained and mean square error in decibels. The computer simulation tests using various power quality disturbances shows that SLT provides a more accurate reconstruction of the original signal than DCT and DWT.

Author details

Gabriel Găşpăresc

"Politehnica" University of Timişoara, Romania

References

[1] Azam, M. S. ., Tu, F. ., Pattipati, K. R. ., & Karanam, R. (2004). A Dependency Model Based Approach for Identifying and Evaluating Power Quality Problems, IEEE Transactions on Power Delivery 19(3), , 1154-1166.

[2] Barrera, Nunez. V. ., Melendez, Frigola. J. ., & Herraiz, Jaramillo. S. (2008). A Survey on Voltage Dip Events in Power Systems, Proceedings of the International Conference on Renewable Energies and Power Quality.

[3] Bollen, M. ., & Gu, I. (2006). Signal Processing of Power Quality Disturbances. John Wiley & Sons.

[4] Dash, P. K. ., Nayak, M. ., Senapati, M. R., & Lee, I. W. C. (2007). Mining for similarities in time series data using wavelet-based feature vectors and neural networks. *Engineering Applications of Artificial Intelligence*, 185-201.

[5] Duda, K. (2008). Lifting Based Compression Algorithm for Power Systems Signals, Metrology and Measurement Systems XV(1), , 69-83.

[6] Dungan, R. C. ., Mc Granaghan, M. F. ., Santoso, S. ., & Beaty, H. W. (2004). Electrical Power System Quality, McGraw-Hill.

[7] Găşpăresc, G. (2010). Data compression of power quality disturbances using wavelet transform and spline interpolation method, Proceedings of the 9th International Conference on Environment and Electrical Engineering.

[8] Găşpăresc, G. (2011). Methodes of Power Quality Analysis, in Power Quality- Monitoring, Analysis and Enhancement, Ed. Ahmed Zobaa, Mario Manana Canteli and Ramesh Bansal, Chapter 6, INTECH., 101-118.

[9] Hamid, E. Y. ., & Kawasasaki, Z. I. (2002). Wavelet-based data compression of power disturbances using the minimum description length criterion, IEEE. Transactions on Power Delivery 17, , 460-466.

[10] Lin, L. ., Huang, N. ., & Huang, W. (2009). Review of Power Quality Signal Compression Based on Wavelet Theory. Proceedings of the International Conference on Test and Measurement.

[11] Littler, T. B. ., & Morrow, D. J. (1999). Wavelets for the analysis and compression of power system disturbances. IEEE Transactions on Power Delivery 14, , 358-364.

[12] Lorio, F. ., & Magnago, F. (2004). Analysis of Data Compression Methods for Power Qualiy Events, Proceedings of the Power Engineering Society General Meeting.

[13] Qian, S. (2002). Time-Frequency and Wavelet Transforms, Prentice Hall PTR.

[14] Panda, G. ., Dash, P. K., Pradhan, A. K., & Meher, S. K. (2002). Data Compression of Power Qualtity Events Using the Slantlet Transform, IEEE Transactions on Power Delivery 17(2), , 662-667.

[15] Ribeiro, M. V. ., Park, S. H. ., Romano, J. M. T. ., & Duque, C. A. (2004). An Improved Method for Signal Processing and Compression in Power Quality Evaluation. *IEEE Transactions on Power Delivery*, 464-471.

[16] Ribeiro, M. V. ., Park, S. H. ., Romano, J. M. T. ., & Mitra, S. K. (2007). A Novel MDL-based Compression Method for Power Quality Applications. IEEE Transactions on Power Delivery 22(1), , 27-36.

[17] Santoso, S. ., Powers, E. J. ., & Grady, W. M. (1997). Power Quality Disturbance Data Compression using Wavelet Transform Methods. IEEE Transactions on Power Delivery 12(3), , 1250-1256.

[18] Selesnick, I. W. (1999). The Slantlet Transform. *IEEE Transactions on Signal Processing*, 1304-1313.

[19] Zhang, M. ., Li, K. ., & Hu, Y. (2011). A High Efficient Compression Method for Power Quality Applications, IEEE Transaction on Power Delivery 60(6), , 1976-1985.

[20] Wang, J. ., & Wang, C. (2005). Compression of Power Quality Disturbance Data Based on Energy and Adaptive Arithmetic Encoding, Proceedings of the TENCON.

[21] Wu, C. J. ., & Fu, T. H. (2003). Data compression applied to electric power quality tracking of arc furnance load, Journal of Marine Science and Technology 11, , 39-47.

Chapter 2

Electric Power Quality Recognition and Classification in Distribution Networks

Soliman Abdel-Hady Soliman and
Rashid Abdel-Kader Alammari

Additional information is available at the end of the chapter

http://dx.doi.org/10.5772/53081

1. Introduction

The power quality problem is now of a great concern to electric utilities of power industry and they are trying hard to supply their customers with a good quality of power especially in the open market. Due to the wide spread use of power electronics in every place in the power industry, the power supplied to the customers are now distorted in either the voltage signal or current signal or both of them. This distortion has a great effect on the sensitive equipments and may cause interruption to such equipments that result in very expensive consequence. It has been reported that 30% voltage sag for very short duration can reset programmable controllers for the entire assembly line. As such an accurate algorithm is needed for identification and measurements of these events

Power quality involves how close the voltage waveform is to being a perfect sinusoid with a constant frequency and amplitude. Historically, it has been the utilities' responsibility to provide a "clean" voltage waveform, and most customers' load did no affect the quality of their power. Today, a new factor, harmonics, has been added to the power quality scenario because utility customers, including residential ones, are using electronic devices that require non-sinusoidal currents, currents rich in harmonics.

The presence of power system harmonics is not a new problem; it has been well known since the first generator was built. However, nowadays due to the widespread use of electronic equipment, arcing devices, such as arc furnaces and equipment with saturated ferromagnetic cores, such as transformers, power engineers pay more attention to power system harmonics. The presence of voltage and current waveform distortion is generally expressed in terms of harmonic frequencies that are integral multiples of the power system nominal frequency. It is

INTECH
open science | open minds

© 2013 Soliman and Alammari; licensee InTech. This is a paper distributed under the terms of the Creative Commons Attribution License (http://creativecommons.org/licenses/by/3.0), which permits unrestricted use, distribution, and reproduction in any medium, provided the original work is properly cited.

a steady phenomenon in a power system, and it is completely different from the distortion results from the transient, such as faults, in a power system.

The subject of harmonics has been deemed so important that there are international conferences dedicated to the subject, working groups and committees in international engineering societies, which deal only with harmonics, and several dedicated books on the topics. Power system, harmonics are become one of the important index to the power quality issues these days. Good power quality means less distortion, less harmonics, in the voltage and current sources.

Good harmonic prediction requires clear understanding of two different but closely related topics. One is the non-linear voltage/ current characteristic of some power system components and its related effect, the presence of harmonic sources. The main problem in this respect is the difficulty in specifying these sources accurately. The second topic is the derivation of the suitable harmonic models of the predominantly linear network components, and of the harmonic flows resulting from their interconnection. This task is made difficult by insufficient information on the composition of the system loads and their damping to harmonic frequencies. Further impediments to accurate prediction are the existence of many distributed non-linearities, phase diversity, the varying nature of the load, etc.

To assess the quality of delivered power, especially in open markets, it is necessary to estimate the harmonic components in a power system. The quality of power delivered necessitates knowledge regarding the magnitude of harmonic components and phase angle of these components. The reduction of harmonics in a system means good system quality. Installing filters at feeding points along the network can do this if the harmonic components magnitudes as well as their phase angles are known in advance.

Over the past 25 years major improvements in the field of signal theory have been achieved and many algorithms and techniques have been published in the literature. In Ref. [1] an algorithm for tracking the voltage envelope based on calculating the energy operator of a sinusoidal waveform is presented. It is assumed that the frequency of the sinusoidal waveform is known and a lead-lag network with unity gain is used. An approach to power quality assessment based on real-time (RT) hardware-in-the-loop (HIL) simulation is proposed in Ref. [2]. The RT-HIL platform is being used for power quality studies of NAVY all-electric ships. Ref. [3] describes an analytical method for evaluating the voltage sag performance of a power-supply distribution network, different evaluation algorithms are proposed to account for each type of sensitivity.

Ref. [4] proposes a fuzzy pattern recognition system for power quality disturbances. It is a two-stage system in which a multi-resolution S-transform is used to generate a set of optimal features vectors in the first stage. In the second stage a fuzzy logic-based pattern recognition system is used to classify the various disturbances waveforms generated due to power quality violations. The Teager energy operator (TEO) and the Hilbert transform (HT) are introduced in Ref. [5] as effective approaches for tracking the voltage flicker levels. It has been found that TEO and HT are capable of tracking the amplitude variations of the voltage flicker and supply frequency in industrial systems with an average error 3%. Root-mean-square (rms) calculation is a popular method adopted in power system parameter classification such as voltage sag classification and relay protection. Ref. [6] studies the characteristics of the rms method, for both the single-frequency and mixed-frequency signals. Analysis is made on its dependence on sampling rate, sampling window size, as well as point-on-wave through strict mathematical deductions.

Ref. [7] presents a control technique for flicker mitigation. This technique is based on the instantaneous tracking of the measured voltage envelope. The ADALINE (ADAptive LINear) neuron algorithm and the Recursive Least Square (RLS) algorithm are introduced for the flicker envelope tracking. Presented in Ref. [8] is a fuzzy-expert system for automated detection and classification of power quality disturbances. The types of concerned disturbances include voltage sags, swells, interruption, switching transients, impulses, flickers, harmonics and notches from a signal that is available in sampled form. Fourier transform and wavelet analysis are utilized to obtain unique features for the waveforms. Ref. [9] and Ref. [10] present concepts based on fuzzy expert system for power quality study, the concept integrates the power system modeling, classifying and characterizing for power quality events, studying equipment sensitivity to the event disturbance, and locating point of event occurrence into one unified frame. Both Fourier and wavelet analyzes are applied for extracting distinct features of various types of events as well as for characterizing the events. Ref. [11] presents a technique for learning power- quality waveforms based adaptive neuro-fuzzy systems to learn power quality signature waveforms. Ref. [12] compared three signal processing tools for power quality analysis; the continues wavelet transform, the multi-resolution analysis and the quadratic transform. It has been concluded that the continues Wavelet transform is as reliable as these methods but has been the advantage of giving directly the magnitude of the 50/60 Hz signal, and is suitable for quantifying power quality when detecting and measuring voltage sags, transients over voltages or flicker. None of these techniques are as adequate to detect and measure the voltage magnitude of harmonic content as the Fourier transform. Reference [15] presents an optimal measurements scheme for tracking the harmonics in power system voltage and current waveforms. The proposed scheme was based on Kalman filtering. Reference [15] implements the well-known LES algorithm for identification and measurement of power system harmonics. The mathematical model for the identification and measurement process is presented in this reference. The samples used in this reference are for one of the three- phase voltage or current signals. Reference [16] presents a comparative study for power system harmonic estimation, where it compares the results obtained using discrete Fourier transform (DFT), the well known least errors square (LES) parameter estimation algorithm and the least absolute value (LAV) parameter estimation algorithm. It has been concluded that the three algorithms produce the same estimate, if the signal under study is free of noise. However, if some data samples are missed, the least absolute value produces better estimates than the DFT and LES algorithms. The algorithms in this reference use the samples of a one-phase voltage or current signal. An approach based on singular value decomposition (SVD) for estimating harmonic components in a power system is presented in Reference [17]. Three different techniques are investigated in this reference; the standard averaged SVD, the total LS and double SVD. Reference [18] implements the neural network in its analogue form for estimation of harmonics. It has been shown that such problem formulation leads to a quadratic objective of the global minimum, which can be found by using simple electronic circuitry in real time. Reference [19] presents an approach for the estimation of harmonic components of a power system using a linear adaptive neuron called Adaline. The proposed estimator tracks Fourier coefficients of signal data corrupted with noise and decaying DC components. Reference [28] develops a fast Newton type solution of the six-pulse rectifier and dc system in the harmonic domain. The nonlinear equations are solved using Newton's method, which employs a Jacobian matrix of partial derivatives. A twelve states Kalman filtering algorithm is applied in Reference [29], using an 8-bit microprocessor, for continues real-time tracking of the harmonics in the voltage or current waveforms of a power system to obtain in real time the instantaneous values for a maximum of six harmonics as well as the existing harmonic distortion. Reference [30] reviews the problems associated with direct application of the Fast Fourier Transform to compute harmonic levels of non-steady state distorted waveforms, and various ways to describe recorded data in statistical terms. Reference [31] presents an approach based on fuzzy linear regression for the measurement of power system harmonic components. The non-sinusoidal voltage or current waveform is written as a linear function. The parameters of this function are assumed to be fuzzy numbers having certain middle and spread value. The problem in this reference is formulated as a linear optimization problem, where the objective is to minimize the spread of voltage or current samples. The on-line digital measurement on power systems for the power quality analysis under non-sinusoidal conditions is considered in Reference [32]. The proposed instrument, in this reference, adopts a floating point digital signal processing (DSP) hosted on a IBM PC and interfaced with a special high-speed data acquisition system (DAS). Voltage and current waveforms are acquired and processed by using a fast recursive least-square (FRLS) measurement algorithm. Using such an algorithm, different quantities are obtained, such as the current and voltage rms values, their harmonic content, the active power, the harmonic active power, the power factor, etc. Reference [33] applies a technique to the computation of individual harmonics in digital protections, where only certain isolated harmonics, rather than the full spectrum, are needed. This leads to O ($\log_2 N$) computations per harmonics. A technique based on modeling and identification method is proposed in Reference [34] using a mathematical model describing the signal in question. The recursive least-square-error identification algorithm is used to identify the harmonic parameters. These are including the frequency, the amplitude and phase angle. Because of the limitations associated with conventional algorithms, particularly under supply-frequency drift and transient situations, an approach based on non-linear least squares parameter estimation has been proposed in Reference [35]. To reduce the computational time the Hopfield type feedback neural networks for real-time harmonic evaluation. The neural network implementation determines simultaneously the supply frequency variation, the fundamental-amplitude/phase variation as well as the harmonics-amplitude variation. Reference [36] considers state estimation of harmonic signals with time varying magnitudes. Harmonic signals are modeled using elliptical set-theoretic methods and an optimal reduced-order estimator, which has one-half the dimension of the state vector, is developed for predicting the unknown time-varying harmonic magnitudes. Reference [37] proposes the optimization of spectrum analysis to reduce the restriction on Fast Fourier Transform (FFT) [38] resulting mismatch the frequency scale with signal characteristics. By using this method

http://dx.doi.org/10.5772/53081

both of the picket-fence effect and the leakage effect are reduced, and it makes the harmonic parameters show on spectrum more accurately. Reference [39] proposes a harmonic model based on Wavelet Transform (WT) for on-line tracking of power system using Kalman filtering. The close relation between the Wavelet and Multi-resolution analysis is utilized to express the harmonic magnitudes and phase angle as a sum of Wavelet and scaling function.voltage sags define as an rms reduction in the AC voltage, at the power frequency, for duration from a half a cycle to a few second. If the voltage drops below normal level for several cycles it would affect the critical load and cause shutdown to these loads. Voltage sags constitute the majority of power line problem representing about 60% of all problems. In addition to these quantities, sags also characterize by unbalance, non-sinusoidal wave-shapes and phase angle shift (phase jump). These factors are important for determining the behavior of ac motor drives during sags [40]. The main sources for the voltage sags are start-ups of large motors, sudden increase in line loads, electrical faults on utility power lines caused by animals, trees, storms, or other objects in contacts with power lines. A majority of faults on a utility system are single-line-to-ground faults (L-G). During L-G faults, the voltage on the faulted phase goes to nearly zero volts at the fault location. The corresponding voltage at a customer bus depends on the system configuration, location of the fault, the impedance of the system upstream of the fault, the feeder impedance, the distance of the fault and the transformer connections between the faulted system and customer bus. Furthermore, Electronic loads that pull large currents such as copy machines, laser printers can cause voltage sags. Further more, loose wiring in the distribution installation can cause voltage sags [14]. To quantify the effect of sensitive equipment to voltage sags, it is necessary to characterize the parameters of voltage sags. Most often, voltage sags characterize by a duration and depth parameter and represent in a two-dimensional rms voltage magnitude versus duration plot. This simplified representation of voltage sag characteristics does not take into account the different in individual phase voltages (voltage asymmetry or unbalance) and the associated phase angle shift during voltage sag. Furthermore, it does not take into account the non-sinusoidal nature of the voltage waveform during the sag. The magnitude of voltage sag can be determined from the rms voltage. As long as the voltage is sinusoidal, it does not matter whether rms voltage, fundamental voltage, or peak voltage used to obtain the sag magnitude. But especially during a voltage sag this is often not the case.[15] Chapter 4 in Ref. [15] explains in details the voltage sag characterization, the different available techniques used to measure the voltage sags at different modes of operation of power systems Fast Fourier Transform used to measure the phase angle jumps using the moving window length technique.

In Reference [42], a method for voltage disturbances such as voltage sags, voltage swells, flicker, frequency change in the utility voltage, and harmonic distortion of a single-phase or poly phase voltage disturbances is explored The algorithm is based on the theory that allows a set of three-phase voltages be represented as dc voltages in a d–q synchronous rotating frame. In this case, the utility input voltages are sensed and then converted to dc quantities in the d–q reference frame. The output of the algorithm is compared with a set of reference voltages, and the error is used as a measure to such disturbances. Reference [43] presents a Monte Carlo based approach to evaluate the maximum voltage sag magnitudes as well as the voltage unbalance in transmission systems. In this context, investigations have been conducted on a system model taking into consideration the uncertainty in several factors associated with the practical operation of a power system.

Reference [44] has discussed the limitations of the conventional sag characterizing method and proposed a new sag characterizing method. The conventional method overestimates the nonrectangular sag and cannot reflect the exact effect of voltage sag according to the voltage tolerance characteristics. The method has approximated the voltage profile during voltage sag using order radical root function. It has modified the voltage sag duration using the known parameters, and which are measured at PQ monitors. With the modified sag duration, the method has evaluated the effect of voltage sag correctly. Moreover, it can be applied to not only rectangular sag but also nonrectangular sag.

Reference [45] proposes the concept of "voltage sag state estimation" and associated algorithms to achieve this goal. The method has the characteristics: 1) It makes use of the radial connection characteristic of a distribution feeder, 2) it is based on a limited number of metering points, and 3) it employs a least-square method to predict the sag profile along a distribution line. The results of sag state estimator can be used to calculate the feeder power quality performance indices such as the System Average RMS Frequency Index (SARFIx). Reference [46] presents a technique for accurate discrimination between transient voltage stability and voltage sag by combining damped sinusoids-based transient modeling with neural networks. Transient modeling is accomplished by energy adapted matching pursuits with an over-complete dictionary of damped sinusoids. In this approach, the information provided by the damped sinusoids-based transient modeling stage is applied to a Neural Network, which determine in a fast and accurate fashion the class to which the waveform belongs. In Reference [47] a power quality assessment method was proposed. Test results show that the method can tolerate highly distorted voltages, significant sudden frequency change, and three-phase voltage sags, but it cannot tolerate certain short-term phase-shifted single-phase voltage sags. In Reference [40] a control algorithm for the (Distributed Generation) DG interface based on the Hilbert transform (HT) is presented. The HT is employed as an effective technique for tracking the voltage flicker levels in distribution systems. The technique can be used on-line tracking of voltage flicker. The accurate tracking of the HT facilitates its implementation for the control of flicker mitigation devices. The HT realized with long filter length provides a minimum error in tracking the voltage envelope but with higher computation cost and a larger delay time than that of the short filter length. The HT can surpass the Kalman Filter in voltage tracking in a sense that it requires less computation effort and avoid the pitfalls of the FFT. In Reference [48], an algorithm for tracking the voltage envelope based on calculating the energy operator of a sinusoidal waveform is presented. This algorithm is used to evaluate the instantaneous changes in the amplitude and so track the envelope of the waveform. The algorithm is fast and robust and uses only a few samples to calculate the energy. It is not sensitive to the noise or the distortion in the waveform. The results show the capability of the algorithm to track different shapes of envelopes associated with high signal-to-noise ratio (SNR). However, there will be a delay between the actual and the tracked envelope, due to using the lead/lag networks

A prototype wavelet-based neural network classifier for recognizing power-quality disturbances is implemented and tested in Reference [50], under various transient events. The discrete wavelet transform (DWT) technique is integrated with the probabilistic neural-network (PNN) model to construct the classifier. First, the multi resolution-analysis technique of DWT and the Parseval's theorem are employed to extract the energy distribution features of the distorted signal at different resolution levels. Then, the PNN classifies these extracted features to identify the disturbance type according to the transient duration and the energy features. Various transient events are tested, such as, momentary interruption, capacitor switching, voltage sag/swell, harmonic distortion, and flicker. The results show that the classifier can detect and classify different power disturbance types efficiently. Reference [51] proposes a wavelet-based extended fuzzy reasoning approach to power-quality disturbance recognition and identification. To extract power-quality disturbance features, the energy distribution of the wavelet part at each decomposition level is introduced and its calculation mathematically established. Based on these features, rule bases are generated for extended fuzzy reasoning. The power-quality disturbance features are finally mapped into a real number, in terms of which different power-quality disturbance waveforms are classified. One prominent advantage of this approach is that the power-quality engineers' knowledge about features of the disturbance waveforms can be easily included into the algorithm using linguistic description.

Reference [52] presents a procedure for stochastic prediction of voltage sags based on the Monte Carlo method. A medium size distribution network is used to analyze, the convergence of the Monte Carlo method, the influence of protective devices and the importance of voltage sag indices is studied. This reference has presented the scope and advantages of an EMTP-based procedure for voltage sag analysis. The aim of a stochastic prediction is not only to deduce the number of voltage sags but also the number of trips of sensitive equipment. Therefore, the representation of equipment sensitivity is also required. In Reference [53] the Recursive Least Square (RLS) algorithm are introduced for the flicker envelope tracking. Both the ADALINE and the RLS algorithms are used to track the voltage envelope. The difference between the estimated envelope and the required voltage level is passed to the controller to obtain the required reactive power to compensate flickers. A fast response, accurate tracking, and robustness of the proposed control system are revealed from the results. In addition, the performance of the ADALINE and the RLS algorithms for the estimation of flicker envelopes are investigated.

Reference [54] presents and verifies a voltage sag detection technique for use in conjunction with the main control system of a DVR. A problem arises when fast evaluation of the sag depth and phase shift is required, as this information is normally embedded within the core of a main DVR control scheme and is not readily available to either user monitoring the state of the grid or parallel controllers. The voltage sag detection method in this reference proposes a matrix method, which is able to compute the phase shift and voltage reduction of the supply voltage much quicker than the Fourier transform or a PLL. DPQ sites have widely dispersed power quality. Many sites have many more sags than other sites. Rural sites have many more sags and momentary interruptions than suburban and urban sites. The three strongest indicators of voltage sags are 1) circuit exposure, 2) lightning, and 3) a term with transformer size and number of feeders. A linear model can predict sags based on a small number of site characteristics. Load density and three-phase circuit exposure most strongly affect momentary [55]. In Reference [56] a potential problem area in using RMS values in power quality assessment are identified and discussed. The RMS can be computed either using a fixed window (s-RMS) or a moving average technique (m-RMS). In both cases, RMS is a function of window length, and is a constant function for periodic signals of fundamental period.

Reference [57] presents an expert system for automatic classification of power quality recordings. The classification procedure is based on segmenting the voltage waveforms in points of sudden changes in the fundamental magnitude. Based on the segmentation results, a set of classification modules is utilized to classify the event. Classification is based on features extracted from the voltage waveforms. The system successfully classifies the largest part of the recordings. The only problems that are found are related with either the failure in detecting very small changes in the voltage magnitude or the time resolution problems of the magnitude estimation and the detection. The expert system enables fast and accurate analysis of large databases and classification of the recordings in terms of the origin. Event classification (instead of disturbance classification) offers the means for better understanding and description of the operation of the system in terms of power quality.

Reference [58] uses the continuous wavelet transform to detect and analyze voltage sags and transients. Recursive algorithm is used and improved to compute the time-frequency plane of electrical disturbances. Characteristics of investigated signals are measured on a time-frequency plane. A comparison between measured characteristics and benchmark values detects the presence of disturbances in analyzed signals and characterizes the type of disturbances. Duration and magnitude of voltage sags are measured; transients are located in the width of the signal. Furthermore, meaningful time and frequency components of transients are measured. Detection and measurement results are compared using classical methods. This algorithm enables very accurate time location and magnitude measurements of voltage sags and meaningful transient identifications. Furthermore, the method enables an accurate classification of transient events to be performed, and characteristics are easily read from the time-frequency plane. Reference [59] shows that it is possible to use sampled voltage and current waveforms to determine on which side of a recording device a disturbance originates. This is accomplished by examining the energy flow and peak instantaneous power for both capacitor energizing and voltage sag disturbances. By examining sampled voltage and current waveforms, it is possible to make a judgment as to which side of a recording device a power quality disturbance event originates. This is accomplished by examining the disturbance power and energy flow and the polarity of the initial peak of the disturbance power. If several recording devices are available in a network, the source of the disturbance may be pinpointed with a higher degree of accuracy.

2. $\alpha\beta$-Transformation for identifying the power quality events (Clarkes transformation)$\alpha\beta$-Transformation (Clarks Transformation) is a dc transformation used to transfer the three-phase ac system of voltage or current to a dc system. This section presents the application of this transformation to identifying the power quality events, such as harmonics, voltage

sages, flicker, swell and transients. This transformation can be implemented for a single phase or three- phase system of voltage. The phasor voltage resulting from the combination of V_α and V_β has a magnitude proportional to the system operating voltage and rotates with a speed of the frequency of the system voltage. The estimated shape of the phasor voltage over the data window size gives the nature of the power quality event. The proposed technique can be implemented in off-line or on-line modes. Simulated results, for different types of events are presented in the text and the results show excellent identification using the proposed technique.

2.1. Modeling the voltage signals [1]

The three phase voltages of the power systems can be written at any sample k, k=1, m, m is the total number of samples available

$$\begin{aligned} v_a(k) &= V_{ma}\sin(\omega k\Delta T + \varphi_a) + \zeta_a(k) \\ v_b(k) &= V_{mb}\sin(\omega k\Delta T + \varphi_b - 120^0) + \zeta_b(k) \\ v_c(k) &= V_{mc}\sin(\omega k\Delta T + \varphi_c + 120^0) + \zeta_c(k) \end{aligned} \tag{1}$$

Where we define the parameters in (1) as:

V_m is the signal amplitude

ω is the signal nominal frequency

ΔT is the sampling time=$\frac{1}{F_s}$,F_s is the sampling frequency

k is the sampling step; $k=1, .., m$

φ is the signal phase angle

$\zeta_a(k)$, $\zeta_b(k)$, $\zeta_c(k)$ are the noise terms which may contain harmonics

The well-known $\alpha\beta$- transformation for the three-phase signal is given as:

$$\begin{bmatrix} v_\alpha(k) \\ v_\beta(k) \end{bmatrix} = \sqrt{\frac{2}{3}} \begin{bmatrix} 1 & -0.5 & -0.5 \\ 0 & \frac{\sqrt{3}}{2} & \frac{-\sqrt{3}}{2} \end{bmatrix} \begin{bmatrix} v_a(k) \\ v_b(k) \\ v_c(k) \end{bmatrix} \tag{2}$$

Equation (2) can be rewritten as:

$$v_{\alpha}(k) = \sqrt{\frac{2}{3}}\left[v_{a}(k) - 0.5v_{b}(k) - 0.5v_{c}(k\right] \tag{3}$$

$$v_{\beta}(k) = \sqrt{\frac{2}{3}}\left[\frac{\sqrt{3}}{2}v_{b}(k) - \frac{\sqrt{3}}{2}v_{c}(k)\right] \tag{4}$$

For m samples of three-phase signals, m samples for $v_{\alpha}(k)$, $v_{\beta}(k)$are obtained. By using this transformation, harmonics of order three and their multiples are suppressed. Equation (3) and (4) gives the transformed voltage $v_{\alpha}(k)$, $v_{\beta}(k)$at any sample k. The complex voltage phasor formed from these two voltages which having the same frequency ω and phase angle φ as the original three phases is:

$$\begin{aligned} v(k) &= v_{\alpha}(k) + jv_{\beta}(k) \\ &= V(k)e^{j(\omega k\Delta T + \varphi)} \end{aligned} \tag{5}$$

Where the amplitude *V (k)* of the complex signal is calculated at any sample *k* as:

$$V(k) = \left[v_{\alpha}^{2}(k) + v_{\beta}^{2}(k)\right]^{\frac{1}{2}} \tag{6}$$

and the rms magnitude is given as:

$$V = \frac{1}{m}\sum_{k=1}^{m}\left[v_{\alpha}^{2}(k) + v_{\beta}^{2}(k)\right]^{\frac{1}{2}} \tag{7}$$

Equation (6) gives the value of phasor voltage at any instant *k*, while equation (7) is used to calculate the rms value of the phasor voltage.

The phase angle of this transformed voltage is:

$$\theta(k) = \tan^{-1}\frac{v_{\beta}(k)}{v_{\alpha}(k)} \tag{8}$$

2.2. Testing the algorithm

The proposed algorithm explained in the previous section is used to identifying the power quality events for a three phase balanced voltage. These are including voltage swell, voltage flicker, momentary interruption, voltage sags, and harmonics and voltage transients due to switching on/off capacitors used in reactive power compensation.

2.2.1. *Voltage swell*

In this test a voltage swell is assumed on the system, over voltage, for a short period. The rms value for the complex phasor is calculated, using equation (7) as well as the phasor phase angle in radian is computed using equation (8)

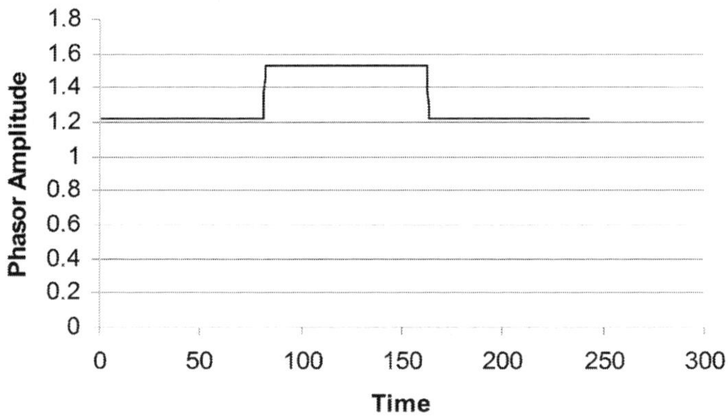

Figure 1. A system of three-phase voltage experiences a voltage swell.

Figure 1 gives the amplitude, rms value and the phase angle of the phasor for a system of three-phase voltage experiences a voltage swell. It can easily be noticed, from the phasor amplitude that the voltage is experienced a voltage swell of magnitude 1.53, normal voltage is 1.224 p. u, during the time that equivalent to sample number 83 to the time that equivalent to sample number 163.

2.2.2. *Voltage flickers*

In this test we assume that the three phase voltages are experienced a flicker the complex phasor of the voltage is extracted from the available samples for the three phase voltages.

Figure 2 gives the results obtained. Examining this figure reveals that this amplitude is modulated amplitude during the data window size under investigation, i.e. looking to this figure one can conclude that a flicker signal is imposed on the nominal voltage signal.

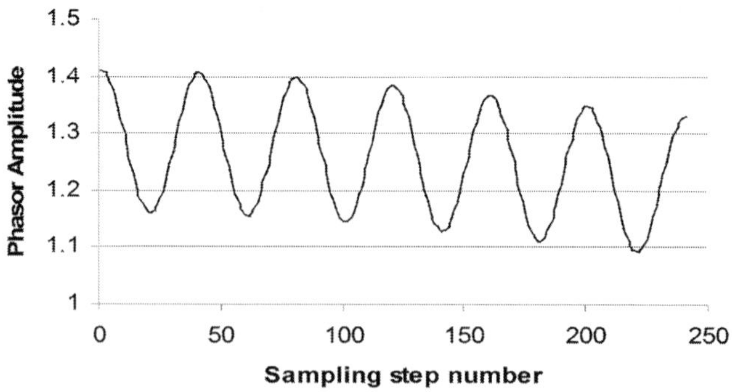

Figure 2. Phasor amplitude for a voltage flicker

2.2.3. Momentary interruption

A momentary interruption for a small period of time is implemented in this test, and then the voltage is restored again. Figure 3 gives the phasor amplitude when the system voltage experienced a momentary interruption. Examining this curve reveals that the proposed algorithm is succeeded in estimating the voltage amplitude and the period of interruption.

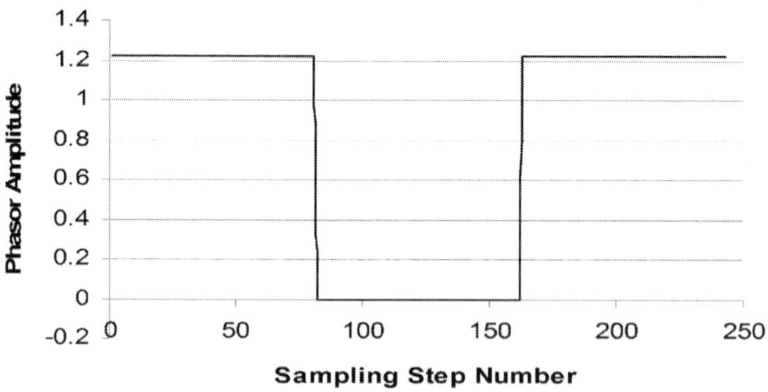

Figure 3. Phasor Amplitude for a momentary interruption

This figure also reveals that the phasor amplitude tends to zero from the time equivalent to the samples between 85 and 162, the time of interruption is (162-85)*sampling time, and after the sample number 162 the system voltage is restored again to nominal voltage.

2.2.4. Voltage sags

In this test voltage is assumed to experience voltage sag, where the voltage is reduced to 0.6 pu, the voltage phasor amplitude of equation 6 is calculated. Figure 4 gives the results obtained, examining this curve reveals that the proposed algorithm is succeeded in estimating the voltage amplitude as well as the sags period.

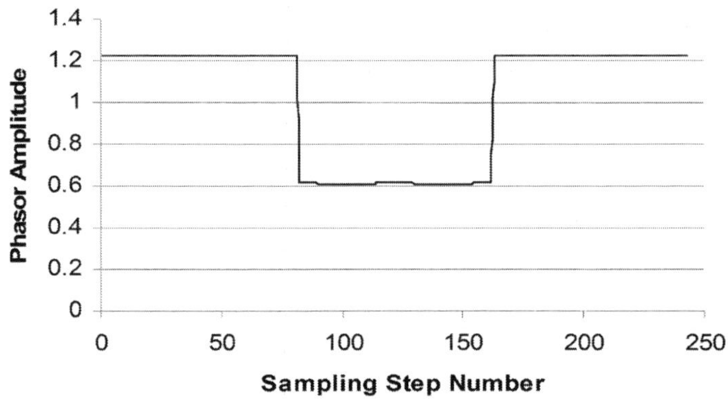

Figure 4. Phasor amplitude for a Voltage experiences voltage sag

2.2.5. Voltage transients

In this test the three-phase voltage signal is experienced voltage transient, this may come from faults and/or lightning. The phasor amplitude is calculated using the proposed algorithm.

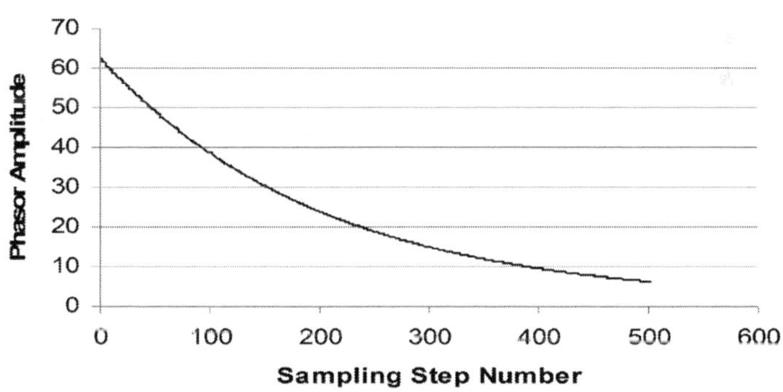

Figure 5. Phasor amplitude for voltage experiences voltage transients

Figure 5 depicts the results obtained. Examining this curve reveals that the proposed transformation is succeeded in estimating the voltage amplitude as well as the transient period.

2.2.6. Harmonics

The voltage signal in this test is assumed to be polluted by harmonics up to the fifth order, and the proposed algorithm is used to estimate the voltage phasor amplitude. Figure 6 depicts the results obtained. If we examine this figure we can conclude that the phasor amplitude is a harmonically polluted one, and is following exactly the system voltage.

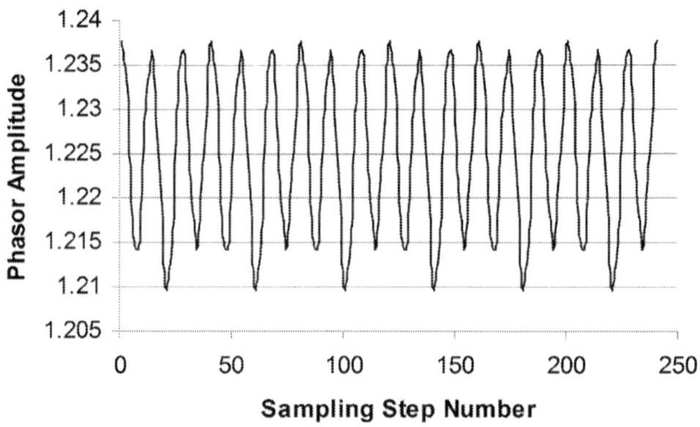

Figure 6. Voltage signal polluted with harmonics

2.3. Conclusion

We present, in this section, the application to $\alpha\beta$-Transformation for identifying and discriminating between different types of power quality events. Simulated examples for these events are presented in the body of the text and more results for the nature of the transformed voltages are given. It has been shown, through extensive runs, that if one of the phase is experienced a power quality event alone while the others are not, the algorithm is succeeded in identifying this events which is not the case of the other earlier algorithms in the past.

3. Kalman filtering algorithm

Power quality involves how close the voltage waveform is to being a perfect sinusoid with a constant frequency and amplitude. In this section, we review the applications of linear Kalman filtering algorithm for on-line electric power quality analysis. These applications include the measurement of harmonics and voltage sags. Mathematical models for each problem are developed to suite the filter and tested using simulated examples.

3.1. Harmonics model

The voltage signal can be written as Fourier series as:

http://dx.doi.org/10.5772/53081

$$v(t) = \sum_{n=0}^{N} \sqrt{2}V_n \cos(n\omega t + \phi_n) \tag{9}$$

A d.c component may exist in the voltage signal, this occurs at n=0. Equation (9) can be written, for a specific number of harmonics, as:

Equation (9) can be rewritten as:

$$\begin{aligned} v(t) = V_o + \sqrt{2}V_1 \cos\phi_1 \cos\omega t - \sqrt{2}V_1 \sin\phi_1 \sin\omega t + \sqrt{2}V_2 \cos\phi_2 \cos 2\omega t - \sqrt{2}V_2 \sin\phi_2 \sin 2\omega t \\ + \sqrt{2}V_3 \cos\phi_3 \cos 3\omega t - \sqrt{2}V_1 \sin\phi_3 \sin 3\omega t \end{aligned} \tag{10}$$

In equation (10) we assume that the voltage signal is contaminated up to the third harmonic. But the incoming steps are generally enough to apply for any number of harmonics.

Define the following states as :

$$\begin{aligned} X_o &= V_o \\ X_1 &= \sqrt{2}V_1 \cos\phi_1 \\ Y_2 &= \sqrt{2}V_1 \sin\phi_1 \\ X_2 &= \sqrt{2}V_2 \cos\phi_2 \\ Y_2 &= \sqrt{2}V_2 \sin\phi_2 \\ X_3 &= \sqrt{2}V_3 \cos\phi_3 \\ Y_3 &= \sqrt{2}V_3 \sin\phi_3 \end{aligned} \tag{11}$$

Moreover, define the following coefficients as:

$h_o = 1$

$h_1(k\Delta t) = \cos 2\pi fk\Delta t$

$h_2(k\Delta t) = -\sin 2\pi fk\Delta t$

$h_3(k\Delta t) = \cos 4\pi fk\Delta t$

$h_4(k\Delta t) = -\sin 4\pi fk\Delta t$

$h_5(k\Delta t) = \cos 6\pi fk\Delta t$

$h_6(k\Delta t) = -\sin 6\pi fk\Delta t$

Equation (10) can be written as:

$$v(k\Delta t) = X_o + X_1 h_1(k\Delta t) + Y_1 h_2(k\Delta t) + X_2 h_3(k\Delta t) + Y_2 h_4(k\Delta t) + X_3 h_5(k\Delta t) + Y_3 h_6(k\Delta t) \tag{12}$$

In equation (12) k=1,2,..,m, and Δt is the sampling time.

Equation (12) can be written as an observation equation as:

$$\begin{bmatrix} v(\Delta t) \\ v(2\Delta t) \\ \cdot \\ \cdot \\ \cdot \\ \cdot \\ v(m\Delta t) \end{bmatrix} = \begin{bmatrix} 1 & h_1(k\Delta t) & h_2(k\Delta t) & h_3(k\Delta t) & h_4(k\Delta t) & h_5(k\Delta t) & h_6(k\Delta t) \\ 1 & h_1(2k\Delta t) & h_2(2k\Delta t) & h_3(2k\Delta t) & h_4(2k\Delta t) & h_5(2k\Delta t) & h_6(2k\Delta t) \\ & & & & & & \cdot \\ \cdot & \cdot & \cdot & \cdot & \cdot & \cdot & \cdot \\ & & & & & & \cdot \\ 1 & h_1(m\Delta t) & h_2(m\Delta t) & h_3(m\Delta t) & h_4(m\Delta t) & h_5(m\Delta t) & h_6(m\Delta t) \end{bmatrix} \begin{bmatrix} X_o \\ X_1 \\ Y_1 \\ X_2 \\ Y_2 \\ X_3 \\ Y_3 \end{bmatrix} \tag{13}$$

Equation (13) can be written in vector form as:

$$\underline{Z} = H\underline{\theta} + \underline{\xi} \tag{14}$$

Where $\underline{Z}$ is an $m \times 1$ samples of voltage, H is an $m \times 7$ measurement matrix which can be calculated off line, $\underline{\theta}$ is 7×1 parameters vector to be identified and $\underline{\xi}$ is $m \times 1$ errors vector to be minimized. Note that the order of H matrix as well as the vector X depends on the number of harmonic expected to contaminate the voltage signal.

We assume that the system state equation is given by

$$\underline{\theta}(k+1) = \phi\,\underline{\theta}(k) + \underline{\upsilon}(k) \tag{15}$$

Where φ is 7×7 state transition matrix, and it is assumed to be identity matrix. Moreover, $\underline{\upsilon}(k)$ is 7×1 errors vector associated with the states when they move from stage k to k+1. Now equations (14) and (15) are suitable for Kalman filter application

3.2. Computer experiments

3.2.1. Harmonics identification

The voltage signal in this test is assumed to be contaminated with harmonics, 3rd, 5th, and 7th order. The voltage signal in this case is given as:

$$v(t)=\sqrt{2}\left[\cos(\omega t+60^{o})+0.25\cos(3\omega t+15^{o})+0.15\cos(5\omega t-10^{o})+0.05\cos(7\omega t)\right]$$

Where $\omega = 2\pi f$, f=50 Hz, and it is assumed to be constant during the estimation process. This signal is sampled at frequency of 2000 Hz, 40 samples per cycle.

Figure 7 shows the estimated fundamental, 3rd, 5th components when the voltage signal is modeled using up to the 7th harmonic. While Figure 8 shows the estimated phase angle for each harmonics component including the fundamental component. The following observation can be concluded from these two Figures:

- Kalman filtering algorithm is succeeded in estimating the magnitude of harmonics components, as well as the phase angle of each component, providing that the voltage signal is accurately modeled
- It has been shown through extensive runs that the sampling frequency and number of samples do not affect the estimates of the magnitudes and phase angles of each harmonics component. Providing that the sampling frequency satisfies the sampling theory.

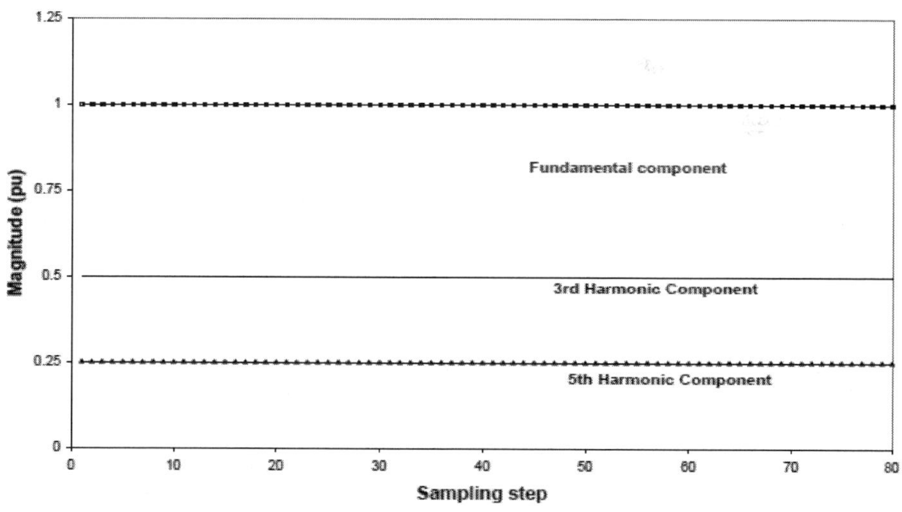

Figure 7. Estimated Magnitudes

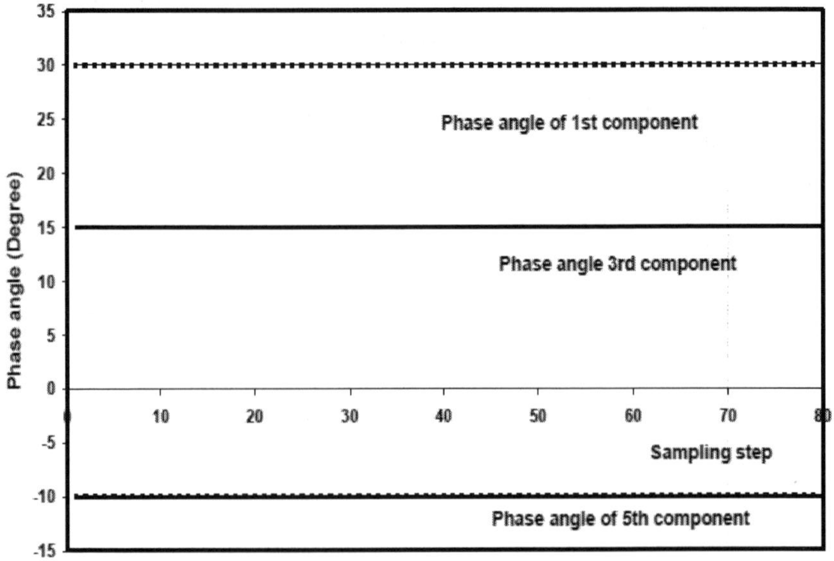

Figure 8. Estimated phase angle

Another test is conducted, where we modeled the voltage signal in the observation equation to take only the fundamental component i.e we assume that the voltage signal does not contaminated with harmonics, but in reality, it does. Figure 9 and 10 gives the estimated fundamental magnitude and its phase angle respectively.

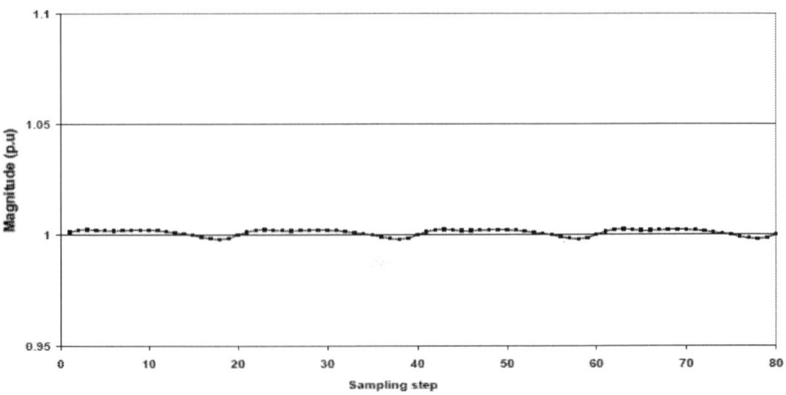

Figure 9. Estimated Fundamental Magnitude

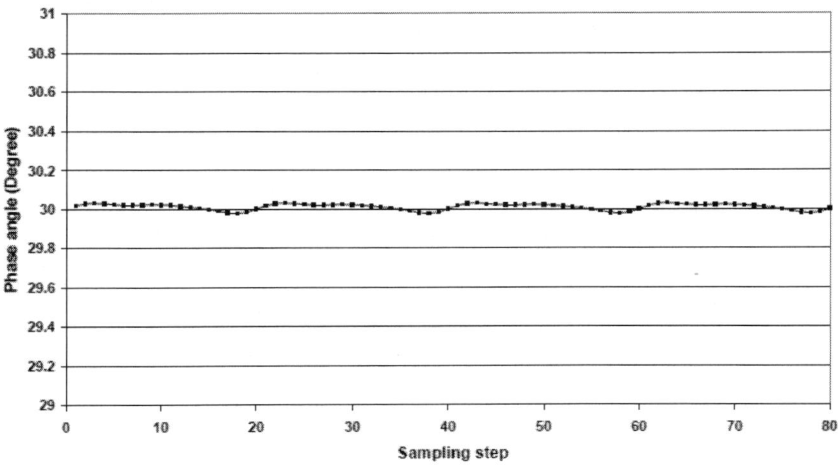

Figure 10. The estimated fundamental phase angle

Examining these curves reveals the following remarks:

- Although the fundamental component is the only one modeled in the waveform, good estimates are obtained for both amplitude and phase angle.
- The proposed algorithm can easily estimate the fundamental component and its phase angle using 20 samples, since the estimates are periodical, i.e a half cycle data window size is enough.

3.2.2. Voltage sags

3.2.2.1. A pure sinusoidal voltage signal

The source voltage in this case takes the following waveform equation:

$$v(t) = \sqrt{2}V_s \cos(\omega_o t + 60^o) \tag{16}$$

Where ω_o is the nominal frequency = $2\pi f_o$, f_o=50 Hz, V_s is the value of the voltage during the period of measurements, presage value, during sage and post sage voltage. In this simulation, we assume the following values for this voltage

$$V_s = \begin{cases} 1.0 & 0 \le t \le T_s \\ 0.2 & T_s \le t \le T_f \\ 1.0 & T_{f \le t \le T} \end{cases}$$

where T_s is the time at which the voltage sag starts, while T_f is the time at which the voltage sage restrain. T is the time at which the measurements end up. Figure 11 shows such type of voltage signal.

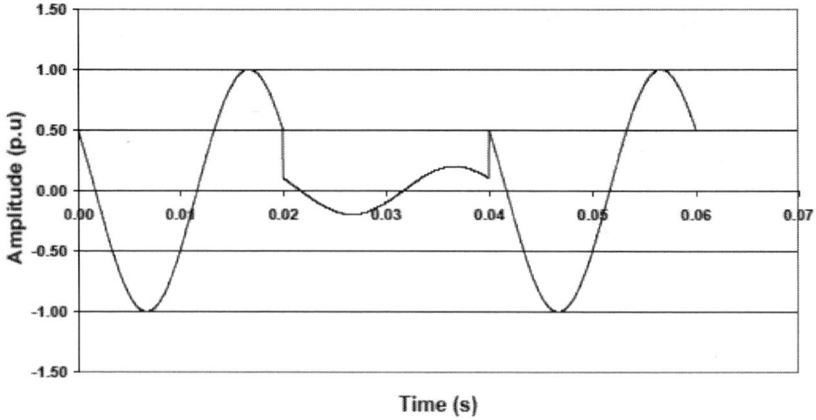

Figure 11. Voltage Sag

This signal is sampled at a sampling frequency of 40 kHz with a one-cycle data window size for the three stages. The proposed technique, Kalman filter algorithm, is implemented to estimate the voltage magnitude as well as the phase angle.

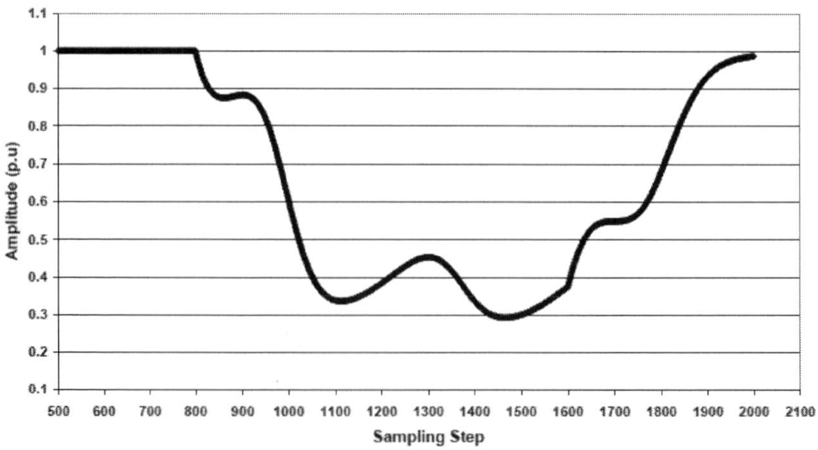

Figure 12. The estimated Voltage Amplitude (p.u)

Figure 12 gives the estimated rms voltage during the three stages, while Figure 13 gives the estimated phase angle. Examining the two curves reveals the following remarks;

- The proposed algorithm estimates the voltage amplitude accurately in the first cycle and with a fair accuracy during the voltage sages and after the voltage sages
- One cycle data wind size during voltage sages is not enough to reach the actual voltage sages at least we need more than four cycles. This will be discussed in the next test

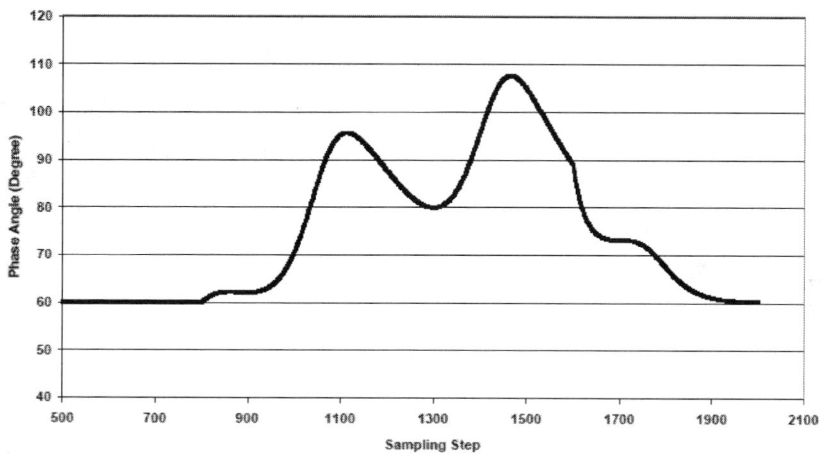

Figure 13. The estimated phase angle

- There is a phase jump during the voltage sags period, and the proposed algorithm estimates the phase angle accurately during pre voltage sages and after voltage sages.

Another test is conducted in the same voltage signal, where we increase the voltage sages' period to three cycles. Figure 14 gives such waveform

The estimated voltage amplitude and the phase angle are given in Figure 15, 16. Examining these two curves reveals that:

- Kalman filter is succeeded in identifying the period, where the voltage sags are occurred, as well as the amplitude of the voltage sags with an acceptable accuracy
- Increasing the number of cycles for the voltage sags makes the filter able to identify pre, during and after sags amplitudes with acceptable accuracy.
- It has been shown, as a common practice, that there is phase jump always occurs during the period of voltage sags

Another test is performed in this study, where we assume that the voltage signal contaminates with third and fifth harmonics with magnitude 0.25 and 0.125 respectively. Figure 17 gives the

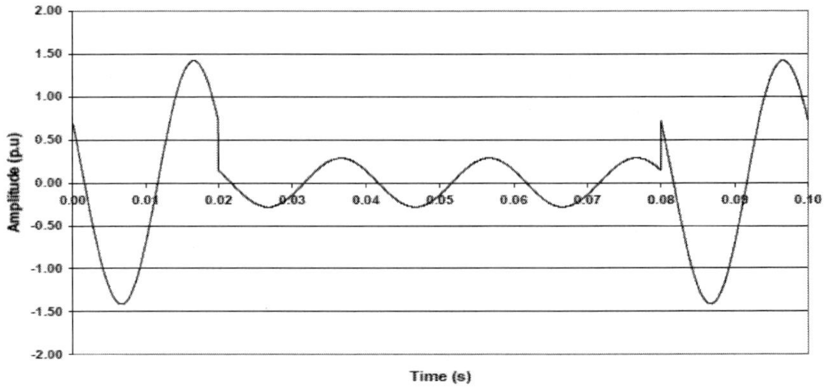

Figure 14. A Voltage sag

waveform of this voltage signal, while Figures 18, 19 give the estimated rms value of the fundamental component and its phase angle.

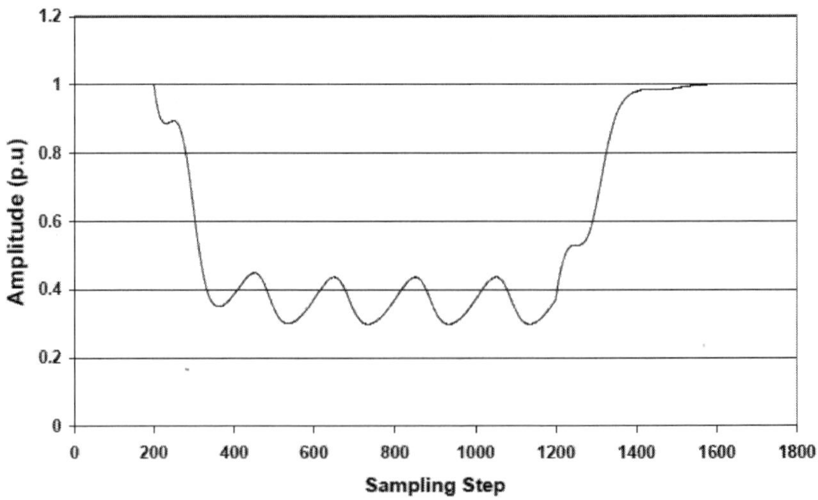

Figure 15. The estimated voltage amplitude

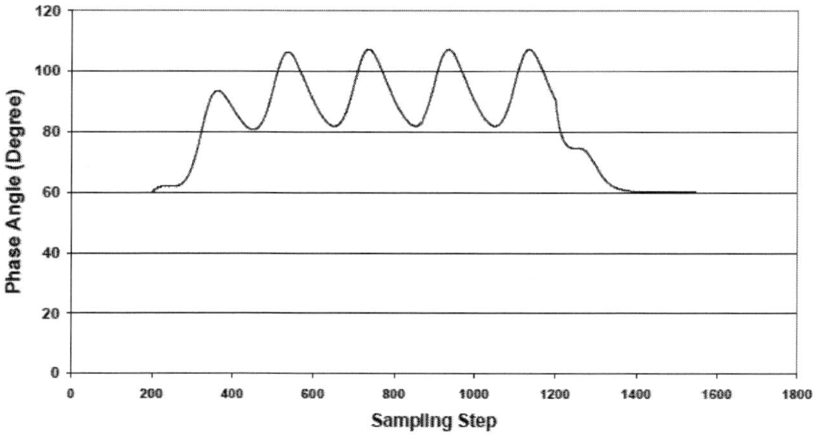

Figure 16. The Estimated Phase Angle

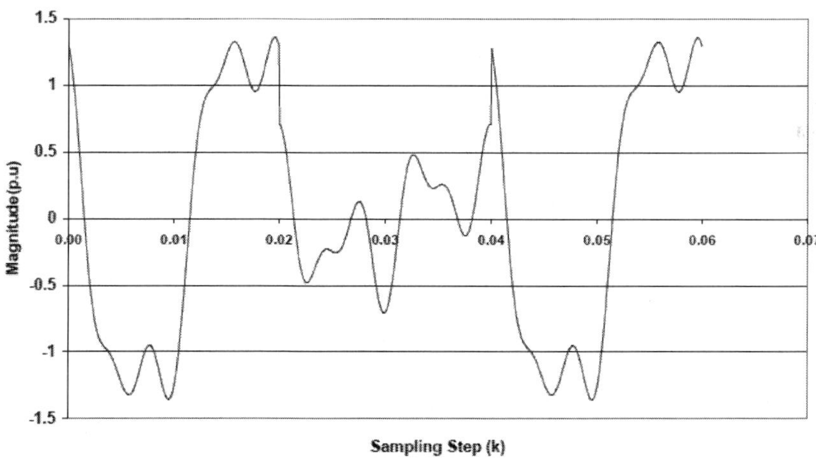

Figure 17. A Non-sinusoidal voltage experienced voltage sag

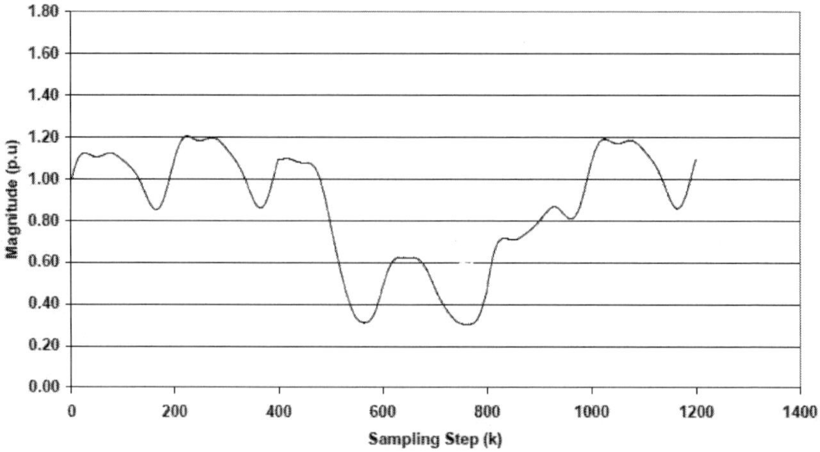

Figure 18. Estimated Fundamental Component

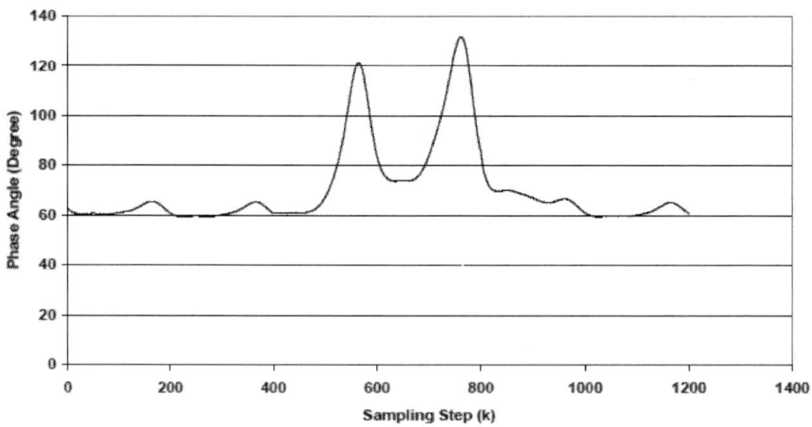

Figure 19. Estimated Fundamental Phase Angle

Examining these curves reveals that:

- Although the voltage signal is contaminated with harmonics, the proposed algorithm is succeeded to estimate the amplitude of the fundamental components in pre, during and after voltage sags with fair accuracy.
- The proposed technique estimates exactly the period of the voltage sags,

- The phase angle jump during the voltage sags, in this case, is greater than that of non-harmonics contamination.

3.3. Conclusions

The main contributions of this section are:

- Kalman filtering algorithm is applied to estimate the harmonic components in a voltage signal of a power system. A full spectrum for identification of the fundamental component is not necessarily. Since the proposed algorithm estimates this component with a good accuracy.
- The proposed algorithm is succeeded in estimating the magnitude of the voltage signal that experienced voltage sag. Good estimates are produced for the voltage signal before, during and after the voltage sags.
- An accurate estimate for the phase angle and the phase angle jump is produced using the proposed algorithm.
- The proposed algorithm produces good estimate for the period of the voltage sags. The starting and ending time of the voltage sags is shown in the Figures within the text.
- The proposed Kalman filtering technique does not affect by the number of samples for harmonics measurements. A small data window size can provide all the information for harmonics components. Furthermore, the sampling frequency has no effect on the estimated components. Providing that the sampling theorem be satisfied.

4. Parks' transformation

Park's transformation is a well known transformation used in the analysis of electric machines, where the three rotating phases abc are transferred to three equivalent stationary dq0 phases (d-q reference frame). In this section this transformation is implemented to recognizing and classifying the power quality events, either for three-phase or single phase circuits. The proposed algorithm transferred the utility signal to a complex phasor. The magnitude of this phasor depends on the magnitude of the utility signal either a three-phase or a single phase signal. This technique produces the complex phasor loci that depend on the power quality event; voltage sags, voltage flickers, voltage swell and harmonics. The time of starting the disturbance is chosen randomly and the length of disturbance is arbitrary. Implementation of this technique is succeeded in recognizing and classifying the power quality events. Simulated results are presented, for three-phase and single phase events. wi

4.1. Identification processes

In the following steps we assume that m samples of the three phase currents or voltage are available at a pre-selected sampling frequency that satisfying the sampling theorem.

The forward transformation matrix

$$P = \sqrt{\frac{2}{3}} \begin{bmatrix} \sin \omega t & \sin(\omega t + 120) & \sin(\omega t + 240) \\ \cos \omega t & \cos(\omega t + 120) & \cos(\omega t + 240) \\ \frac{1}{\sqrt{2}} & \frac{1}{\sqrt{2}} & \frac{1}{\sqrt{2}} \end{bmatrix} \quad (17)$$

The matrix given in equation (17) can be computed off line if the frequency of the voltage and/or current signals as well as the sampling frequency and the number of samples are known in advance. If the matrix given in equation (17) is multiplied digitally by the samples of the three-phase voltage that are sampled at the same sampling frequency of matrix (17), a new set of three -phase samples are obtained, we call this set a *dqo* set (reference frame). If we use only the samples for the two perpendicular phases d and q, then the resulting phasor at a sample *k* is given by;

$$V(k) = V_d(k) + jV_q(k) \quad (18)$$

The magnitude of this phasor is given by

$$V(k) = \sqrt{V_d^2(k) + V_q^2(k)} \quad (19)$$

While the phase angle is given by

$$\phi(k) = \tan^{-1}\frac{V_q(k)}{V_d(k)} = \omega\, k\Delta T + \delta(k) \quad (20)$$

In this study we may assume, during the power quality events, that the frequency of the voltage signal is constant and equals to the nominal frequency of the system, 50/60 Hz. However if this frequency is not constant, it can be calculated from equation (20) using the least error squares estimation algorithm. Equation (19) is a good indicator to the status of the system signal. Where *V(k)*is plotted with against the time, and every event has a form that different from other events. If the relation between $V_d(k)$ and $V_q(k)$ a loci is produced that describes the event. In the next section a computer simulation is carried out for different types of power quality events.

4.2. Computer simulation

In this study we assume that the voltage signal frequency is constant at 50Hz during the events, and a sampling frequency of 10000 Hz is used (($\Delta T = 0.1ms$), 200 samples per cycle and 500 samples are used. Before the event 200 samples, 1cycles, are used as pre-estimated period and 100 samples, 0.5 cycles as the event period, and finally 200 samples, 1 cycles, are used as after event period. Different types of power quality events are simulated, these including voltage flickers, voltage sags, voltage swell, momentary interruption and finally voltage harmonics.

4.2.1. Voltage sags

In the first part of this study we assume that one of the phases has only voltage sag, where the voltage is dropped from 1 p.u to 0.25 p.u, single- phase sag. While in the second study, the three phases have the same amount of sags, where the voltage in the three phases is dropped from 1.0 p.u to 0.25 p.u. Figure 20 to 22 gives the results obtained for the single phase sag.

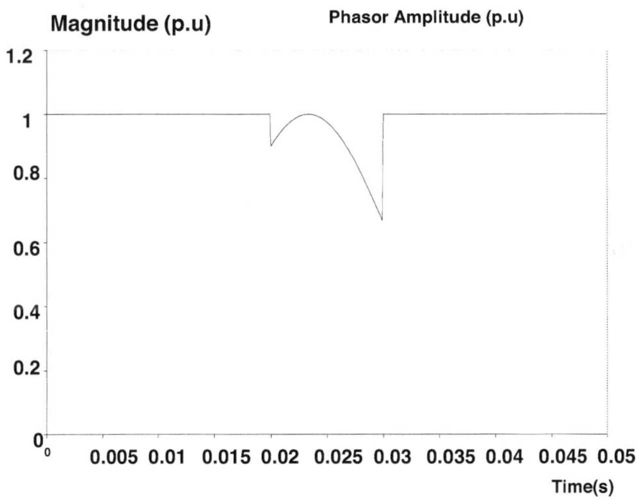

Figure 20. Variation of the phasor magnitude Single Phase voltage sags,

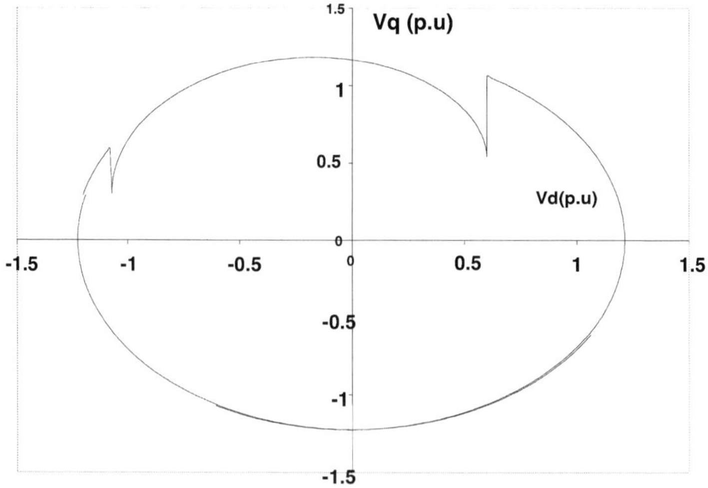

Figure 21. Locus of V_d and V_q for a single phase sag

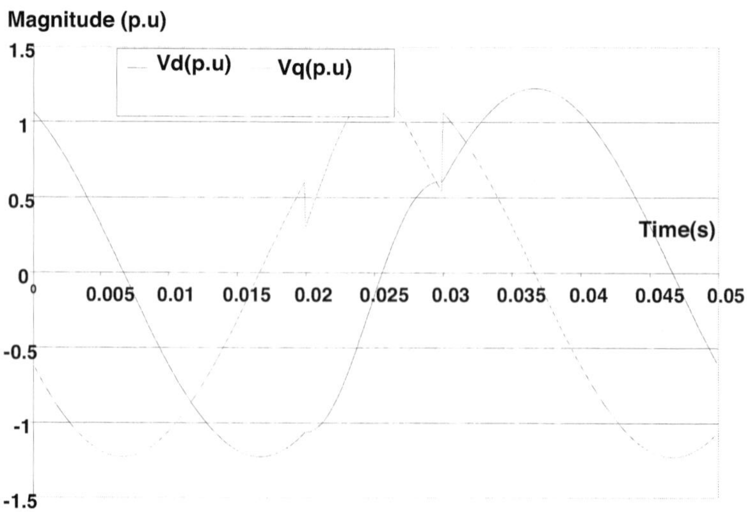

Figure 22. Waveforms for V_d and V_q for single phase sag.

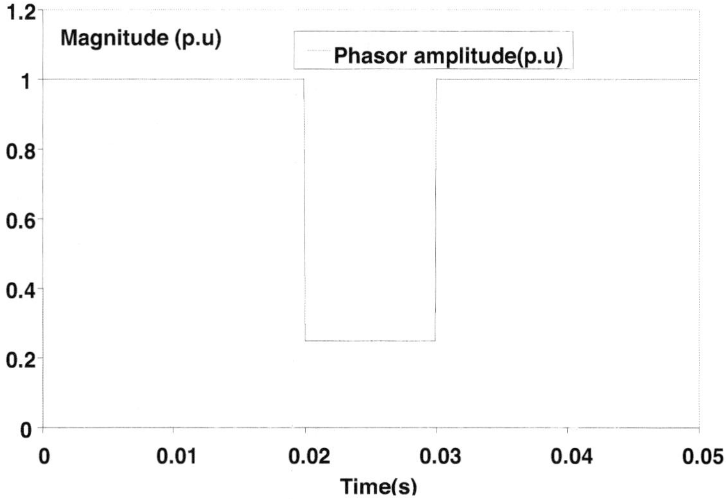

Figure 23. Variation of the phasor magnitude three- Phase voltage sags,

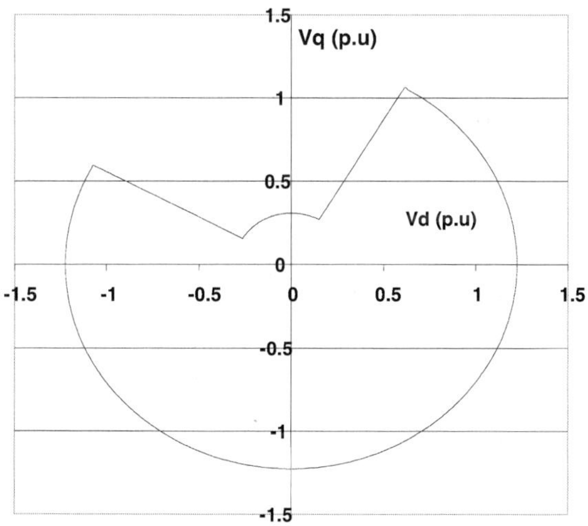

Figure 24. Locus of V_d and V_q for three- phase sag

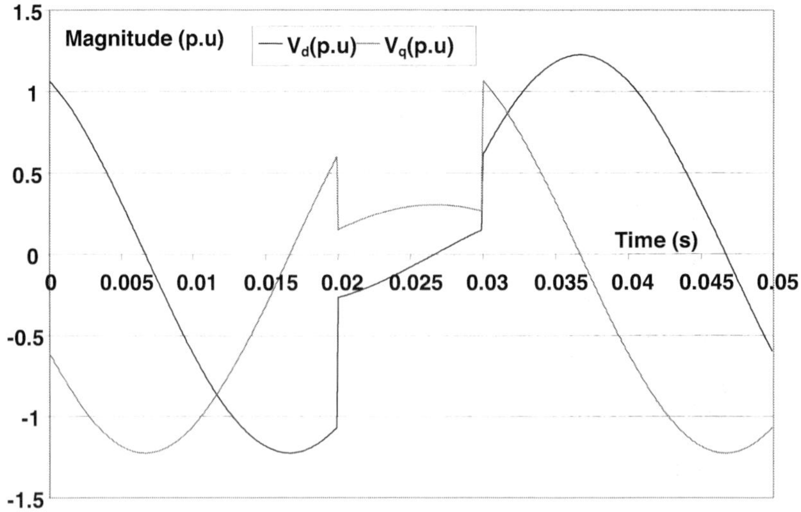

Figure 25. Waveforms for V_d and V_q for three-phase sag.

Examining these figures reveals:

- For the single phase sage, during the sage period the rms value of the voltage is dropped to a value that does not equal to the voltage magnitude at that period, but it does for three -phase voltage sag. Figures 20 and 26. experienced a voltage sage, while in the This is due to the nature of the transformation, where in the single pahse sags one of the phases is only three phase, the three phases are experienced the voltage sags. It does mean that during the sage period, for single phase, two phases are balanced while the third one is distorted, while in the three- phase the three phases are distorted. This is clearly indicated in Figures 21 and 23
- Figures 22 and 25 indicate that the variation of the two voltages V_d *and* V_q in the time domain is not the same for single phase sags, but they are the same for the three phases. In the single phase V_q is a little bit distorted during the sag period while for three -phase sag V_d*and* V_q are large distorted during the sag period.

4.2.2. Voltage flicker

In the first part of this study we assume that one of the phases has only voltage flicker, single phase flicker. While in the second study, the three phases are experienced the same voltage flicker, Figure 26 to 9 gives the results obtained for the single phase flicker while Figures from 19 to 21 give the results obtained for the three phase voltage flicker, where the three phases are assumed to experience the same voltage flicker. Examining these curves reveals the following remarks:

- The phasor amplitude in figure 26 has an amplitude modulation during the data window size
- The locus of V_d and V_q in Figure 27 is not a pure circle as it should be, but it is an ellipse with distorted amplitude.
- The signals waveform for the two voltages V_d and V_q are distorted signals.

Indeed looking to Figure 26, one can notice that the system experiences a voltage flicker

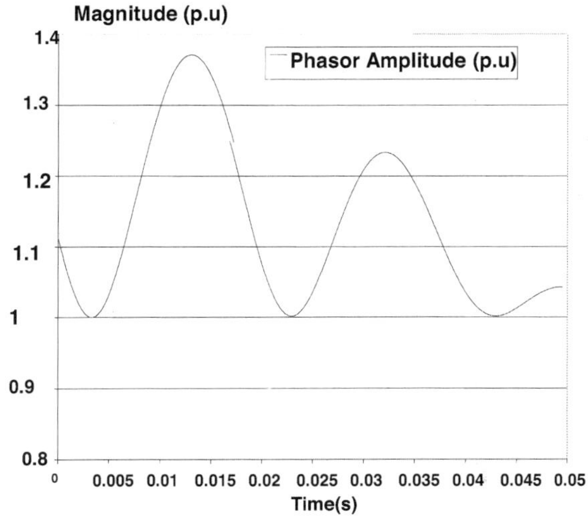

Figure 26. The phasor voltage in the Time domain for a single phase flicker

Examining figures 29 to 31, one can conclude the following remarks:

- The phasor amplitude decreases down during the data window size, and looking to this figure one can notice that the system experiences voltage flicker.
- The locus of V_d and V_q is an open ellipse despite of the single phase voltage flicker where this locus is a distorted closed ellipse
- The wave form of the voltage V_d and V_q is an amplitude modulated wave.

General speaking, for voltage flicker, the amplitude of the transformed signals is demodulated amplitude with amplitude greater than one per unit.

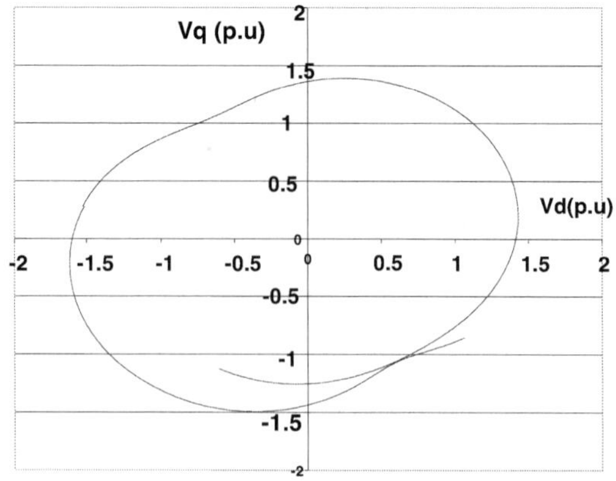

Figure 27. Locus of V_d and V_q for a single phase flicker

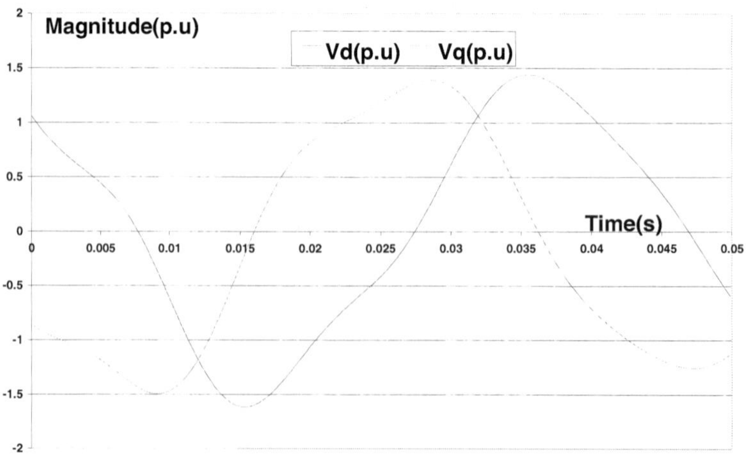

Figure 28. V_d and V_q in the time domain for a single phase flicker

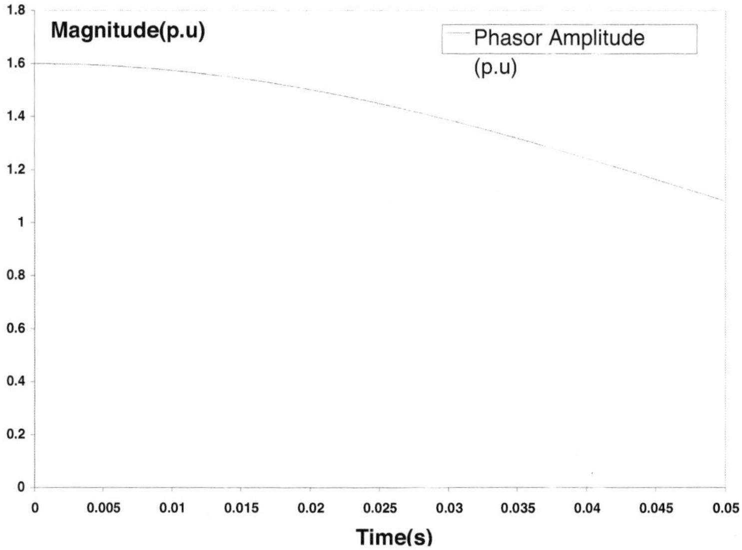

Figure 29. Phasor magnitudes in the time domain for a three phase flicker

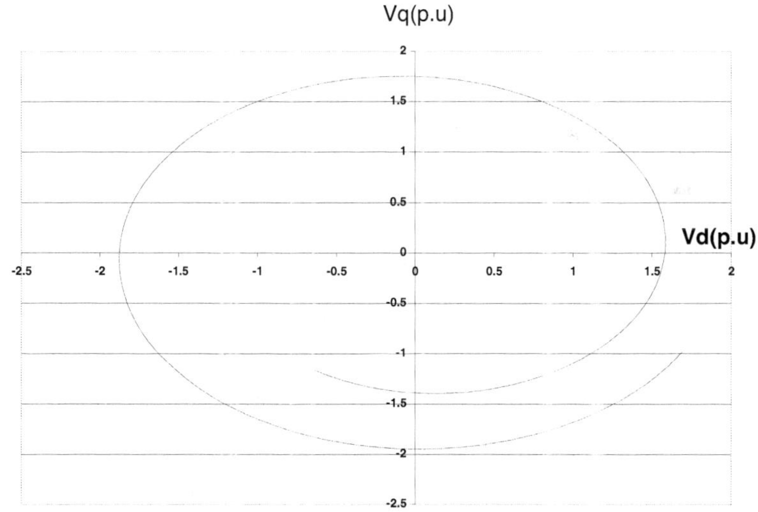

Figure 30. Locus V_d and V_q for a three phase flicker

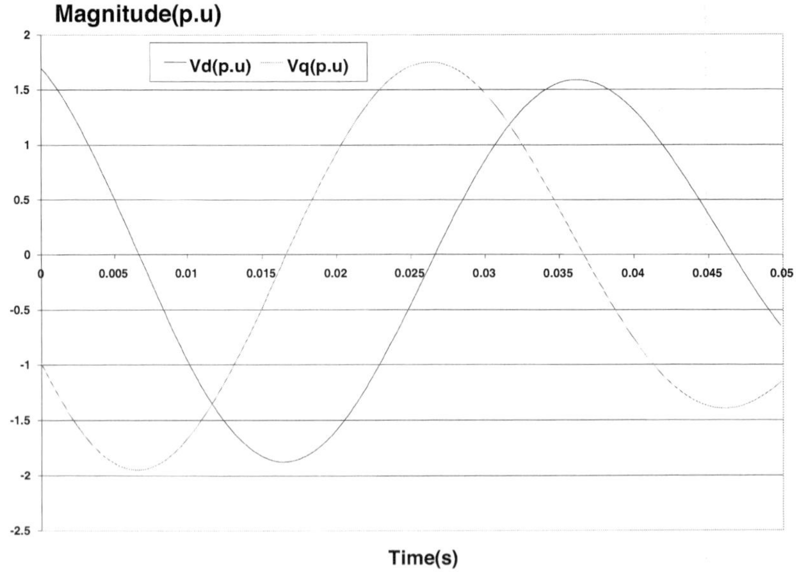

Figure 31. V_d and V_q in the time domain for a three-phase flicker

4.2.3. Voltage swell

In this test a single phase swell is implemented on the voltage signal, where the voltage is increased 50 percent more for a short period, about 10ms. Figures 32, 33 and 34 give the results obtained. Examining these figures one can reveal the following remarks:

- Examining Figure 32, the phasor amplitude equal to 1 per unit in the pre-swell period, then it increases during swell period and then comes back again to a value of one per unit.
- The locus of V_d and V_q is not a pure circle, but it is a distorted circle during the time of voltage swell.
- The signal for V_d and V_q is not pure sinusoids, especially during the time of swell.

In the second part of the test an equal swell is implemented for the three phase signals, and V_d and V_q are calculated. Figures 35, 36 and 37 give the results obtained. Examining these figures reveals the following

- Looking to Figure 35, immediately one can notice that the voltage has a swell with a magnitude of 1.5 p.u during the swell period
- The locus of V_d and V_q is not a pure circle, but it is a distorted circle during the time of voltage swell.

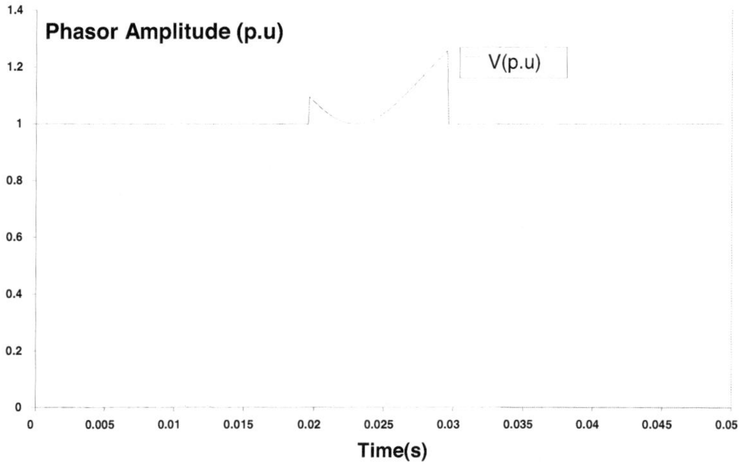

Figure 32. Phasor Magnitude for single phase swells

- The signal for V_d and V_q is not pure sinusoids, especially during the time of swell.

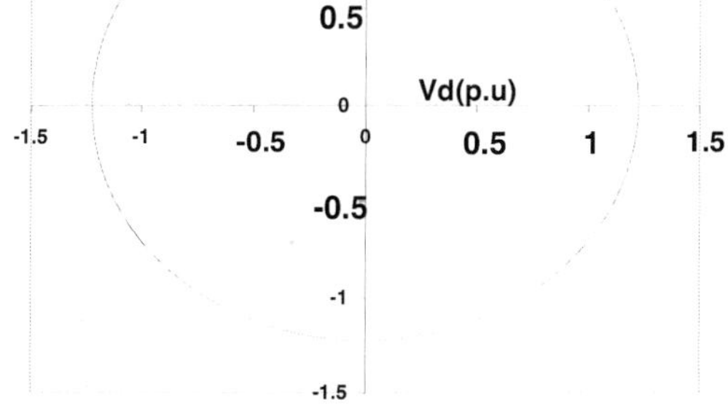

Figure 33. Locus diagram for a single phase swell

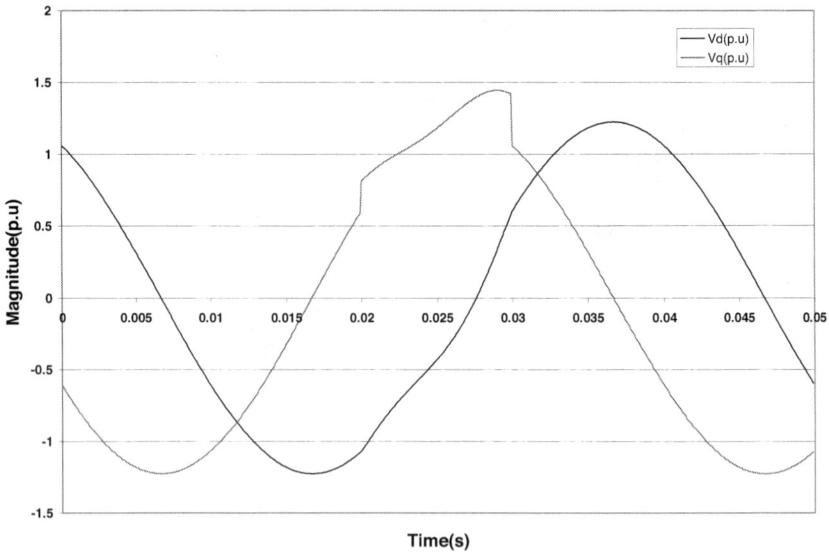

Figure 34. V_d and V_q in the time domain for a single phase swell

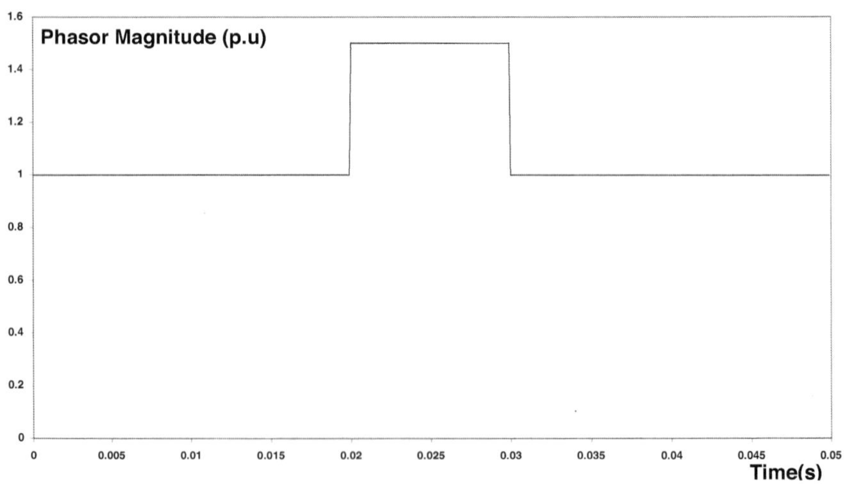

Figure 35. Phasor magnitudes in the time domain for a three phase swell

http://dx.doi.org/10.5772/53081

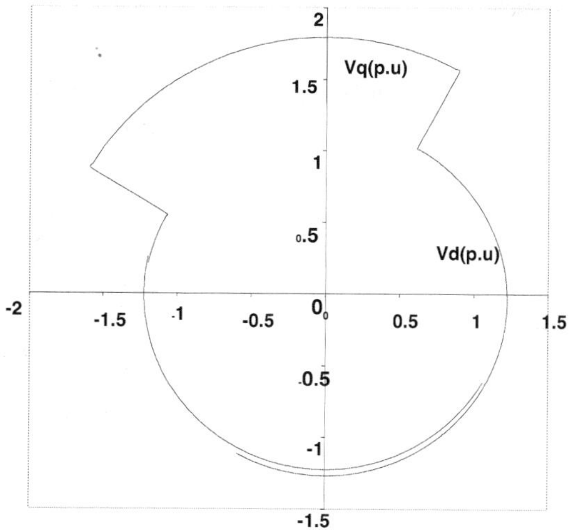

Figure 36. Locus diagram for a three phase swell

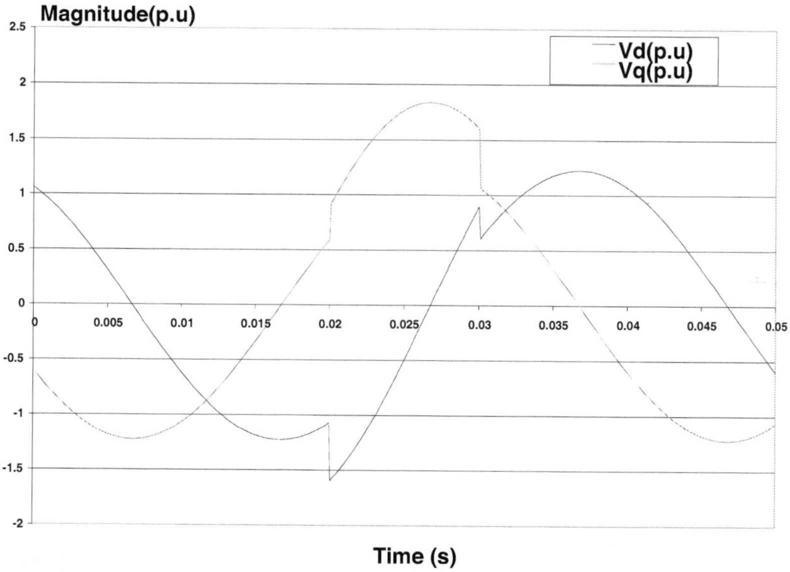

Figure 37. V_d and V_q in the time domain for a three phase swell

4.2.4. Voltage harmonics

In this test the voltage signal is contaminated with third harmonics only. In the first part of the test we assume that one phase is only contaminated with harmonics, while the other two phases are not. Figures 38 to 40 gives the results obtained. Examining this curves one can notice the following:

- The phase magnitude in the time domain is a time varying magnitude and it is not a pure sinusoidal.
- The locus diagram is a distorted locus not a pure circle as it should be for uncontaminated phase.
- The voltages V_d and V_q in the time domain are not sinusoidal signals, but they are distorted.

In the second part of this test, we assume that the three phases are contaminated with the same order of harmonics, a balanced three phase harmonics contaminated system. Figures 41 to 43 give the results obtained. Examining these curves reveals the following remarks:

- Phasor magnitudes in the time domain for three- phase harmonics contamination is a pure sinusoidal with amplitude greater than one per unit, maximum value is 1.2447 per unit and the minimum value is 0.75 per unit and the average value between the two peaks is about one per unit.
- The locus diagram for three- phase harmonics contamination is a symmetrical shape about both axes
- Vd and Vq in the time domain for three- phase harmonics contamination are harmonics contaminated signals, distorted signals

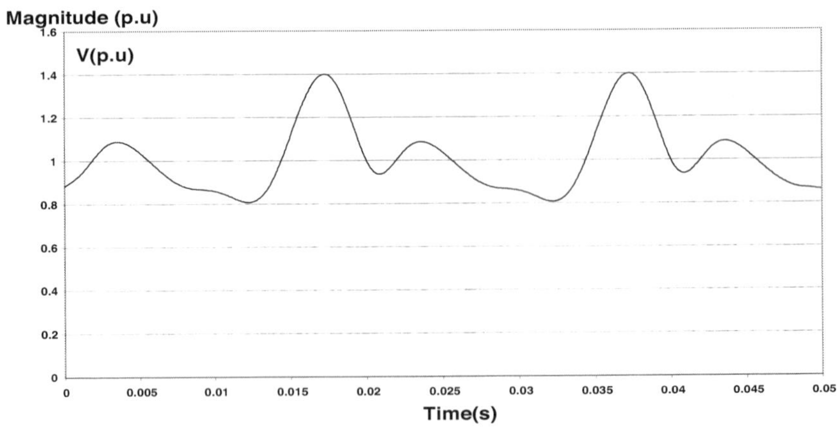

Figure 38. Phasor magnitudes in the time domain for a single phase harmonics contamination

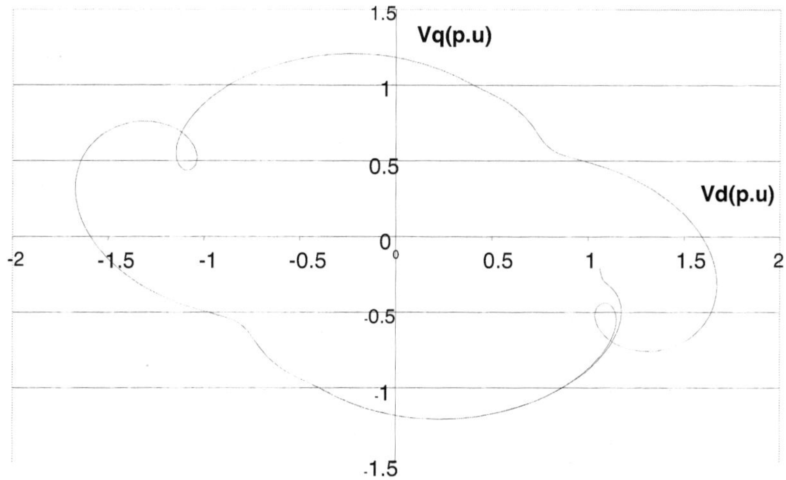

Figure 39. Locus diagram for a single phase harmonics contamination

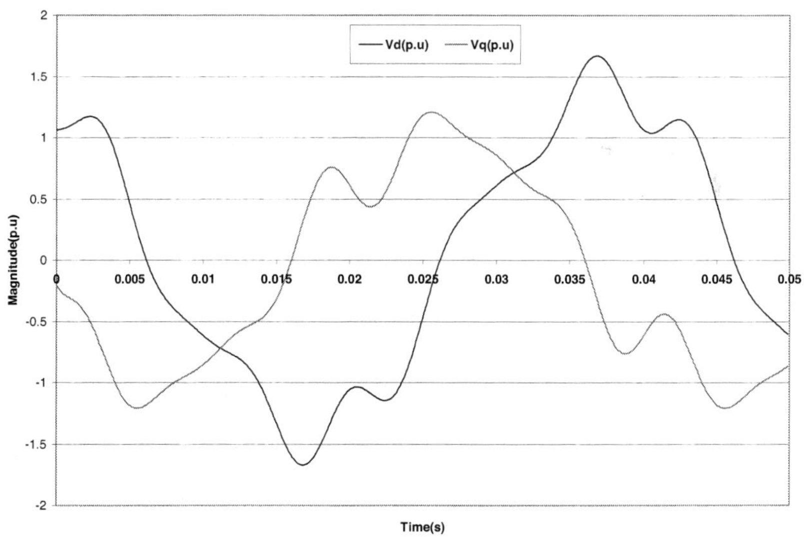

Figure 40. V_d and V_q in the time domain for a single phase harmonics contamination

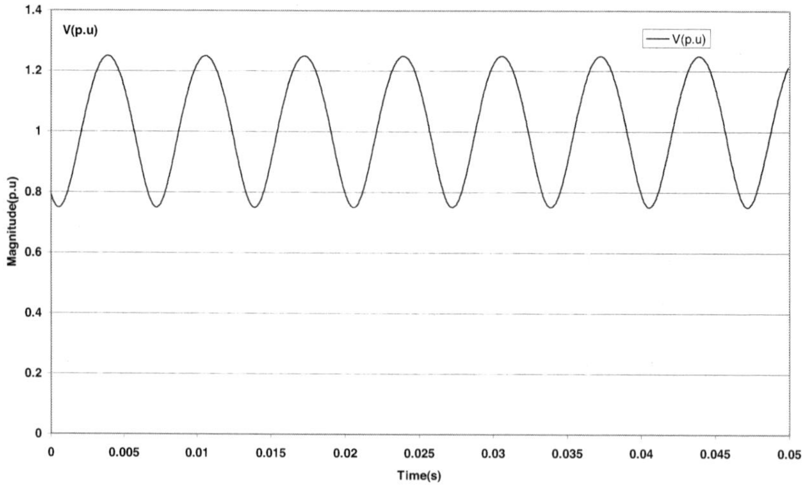

Figure 41. Phasor magnitudes in the time domain for three- phase harmonics contamination

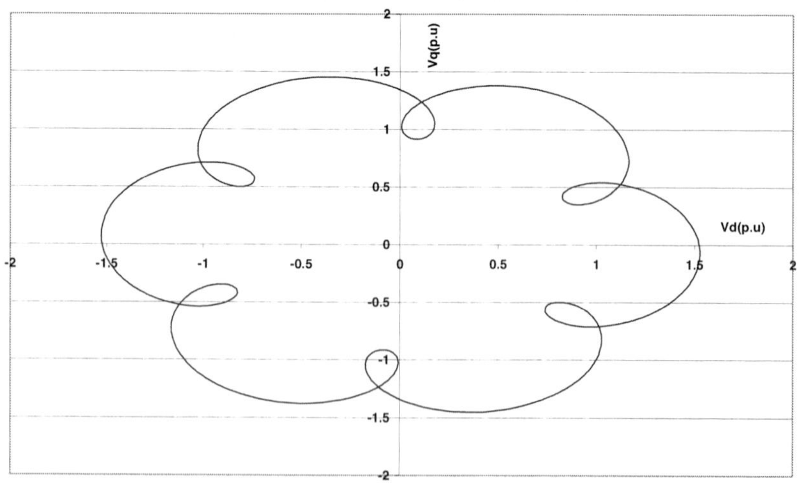

Figure 42. Locus diagram for three- phase harmonics contamination

http://dx.doi.org/10.5772/53081

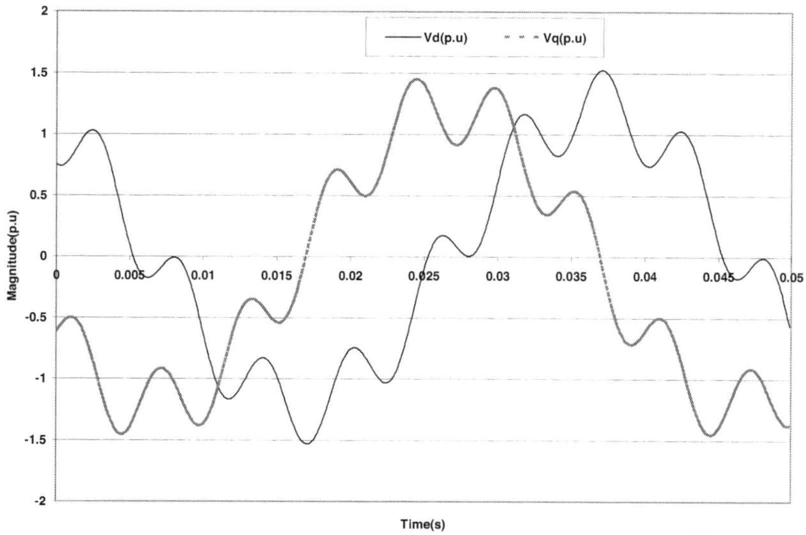

Figure 43. V_d and V_q in the time domain for three- phase harmonics contamination

4.3. Conclusions

The main contributions of this section are:

- Park's transformation is implemented to recognizing and classifying the power quality events; voltage sags, voltage flickers, voltage swell and harmonics, either for three-phase or single phase circuits.
- The proposed algorithm is succeeded in identification and classification of power quality events.
- The utility signal is transformed to a complex phasor. The magnitude of this phasor depends on the magnitude of the utility signal. The proposed technique produces the complex phasor loci that depend on the power quality event. Computer simulation results are presented within the text, these results indicate that the proposed technique is efficient, and one can recognize the type of event just by looking to the locus diagram of the two voltages V_d and V_q.
- No metering instruments are needed to measure these events, since it is hard to find an instrument that can be used to measure such events.
- Classifying these events can help in making a decision to install the suitable power conditioning devices to suppress these events or part of them.
- It has been shown that the starting time of disturbance as well as the length of the disturbance has no effect on the Park's phasor.

Author details

Soliman Abdel-Hady Soliman[1] and Rashid Abdel-Kader Alammari[2]

1 Misr University of Science and Technology, Electrical Power and Machines Department, Giza, Egypt

2 Qatar University, Electrical Engineering Department, Doha, Qatar

References

[1] Eldery, M. A, Saadany, E. F, & Salama, M. M. A Novel Operator Algorithm for Voltage Envelope Tracking" IEEE Transactions on Power Systems, (2005). , 20(1), 510-512.

[2] Liu, Y, Steurer, M, & Riberio, P. A Novel Approach to Power Quality Assessment: Real Time hardware-in-the-Loop Test Bed", IEEE Transactions on Power Delivery, (2005). , 20(2), 1200-1201.

[3] Wang, J, Chen, S, & Lie, T. T. System Voltage Sag Performance Estimation", IEEE Transactions on Power Delivery (2005). , 20(2), 1738-1747.

[4] Chilukuri, M. V, & Dash, P. K. Multi-resolution S-Transform-Based Fuzzy Recognition System for Power Quality Events", IEEE Transactions on Power Delivery (2004). , 19(1), 323-330.

[5] Abdel-galil, T. K, Saadany, E. F, & Salama, M. M. Online Tracking of Voltage Flicker Utilizing Energy Operator and Hilbert Transform" IEEE Transactions on Power Delivery (2004). , 19(2), 861-867.

[6] Wang, F, & Bollen, M. H. J. Frequency-Response Characteristics and Error Estimation in RMS Measurement", IEEE Transactions on Power Delivery, (2004). , 19(4), 1569-1578.

[7] Marie, M. I, Saadany, E. F, & Salama, M. M. Envelope Tracking Techniques for Flicker Mitigation and Voltage Regulation", IEEE Transactions on Power Delivery, (2004). , 19(4), 1854-1861.

[8] Liao, Y, Lee, J. B, & Fuzzy- Expert, A. System for Classifying Power Quality Disturbances" Electrical Power and Energy System, (2004). , 26, 199-205.

[9] Ibrahim, W. R, & Morcos, M. M. A Power Quality Perspective to System Operational Diagnosis Using Fuzzy Logic and Adaptive Techniques" IEEE Transactions on Power Delivery, (2003). , 18(3), 903-909.

[10] Kezunovic, M, & Liao, Y. A Novel Software Implementation Concept for Power Quality Study", IEEE Transactions on Power Delivery, (2002). , 17(2), 544-549.

[11] Anis, W. R. Ibrahim, and M. M. Morcos, "An Adaptive Fuzzy Technique for Learning Power- Quality Signature Waveforms", IEEE Power Engineering Review, January (2001). , 56-58.

[12] Poisson, O, Rioual, P, & Meunier, M. New Signal Processing Tools Applied to Power Quality Analysis" IEEE Transactions on Power Delivery, (1999). , 14(2), 561-566.

[13] Gaouda, M, Salama, M. M, Sultan, M. R, & Chikhani, A. Y. Power Quality Detection and Classification Using Wavelet- Multi-resolution Signal Decomposition" IEEE Transactions on Power Delivery, (1999). , 14(4), 1469-1476.

[14] Girgis, A. A, Chang, W. B, & Makram, E. B. A Digital Recursive Measurement Scheme for on-line Tracking of Power System Harmonics", IEEE Transaction on Power Delivery, (1991). , 6(3), 1153-1160.

[15] Soliman, S. A, Christensen, G. S, & Naggar, K. M. A State Estimation Algorithm for Identification and Measurement of Power System Harmonics" Electric Power System Research Jr. (1990). , 19, 195-206.

[16] Soliman, A. M. A. L-K. a. n. d. a. r. i, S. A. and K.EL-Naggar, "Digital Dynamic Identification Of Power System Sub-Harmonics Based On The Least Absolute Value," Electric Power Systems Research, Jr. (1993). , 28, 99-104.

[17] Abu, E. A, & Al-feilat, I. El-Amin, and M. Bettayeb," Power System Harmonic Estimation: A Comparative Study", Electric Power System Research Jr. (1994). , 29, 91-97.

[18] Soliman, S. A, & Hawary, M. E. Measurement of Power Systems Voltage and Flicker Level for Power Quality Analysis", Int. Jr of Electrical Power & and Energy Systems, (2000). , 22, 447-450.

[19] Soliman, S. A, Alammari, R. A, & Hawary, M. E. Frequency and Harmonics Evaluation in Power Networks Using Fuzzy Regression Technique," Electric Power Systems Research Jr., (2003). , 66, 171-177.

[20] Soliman, S. A, Alammari, R. A, Mantawy, A. H, & Hawary, M. E. On the Application of Park's Transformation for Power System Harmonics Identification and Measurements" Electric Power Systems Components Journal (Formally Electric Machines and Power Systems Journal), (2003). , 31(8)

[21] Soliman, S. A, Mantawy, A. H, & Hawary, M. E. Simulated Annealing Algorithm for Power System Quality Analysis", International Journal of Electrical Power and Energy Systems, (2004). , 26, 31-36.

[22] Al-kandari, A. M, Soliman, S. A, & Alammari, R. A. Power Quality Analysis Based on Fuzzy Estimation Algorithm: Voltage Flicker Measurements" Int. Jr. of Electrical Power and Energy Systems, (2006). , 28, 723.

[23] Al-kandari, A. M, Soliman, S. A, & Measurement, O. F. A POWER SYSTEM NOMINAL VOLTAGE, FREQUENCY AND VOLTAGE FLICKER PARAMETERS" Electri-

cal Power and Energy Systems (2009). http://www.sciencedirect.com/science/article/pii/S0142061509000143, 31(2009)

[24] Osowski, S. Neural Network for Estimation of Harmonic Components in a Power System," IEE Proc. C, Gener. Trans. Distrib. (1992). , 139(2), 129-135.

[25] Osowski, S. SVD Technique for Estimating of Harmonic Components in a Power System: A Statistical Approach", IEE Proc. Gener. Trans. Distrib. Vol.5, (1994). , 141, 473-479.

[26] Dash, P. K, Swain, D. P, Liew, A. C, & Rahman, S. An Adaptive Linear Combiner for on-line Tracking of Power System Harmonics", IEEE Transactions on Power Systems, (1996). , 11(4), 1730-1735.

[27] Smith, B. C, Watson, N. R, Wood, A. R, & Arrillaga, J. A Newton Solution for the Harmonic Phasor Analysis of AC/DC ", IEEE Trans. on Power Delivery, April (1996). , 11(2), 965-971.

[28] Moreno, V. M. Saiz, J. Barros Guadalupe, "Application of Kalman Filtering for Continuous Real-Time Tracking of Power System Harmonics", IEE Proced on Generation, Transmission and Distribution, Part C Jan.(1997). , 144(1), 13-20.

[29] "Time- Varying Harmonics: Part I- Characterizing Measured Data"Probabilistic Aspect Task Force of the Harmonics Working Group, IEEE Trans on Power Delivery, (1998). , 13(3), 938-944.

[30] Soliman, S. A, Helal, I, & Al-kandari, A. M. Fuzzy Linear Regression for Measurement of Harmonic Components in a Power System", Electric Power Systems Research (1999). , 50, 99-105.

[31] Bucci, G, & Landi, C. On-Line Digital Measurement for the Quality Analysis of Power Systems Under Non-sinusoidal Conditions" IEEE Trans. on Instrumentation and Measurement, N0. 4, (1999). , 48, 853-857.

[32] Exposito, A. G. J. A. Rosendo Macias, " Fast Harmonic Computation for Digital Relaying" IEEE Transaction on Power Delivery, (1999). , 14(4), 1263-1268.

[33] Kim, F, Enbouch, L, Chaouiand, A, & Michel, M. Novel Identification-Based Technique for Harmonics Estimation", Can. J.Elect. & Comp.Eng. (1999). , 24(4), 149-154.

[34] Lai, L. L, Chan, W. L, & Tse, C. T. Real-Time Frequency and Harmonic Evaluation Using Artificial Neural Networks", IEEE Transaction on Power Delivery, (1999). , 14(1), 52-59.

[35] Andreon, S, Yaz, E. E, & Olejniczak, K. J. Reduced-Order Estimation of Power System Harmonics Using Set Theory", Proceeding of the 1999 IEEE International Conference on Control Application, Hawai'I, USA, August (1999). , 22-27.

[36] Tsao, T. P, Wu, R. C, & Nig, C. C. The Optimization of Spectral Analysis for Signal Harmonics", IEEE Transaction on Power Delivery, April (2001). , 16(2), 149-153.

[37] Lin, H. C, & Lee, C. S. Enhanced FFT-based Parametric Algorithm for Simultaneous Multiple Harmonics Analysis", IEEE Proc.- Gener. Transm. Distrib., May (2001). , 148(3), 209-214.

[38] Liu, Y. Z, & Chen, S. A Wavelet Based Model for On-Line Tracking of Power System Harmonic Using Kalman Filtering", IEEE Summer Power Meeting, Vancouver, B. C. July (2001). , 17-19.

[39] Skvarenina, T, & Dewitt, W. Electrical Power and Control", Prentice Hall. Ins., New Jersey, U. S. A, (2001).

[40] Bollen, M. H. J. Understanding Power Quality", IEEE Press Series on Power Engineering, Piscataway, NJ. (2000).

[41] Bollen, M. H. J. Understanding Power Quality", IEEE Press Series on Power Engineering, Piscataway, NJ. (2000).

[42] Oscar, C. Montero-Hernández, and Prasad N. Enjeti, "A Fast Detection Algorithm Suitable for Mitigation of Numerous Power Quality Disturbances', IEEE TRANSACTIONS ON INDUSTRY APPLICATIONS, NOVEMBER/DECEMBER (2005). , 41(6)

[43] Sherif Omar FariedRoy Billinton, and Saleh Aboreshaid," Stochastic Evaluation of Voltage Sag and Unbalance in Transmission Systems" IEEE TRANSACTIONS ON POWER DELIVERY, OCTOBER (2005). , 20(4)

[44] Dong-Jun WonSeon-Ju Ahn, and Seung-Il Moon, "A Modified Sag Characterization Using Voltage Tolerance Curve for Power Quality Diagnosis", IEEE TRANSACTIONS ON POWER DELIVERY, OCTOBER (2005). , 20(4)

[45] Bin WangWilsun Xu, and Zhencun Pan,"Voltage Sag State Estimation for Power Distribution Systems"' IEEE TRANSACTIONS ON POWER SYSTEMS, MAY (2005). , 20(2)

[46] Nicolás Ruiz-ReyesPedro Vera-Candeas, and Francisco Jurado, "Discrimination between Transient Voltage Stability and Voltage Sag Using Damped Sinusoids-Based Transient Modeling IEEE TRANSACTIONS ON POWER DELIVERY, OCTOBER (2005). , 20(4)

[47] Liu, Y, Steurer, M, & Ribeiro, P. A Novel Approach to Power Quality Assessment: Real Time Hardware-in-the-Loop Test Bed" IEEE TRANSACTIONS ON POWER DELIVERY, APRIL (2005). , 20(2)

[48] Mostafa, I. Marei, Tarek K. Abdel-Galil, Ehab F. El-Saadany, and Magdy M. A. Salama,"Hilbert Transform Based Control Algorithm of the DG Interface for Voltage Flicker Mitigation" IEEE TRANSACTIONS ON POWER DELIVERY, APRIL (2005). , 20(2)

[49] Eldery, M. A, Saadany, E. F, & Salama, M. M. A. A Novel Energy Operator Algorithm for Voltage Envelope Tracking", IEEE TRANSACTIONS ON POWER SYSTEMS, FEBRUARY (2005)., 20(1)

[50] Zwe-Lee GaingWavelet-Based Neural Network for Power Disturbance Recognition and Classification", IEEE TRANSACTIONS ON POWER DELIVERY, OCTOBER (2004)., 19(4)

[51] Zhu, T. X, Tso, S. K, & Lo, K. L. Wavelet-Based Fuzzy Reasoning Approach to Power-Quality Disturbance Recognition" IEEE TRANSACTIONS ON POWER DELIVERY, OCTOBER (2004)., 19(4)

[52] Juan, A. Martinez, and Jacinto Martin-Arnedo, "Voltage Sag Stochastic Prediction Using an Electromagnetic Transients Program," IEEE TRANSACTIONS ON POWER DELIVERY, OCTOBER (2004)., 19(4)

[53] Mostafa, I. Marei, Ehab F. El-Saadany, and Magdy M. A. Salama, "Envelope Tracking Techniques for Flicker Mitigation and Voltage Regulation", EEE TRANSACTIONS ON POWER DELIVERY, OCTOBER (2004)., 19(4)

[54] Chris FitzerMike Barnes, and Peter Green,"Voltage Sag Detection Technique for a Dynamic Voltage Restorer", IEEE TRANSACTIONS ON INDUSTRY APPLICATIONS, JANUARY/FEBRUARY (2004)., 40(1)

[55] Tom, A. Short, Arshad Mansoor, Wes Sunderman, and Ashok Sundaram, "Site Variation and Prediction of Power Quality" IEEE TRANSACTIONS ON POWER DELIVERY, OCTOBER (2003)., 18(4)

[56] Mihaela Albu and GT. Heydt, "On the Use of RMS Values in Power Quality Assessment" IEEE

[57] TRANSACTIONS ON POWER DELIVERYOCTOBER (2003)., 18(4)

[58] Emmanouil StyvaktakisMath H. J. Bollen, and Irene Y. H. Gu, "Expert System for Classification and Analysis of Power System Events" IEEE TRANSACTIONS ON POWER DELIVERY, APRIL (2002)., 17(2)

[59] Olivier PoissonPascal Rioual, and Michel Meunier, 'Detection and Measurement of Power Quality Disturbances Using Wavelet Transform" IEEE TRANSACTIONS ON POWER DELIVERY, JULY (2000)., 15(3)

[60] Anthony, C, & Parsons, W. Mack Grady, Edward J. Powers, and John C. Soward, "A Direction Finder for Power Quality Disturbances Based Upon Disturbance Power and Energy" IEEE TRANSACTIONS ON POWER DELIVERY, JULY (2000)., 15(3)

Chapter 3

Power Quality Enhancement Through Robust Symmetrical Components Estimation in Weak Grids

António Pina Martins

Additional information is available at the end of the chapter

http://dx.doi.org/10.5772/53138

1. Introduction

Power conditioning systems like active filters, universal power flow controllers, static compensators, and dynamic voltage restorers need accurate control and synchronization circuits capable of dealing with very different grid voltage perturbations. Phasor estimation and symmetrical components estimation is a fundamental task in power systems relaying and power quality characterization. Voltage sags and swells, harmonics, frequency variation, phase steps, DC components and noise, are phenomena that can cause severe malfunction in the control or supervision circuits of such power systems. Additionally, they must be identified, measured and quantified for power quality purposes. Moreover, some protection systems are sensitive to these parameters and can develop erroneous actions. Instantaneous symmetrical components, required for power quality analysis and power electronics converter control, are robust estimated by the Discrete Fourier Transform algorithm in normal conditions. However, strong grid voltage perturbations, occurring in weak grids, especially in the grid connection of renewable energies, are more challenging. The chapter accurately describes the DFT errors under frequency variation, decaying DC components and harmonics. The error analysis allows the extension of the DFT algorithm to manage these known but more frequent and severe phenomena. The method is capable of handling a large variety of grid voltage perturbations, maintaining a good dynamic response and accuracy. These are the main reasons why the DFT algorithm is widely used in Phasor Measurement Units (PMUs) and power quality analyzers [1-2].

INTECH
open science | open minds

© 2013 Pina Martins; licensee InTech. This is a paper distributed under the terms of the Creative Commons Attribution License (http://creativecommons.org/licenses/by/3.0), which permits unrestricted use, distribution, and reproduction in any medium, provided the original work is properly cited.

2. The DFT environment

The Discrete Fourier Transform (DFT) has been widely used in the analysis of the fundamental component and harmonics of the electric grid voltages and currents, namely in PMUs and protection relays [3-4]. The temporal information loss caused by the transformation is recovered by analyzing the signal in just one temporal window with duration of one or a multiple of the fundamental period of the analyzed waveform. The DFT method gives accurate results when the sampled period is equal to the fundamental period.

In case of input frequency variations there is a phase shift between the input and output signals, as well as spectral leakage. With unknown input frequencies the signal components are projected to the presumed frequency components, multiples of the input frequency. During fast transients or faults, the grid voltage is characterized by being a non-periodic signal containing fast oscillations, exponential decaying components and harmonics, among other possible disturbances. Fundamental component extraction by conventional DFT is affected by an important error due to the limited temporal resolution of the analyzed window.

The phase difference resulting from the difference between the presumed frequency and the real frequency can be compensated in different ways: by imposing that the analyzed interval should be equal to the grid period, [5], or by adding a phase offset to cancel the phase difference. The second method is preferable since it does not imply a change in the execution frequency of the digital algorithm that can be embedded in control task with other algorithms that need to be executed at a fixed frequency. The so-called Smart Discrete Fourier Transform, [6], measures the input phasor signal and estimates the frequency with high precision, superior to the conventional DFT, showing robustness and being implemented in a recursive manner. The estimation precision is robust in the presence of noise and is higher if it is considered high order harmonics but implying a more complex algorithm and pre filtering, [7].

Different methods have been proposed to allow the DFT algorithm to deal with variable frequency input signals, exponential decaying components and noise immunity. Adaptive variation of the temporal window, adaptive change of the sampling frequency, phase and amplitude correction and input data modification are the main proposed methods to increase the DFT algorithm performance.

Input low pass anti-aliasing filters can eliminate the high frequency components but can not remove decaying dc components and reject low frequency components. Under these conditions, the DFT-based phasor estimation is more difficult and slower and affects the performance of converters synchronization and digital relaying.

Absolute and recursive DFT modified algorithms can be applied to some of the above mentioned non-nominal operating conditions. However, a more realistic list of abnormal field operating conditions includes:

Amplitude variation (sags and swells). If voltage swells are important because they can cause serious damage in electrical machines and transformers voltage sags are becoming more demanding since there is a need to maintain some important grid connected system in operation even under high amplitude sags, [8].

Harmonics. The presence of high power nonlinear loads, deregulation rules and increased power flow allowed the increase of harmonics presence in the grid voltage.

Spikes and notches. Caused by power devices switching and capacitors commutation they can severely affect zero crossing detection and generate low frequency harmonics.

Frequency variation (step and continuous). High active power variations, generators failure and power transfers between large connected areas can cause frequency variations. As voltage sags, frequency deviations from the nominal value are imposing new and demanding conditions in the new energy generation era.

Phase steps. Connecting and disconnecting large loads, especially in weak grids, are the main origin of phase steps occurrence. Having an extremely large frequency spectrum they cause important transient phenomena as in amplitude measurement or frequency estimation as in control systems.

Exponential decaying components. Generated by different fault types or grid connection of high power electrical machines they constitute with phase steps the more important disturbances that appear in a grid voltage system.

Noise. Always present and generated by very different sources.

Phasor estimation under different grid voltage perturbations and symmetrical components calculation for power conditioning converters control are the parameters to be analyzed in the chapter. Performance optimization for the all conditions should be a compromise.

3. Symmetrical components estimation with the DFT

Apart from noise and harmonics, which will be considered later, the main components of a voltage signal from the phasor detection point of view can be expressed as:

$$x(t) = X\cos(\omega t + \varphi) + Ae^{-t/\tau} \tag{1}$$

where A is the initial amplitude of a decaying DC component being τ its time constant, X is the amplitude of the voltage signal, ω the fundamental frequency, and ϕ the phase angle of the voltage signal.

Assume that $x(t)$ is sampled at a sampling rate f_s, multiple of the nominal frequency, f_o:

$$f_s = \frac{1}{T_s} = f_o N \tag{2}$$

Being f_o the nominal frequency and N the number of samples per fundamental nominal period, the sampling produces a data sequence $x(kT_s)$, or $x(k)$:

$$x(k) = X\cos\left(\omega\frac{k}{f_o N} + \varphi\right) + Ae^{-kT_s/\tau} \tag{3}$$

Using a phasor representation with

$$\bar{x} = Xe^{j\varphi} = X\cos\varphi + jX\sin\varphi \tag{4}$$

The signal $x(k)$ can be represented by

$$x(k) = \frac{\bar{x}e^{j\omega t_k} + \bar{x}^{*}e^{-j\omega t_k}}{2} + Ae^{-kT_s/\tau} \tag{5}$$

where * denotes complex conjugate. The fundamental frequency component, X_1, (at instant k and with nominal frequency, $f=f_o$) given by the DFT algorithm is

$$X_1(k) = \frac{2}{N}\sum_{i=0}^{N-1} x(k+i-N)e^{-2\pi f_o i T_s} \tag{6}$$

It is important to note that at instant k, the data window used to compute the DFT goes from k-N to k-1. Taken frequency variation, $f=f_o+\Delta f$, into consideration and substituting the signal phasor representation in the DFT expression, the fundamental component is

$$\begin{aligned} X_1(k) = {} & \frac{\bar{x}}{N}\sum_{i=0}^{N-1} e^{j2\pi(f_o+\Delta f)\frac{k+i-N}{f_o N}} \cdot e^{-j\frac{2\pi}{N}i} + \frac{\bar{x}^{*}}{N}\sum_{i=0}^{N-1} e^{-j2\pi(f_o+\Delta f)\frac{k+i-N}{f_o N}} \cdot e^{-j\frac{2\pi}{N}i} \\ & + \frac{2A}{N}\sum_{i=0}^{N-1} e^{-\frac{k+i-N}{f_o N\tau}} \cdot e^{-j\frac{2\pi}{N}i} \end{aligned} \tag{7}$$

With some algebraic manipulation, the expression can be given by

$$\begin{aligned} X_1(k) = {} & \frac{\bar{x}}{N}e^{j\frac{2\pi}{N}k} \cdot \frac{\sin(N\theta_1)}{\sin(\theta_1)} \cdot e^{j\frac{\pi}{N}\frac{\Delta f}{f_o}(2k-N-1)} + \frac{\bar{x}^{*}}{N}e^{-j\frac{2\pi}{N}(k-1)} \cdot \frac{\sin(N\theta_2)}{\sin(\theta_2)} \cdot e^{-j\frac{\pi}{N}\frac{\Delta f}{f_o}(2k-N-1)} \\ & -\frac{2A}{N} \cdot \frac{e^{\frac{1}{f_o\tau}} - 1}{e^{-\frac{1}{f_o N\tau} - j\frac{2\pi}{N}} - 1} \cdot e^{-\frac{k}{f_o N\tau}} \end{aligned} \tag{8}$$

The frequency deviation dependent angles θ_1 and θ_2 are given by

$$\theta_1 = \frac{\pi}{N}\frac{\Delta f}{f_o}; \theta_2 = \frac{\pi}{N}\left(2 + \frac{\Delta f}{f_o}\right) \tag{9}$$

Harmonics presence in the grid voltage when there is a frequency deviation creates additional errors. It can be shown that the existence of m harmonics causes an error in the fundamental component that is given by:

$$\Delta X_1(k) = \frac{1}{N}\sum_{\substack{h=-m \\ h\neq -1,+1}}^{m} X_h e^{j\varphi_h} \cdot \frac{\sin(N\theta_3)}{\sin(\theta_3)} \cdot e^{j\frac{\pi}{N}\left[h(1+\frac{\Delta f}{f_o})(2k-N-1)-N+1\right]} \tag{10}$$

where

$$\theta_3 = \frac{\pi}{N}\left(h - 1 + h\frac{\Delta f}{f_o}\right) \tag{11}$$

Eq. (8) and (10) clearly show the behaviour of the DFT algorithm under abnormal conditions. In what respects to frequency deviation the resulting fundamental component has two types of errors: amplitude and phase. The positive direction rotating phasor presents a frequency deviation dependent amplitude and phase; the negative direction rotating phasor has also variable amplitude and rotates at a double frequency. Harmonics have a similar but much smaller contribution to the referred errors. The presence of an exponential decay component introduces a complex error, very unfavourable in phasor detection.

The DFT algorithm with these two conditions, variable frequency and exponential component, can be used if the resulting errors are correctly handled.

3.1. Correction for variable frequency

There are some approaches to variable frequency operation: analytical correction based on absolute DFT calculation [6, 9], recursive DFT algorithm [5, 10-11], and high frequency filtering [12].

In [9], frequency estimation is based on the phase variation given by two DFT calculations and approximate expressions so containing a frequency deviation dependent error. The method presented in [6] is simple in ideal conditions and can be recursive, but is susceptible to harmonics; its consideration highly increases the method complexity. Recursive DFT approaches share the same initial simplicity. When dealing with variable frequency the correction methods are very different, essentially depending on the measurement purpose.

In [10] it is analyzed only phase correction, not amplitude, which is important for phasor estimation. Sampling frequency variation, proposed in [5], is not a feasible method; also, linear correction of the measured phase presents good results but is not tested in all conditions. Recursive DFT calculation with phase and amplitude correction, as presented in [11], works well. However, DC decaying components cause significant perturbations in phasor estimation. Filtering the high frequency components present in the amplitude and phase values returned by the DFT is limited to small frequency deviations [12]. Also, filtering introduces an additional delay in the phasor estimation.

In general, it can not be guaranteed the absence of even harmonics, so any method based on the half cycle DFT is not considered. The relative slower dynamics of the full cycle DFT must be assumed.

3.2. Decaying DC component compensation

Accurate elimination of the exponentially decaying DC component from the fundamental phasor calculated by the DFT is treated by different methods: mimic filter [13], input data correction [14-15], and analytical calculation with variable data window [16].

The mimic filter just uses an average value of the presumed decay component time constant, amplifies high frequency components and introduces a phase advance in the fundamental component, so making it unsuitable for accurate instantaneous phasor detection. Also, operation under variable frequency introduces amplitude errors. Input data correction, so eliminating the decay component, can be made just one cycle after the occurrence of a fault. Naturally, there is a need to detect the fault; the method is tailored for fault occurrence.

The correct fundamental component phasor is only obtained after N samples [15] or $N+4$ samples [17]. Of course, this is much better than the simple DFT algorithm that originates an amplitude overshoot and settling time dependent on the time constant but also is done with high complex calculations. In practical applications this complexity results in two very important aspects: execution time and run time errors, due to trigonometric operations and possible divide by zero operations or square root calculations, respectively.

The variable data window method in [16] is not so complex and is also fast but does not consider frequency deviation. Additionally, its main algorithm is tailored for fault detection, not for permanent operation like grid voltage feature extraction or control purposes like synchronization or current control.

3.3. Symmetrical components for power conditioning devices

Single phase phasor estimation and three-phase symmetrical components estimation are a very useful tool in power systems. During unbalanced disturbances their values change significantly. Although the symmetrical components concept is a frequency domain one, it is extended to the time domain, [18]. The estimation of symmetrical components from measured signals can be used for efficient control, supervision and protection in electrical power systems especially in unbalanced ones [19] or for instantaneous phasor detection [20-21].

The digital implementation of three-phase symmetrical components is based on two approaches: by the definition, with the help of a digital time delay, or by the decomposition of the three-phase signals in orthogonal components followed by a complex digital filtering [22]. The former is more general and will be used here.

Being $v_a(t)$, $v_b(t)$, and $v_c(t)$ three-phase instantaneous voltages the instantaneous symmetrical components are defined by:

$$\begin{bmatrix} v_a^0(t) \\ v_a^p(t) \\ v_a^n(t) \end{bmatrix} = \frac{1}{\sqrt{3}} \begin{bmatrix} 1 & 1 & 1 \\ 1 & a & a^2 \\ 1 & a^2 & a \end{bmatrix} \cdot \begin{bmatrix} v_a(t) \\ v_b(t) \\ v_c(t) \end{bmatrix} \tag{12}$$

where a=exp(2π/3).

The instantaneous positive and negative sequence components defined by (12) are in general complex signals. Another definition yielding real signals can be obtained through the substitution of the complex operator a by a 120º phase shift in the time domain as follows:

$$\begin{bmatrix} v_a^0(t) \\ v_a^p(t) \\ v_a^n(t) \end{bmatrix} = \frac{1}{3} \cdot \begin{bmatrix} v_a(t) + v_b(t) + v_c(t) \\ v_a(t) + S_{120} v_b(t) + S_{240} v_b(t) \\ v_a(t) + S_{240} v_b(t) + S_{120} v_c(t) \end{bmatrix} \tag{13}$$

The negative sequence instantaneous symmetrical component is of no interest, because it is the complex conjugate of the positive sequence instantaneous symmetrical component.

Since the DFT algorithm extracts the phasor information, the symmetrical components can be easily obtained through the phase shift operator and algebraic processing. Furthermore, using only the fundamental component returned by the DFT it can be readily obtained the fundamental positive and negative instantaneous components.

The positive sequence instantaneous component has the following expression:

$$\begin{bmatrix} v_a^p(t) \\ v_b^p(t) \\ v_c^p(t) \end{bmatrix} = \frac{1}{3} \begin{bmatrix} 1 & a & a^2 \\ a^2 & 1 & a \\ a & a^2 & 1 \end{bmatrix} \cdot \begin{bmatrix} v_a(t) \\ v_b(t) \\ v_c(t) \end{bmatrix} \tag{14}$$

The application of the DFT algorithm to extract the symmetrical sequence components was discussed in several works [12, 23-24]. The algorithm, with the appropriate and presented counter measures is capable of dealing with non-stationary signals and variable frequency. Symmetrical components estimation in conjunction with amplitude, phase and frequency detection can be made according to the diagram in Figure 1. With the sampling theorem satisfied by fast A/D converters, enough bandwidth is available for fast dynamics, namely for including low frequency harmonics detection.

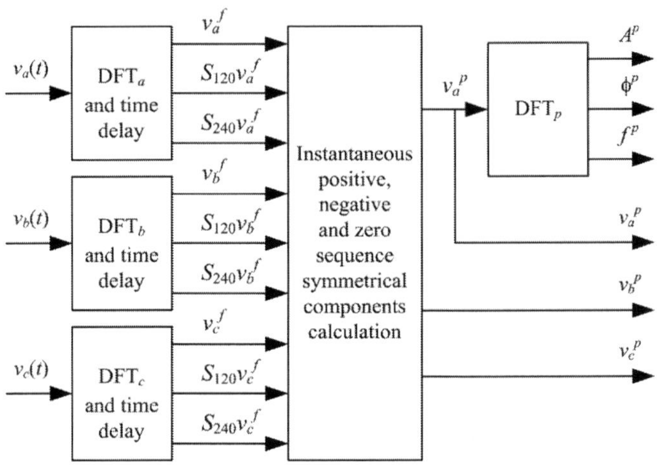

Figure 1. DFT-based three-phase symmetrical components calculation with instantaneous amplitude, phase and frequency estimation.

4. Handling grid perturbations

The proposed method is focussed for power converters control and protection. It is intended for operating under all input voltage conditions including the above referred ones. The most demanding are the decaying DC components and the frequency variations. With these conditions present it was made an extensive study on the performance of a DFT-based method for operating under a general and unknown electrical environment.

Different criteria have been used to assess the performance of a particular method. Operation in all mandatory conditions must be a compromise between dynamics and precision: dynamics in order to efficiently control the converter currents and correctly measure the amplitude value with no over or undershoots; precision to accurately compute the parameters of interest and the control algorithms output values.

4.1. General purpose method

An online DFT-based phasor estimation method with decaying DC component correction and frequency deviation compensation is presented. Among the different possibilities for minimizing the effects of a decaying DC component in DFT fundamental component phasor estimation, the partial sums approach will be used. Being excellent in the case of a fault occurrence, it deteriorates the DFT performance under other transients like voltage sags and swells, and phase steps. Also, it should be considered that a frequency deviation creates an error in the corrected data, so affecting its performance. The partial sums are defined by:

$$PS_1 = x(1) + x(3) + \ldots + x(N-1) \tag{15}$$

$$PS_2 = x(2) + x(4) + \ldots + x(N) \tag{16}$$

In the absence of a decaying DC component and with nominal frequency the two sums are zero and the acquired data are not changed. Frequency deviation introduces an error that must be considered in real-time phasor estimation. The partial sums then result in:

$$PS_1(k) = A \cdot \frac{b(b^N - 1)}{b^2 - 1} + e_1(k) \tag{17}$$

$$PS_2(k) = A \cdot \frac{b^2(b^N - 1)}{b^2 - 1} + e_2(k) \tag{18}$$

where b=exp(-1/($f_o N\tau$)) and the errors are given by:

$$e_1(k) = X \sin\left[\frac{2\pi}{N}(1 + \Delta f / f_o)(k - N/2)\right] \cdot \frac{\sin\left[\pi(1 + \Delta f / f_o)\right]}{\sin\left[2\pi / N(1 + \Delta f / f_o)\right]} \tag{19}$$

$$e_2(k) = X \sin\left[\frac{2\pi}{N}(1 + \Delta f / f_o)(k - N/2 + 1)\right] \cdot \frac{\sin\left[\pi(1 + \Delta f / f_o)\right]}{\sin\left[2\pi / N(1 + \Delta f / f_o)\right]} \tag{20}$$

The errors are dependent on the fundamental component amplitude, X, the frequency deviation, Δf, the sampling frequency, Nf_o, and the actual instant, k. At each sampling instant, the partial sums must be calculated according to (17) and (18); then A and b are determined by simple algebra, [15].

The data correction made at each sampling instant by the DC component compensation algorithm does not allow using the recursive DFT method. The approach to deal with the frequency deviation is based on the method presented in [11] but modified to the absolute version of the DFT algorithm. Some increase in the computational needs is the consequence of a more general algorithm.

The algorithm operates iteratively under the flow diagram shown in Figure 2. When a new sample occurs, the partial sums are calculated according to (15) and (16); the new b and A parameters are determined with (19) and (20) using the errors e_1 and e_2 determined with (17) and (18), with the frequency deviation of the previous sampling; the data window is corrected; the absolute DFT is computed; the phase, frequency and amplitude errors are calculated; the phasor parameters are outputted. Simultaneously, samples for $v_a^f(k)$, $S_{120}v_a^f(k)$ and $S_{240}v_a^f(k)$ are generated in order to compute the instantaneous positive, negative and zero sequence symmetrical components.

4.2. Simulation results

As referred in the Introduction, the grid voltage is subjected to very different phenomena. So, a phasor and symmetrical components estimation method must be tested against conditions like the ones that will be faced in field operation.

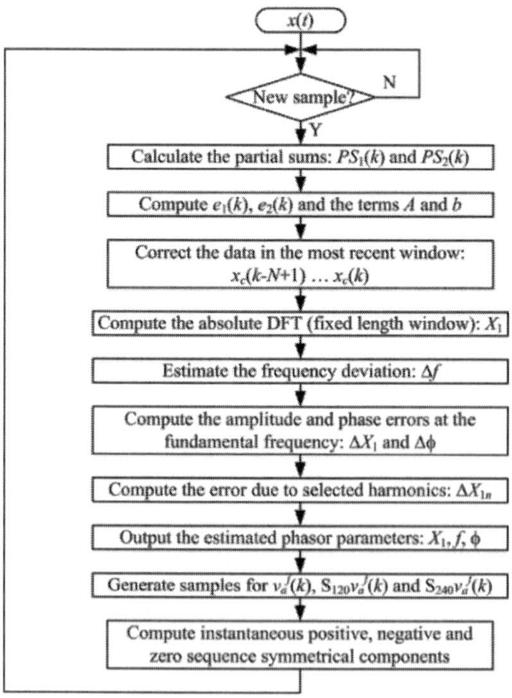

Figure 2. Flow diagram for the DFT-based symmetrical components estimation method.

The simulated conditions can be divided into three categories: voltage perturbations, frequency deviations and harmonics and noise rejection. In the voltage perturbations category

http://dx.doi.org/10.5772/53138

it is considered balanced voltage sags and swells, and decaying DC components; in the frequency deviations it is analyzed the behaviour under frequency variations and phase steps; random noise and low frequency harmonics presence in association with frequency estimation precision are considered in the last group. The main simulation parameters are: the nominal voltage is 1 p.u., the nominal frequency is 50 Hz, with 32 samples per period.

Stationary symmetrical components in different conditions are extracted in Figure 3, where phase *a* has a 20% voltage sag (unbalanced sag) during the time interval *t*=[0.2 s, 0.4 s] and, at *t*=0.6 s, phases *b* and *c* have a phase jump of +20º.

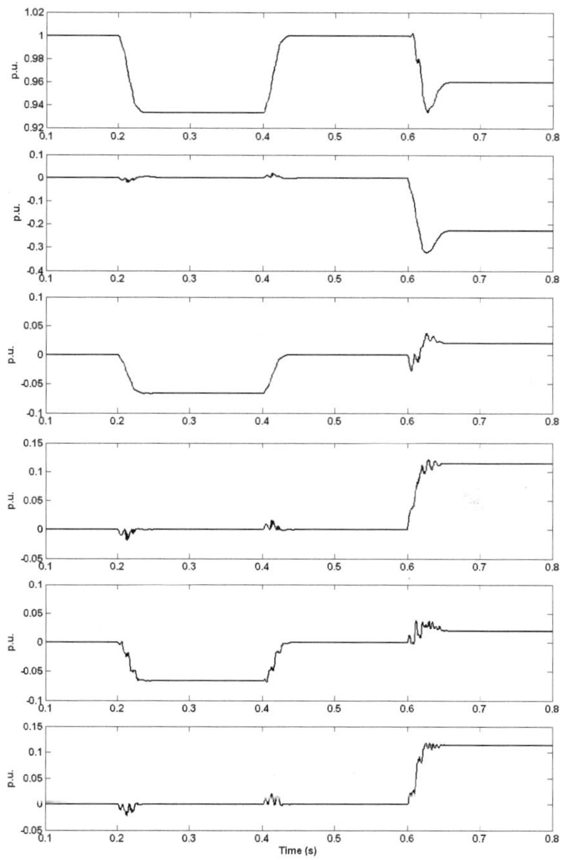

Figure 3. Stationary symmetrical components estimation during a voltage sag in phase *a* in the interval *t*=[0.2 s, 0.4 s] and a phase jump of +20° at *t*=0.6 s, in phases *b* and *c*. Traces from top: real and imaginary parts of the positive sequence, real and imaginary parts of the negative sequence, real and imaginary parts of the zero sequence.

Due to a deregulated market and a high penetration level of renewable power sources strong voltage perturbations are expected in the near future. The amplitude voltage perturbations are presented in Figures 4, 5 and 6. For all the three conditions, several tests have been made. In all of them, the dynamics of the transient response is dependent on the instant when the perturbation occurred. The presented ones show the worst situations.

Figure 4 shows, between *t*=0.2 s and *t*=0.3 s, the collapse of the voltage down to 20% (balanced sag), the instantaneous positive sequence voltage estimation, and the positive and negative sequence amplitudes. Only one and a half cycle is required to reach the steady-state condition.

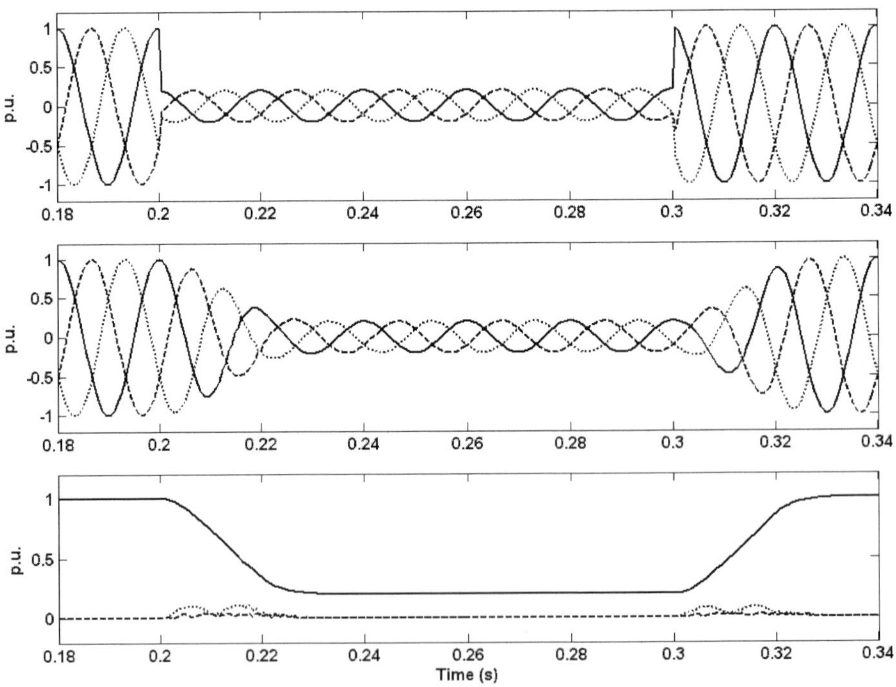

Figure 4. Three-phase balanced sag. Traces in upper window: three-phase input voltages. Traces in middle window: three-phase instantaneous positive sequence voltages. Traces in lower window: positive (—); negative (·····); and zero sequence amplitude (----).

http://dx.doi.org/10.5772/53138

Figure 5 shows the same waveforms but for a voltage swell of 180%, between t=0.4 s and t=0.5 s.

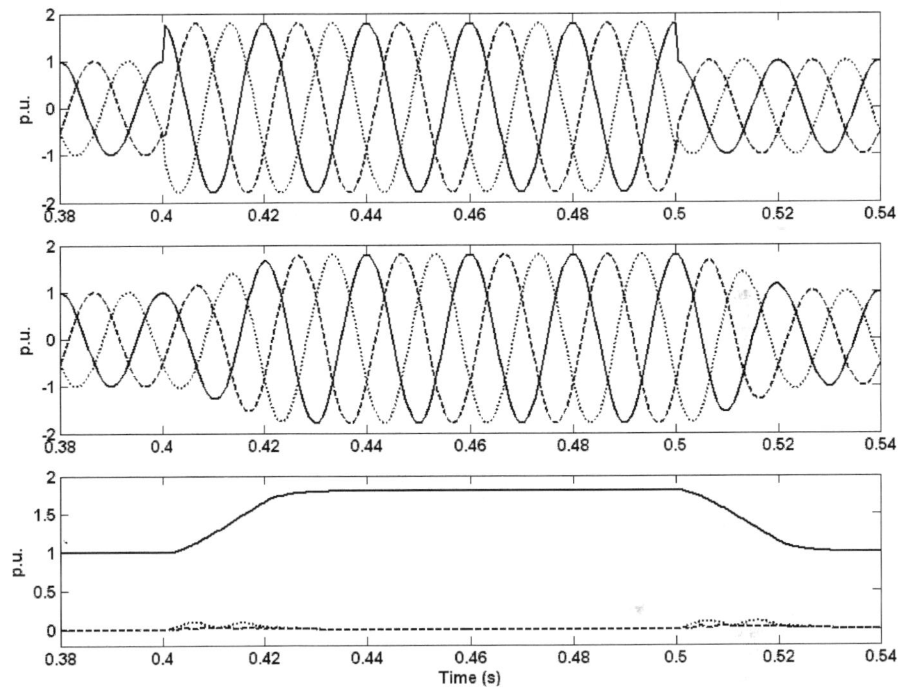

Figure 5. Three-phase balanced swell. Traces in upper window: three-phase input voltages. Traces in middle window: three-phase instantaneous positive sequence voltages. Traces in lower window: positive (—); negative (······); and zero sequence amplitude (----).

Exponentially decaying DC components are caused by several fault types including the recovering of short circuits and overload conditions.

Figure 6 shows such an example: at t=0.24 s the voltage collapses to zero; then, at t=0.3 s the voltage raises again, with an exponential decaying component with A=1 p.u. and a time constant of 30 ms.

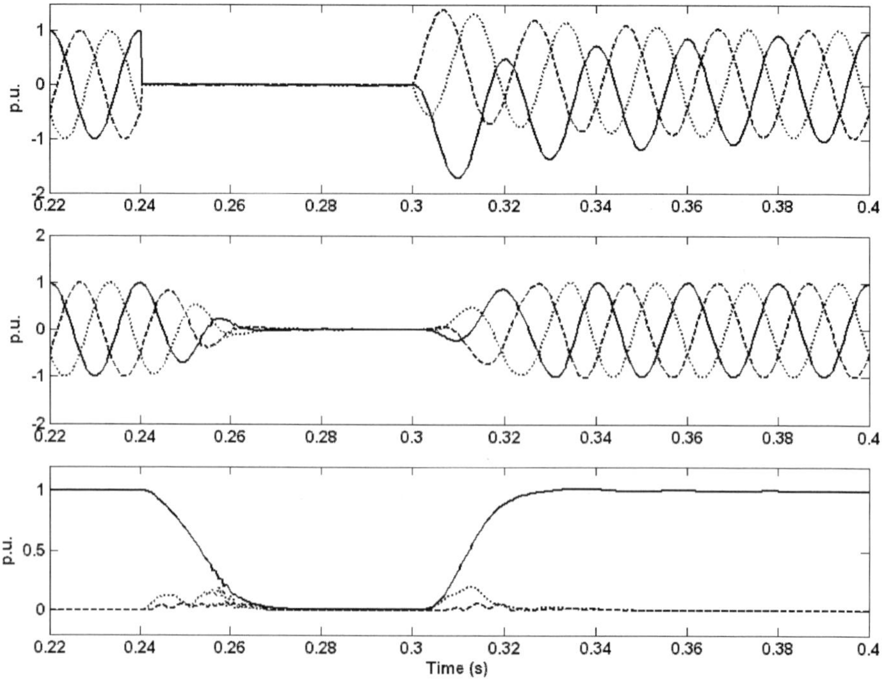

Figure 6. Voltage collapse and recovering with a DC decaying component. Traces in upper window: three-phase input voltages. Traces in middle window: three-phase instantaneous positive sequence voltages. Traces in lower window: positive (—); negative (·····); and zero sequence amplitude (----).

Balanced voltage sags and swells have one and a half cycle time response. Decaying DC components generate a not so small perturbation in the negative component but are efficiently handled in the instantaneous positive sequence.

Different fault types can cause unbalanced voltage sags, possibly with phase steps. Frequency perturbations come from generation-consumption unbalance in static and dynamic conditions. Frequency deviation steps are not usual in strong grids but can occur in weak systems; continuous frequency deviations and phase steps are much more common. Three conditions are presented in Figures 7, 8 and 9. In Figure 7, frequency goes from 50 Hz to 48 Hz at t=0.3 s. With a time response of two cycles the frequency is correctly tracked with no amplitude errors.

An unbalanced voltage sag is presented in Figure 8, during the interval t=[0.2 s, 0.3 s]. In this case, phase a maintains its amplitude and phase while phase b decreases to an amplitude of 0.577 with a phase jump of -30º and phase c decreases to the same amplitude of phase b with a phase jump of +30º. This condition generates a negative sequence component but not a zero sequence one.

Another unbalanced condition, as referred in IEEE 1159 [25], is shown in Figure 9, during the time interval t=[0.4 s, 0.5 s]. While phase a maintains its amplitude and phase, phase b decreases to an amplitude of 0.1 with a phase jump of -55º and phase c decreases to an amplitude of 0.5 with a phase jump of -20º, in a three-phase four-wire system. The three components, positive, negative and zero, are now present in the three-phase system and are efficiently detected.

Frequency deviation and severe phase steps cause severe perturbations at different levels. As in symmetrical components estimation as in amplitude detection and frequency estimation the perturbations are important but are correctly handled by the DFT-based estimation method.

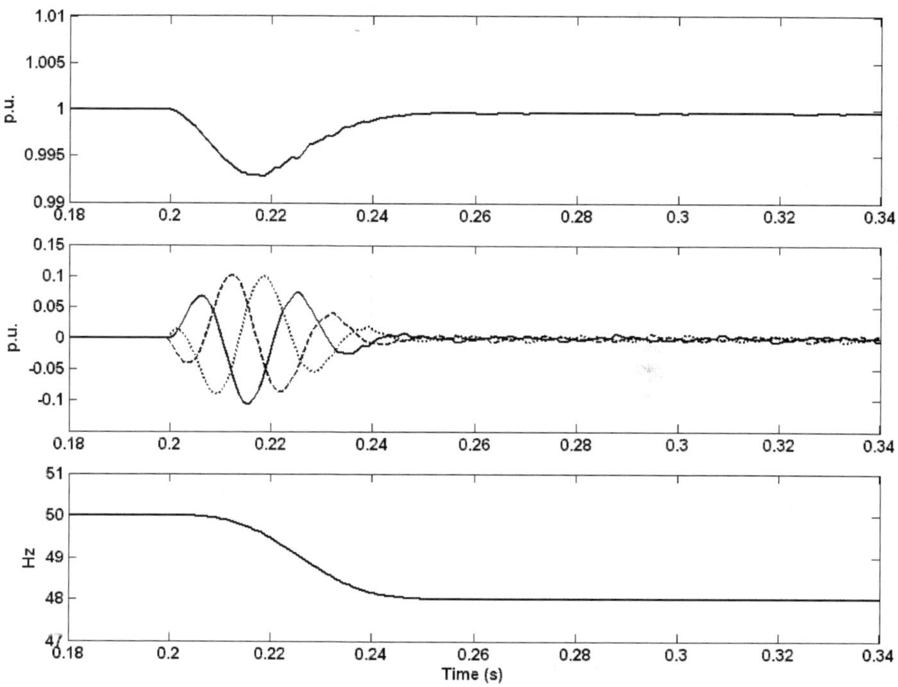

Figure 7. Frequency change from 50 to 48 Hz. Upper window: real part of the positive sequence component. Middle window: errors in the estimation of the instantaneous positive three-phase components. Lower window: frequency estimation.

Quite often there is the occurrence of harmonics in power systems: nonlinear loads generate harmonic currents and the associated voltage drops in the line impedances create voltage harmonics. These degrade the overall quality of the delivered power and can also severely affect the operation of grid-connected equipment. The DFT algorithm can easily extract the harmonics present in the three-phase voltages; in fact it is a common feature of any power quality analyzer.

Figure 10 demonstrates the dynamic operation of the DFT algorithm in the estimation of the three-phase instantaneous positive sequence and its magnitude in the following conditions: between t=0.1 s and t=0.2 s the voltage signal contains the following harmonics: 2nd with 1%, 3rd with 5%, 5th with 10%, and 7th with 5%, and arbitrary phase; the frequency is maintained in 50 Hz and the S/N ratio is 30 dB. As expected there is a very small disturbance in the estimation of the symmetrical components magnitudes; the instantaneous positive sequence is almost undisturbed.

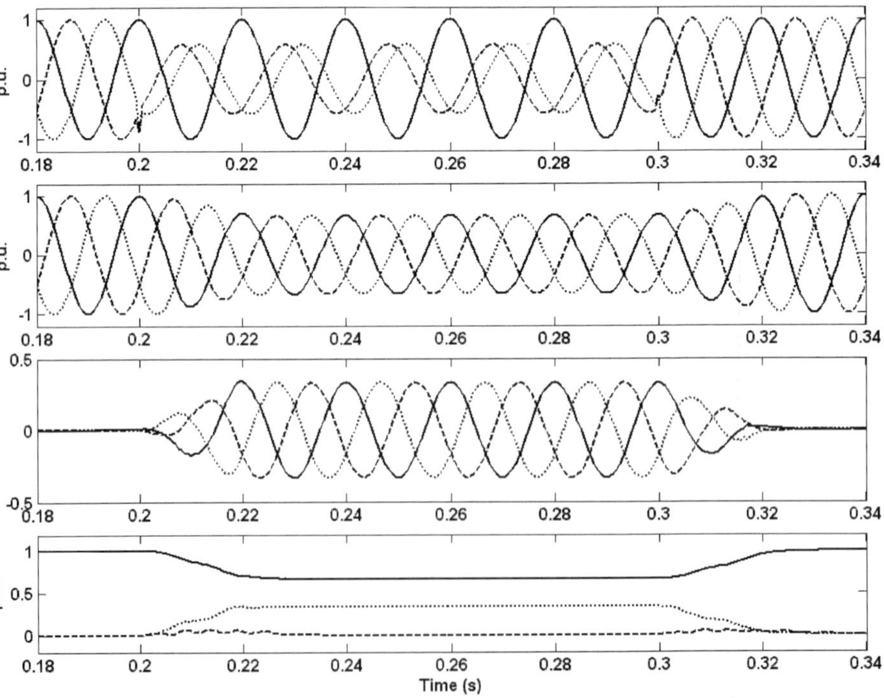

Figure 8. Unbalanced voltage sag without zero sequence component. Upper window: three-phase input voltages. Second window: three-phase instantaneous positive sequence voltages. Third window: three-phase instantaneous negative sequence voltages. Lower window: positive (—); negative (·····); and zero sequence amplitude (----).

http://dx.doi.org/10.5772/53138

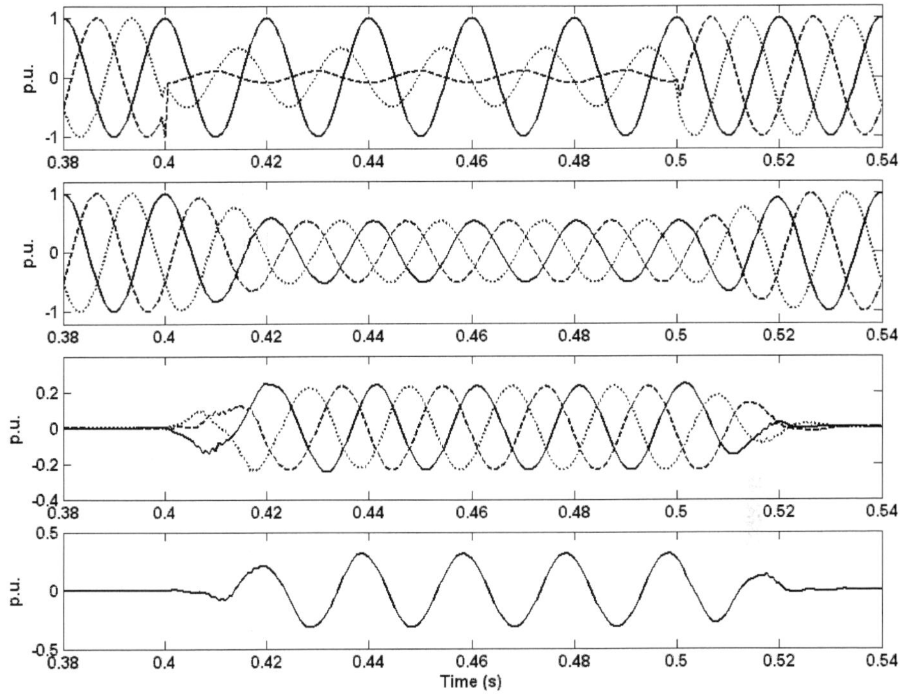

Figure 9. Unbalanced voltage sag with zero sequence component. Upper window: three-phase input voltages. Second window: three-phase instantaneous positive sequence voltages. Third window: three-phase instantaneous negative sequence voltages. Lower window: zero sequence component.

In low-voltage grids another important phenomenon already referred is the occurrence of exponentially decaying DC components: if associated with low-frequency harmonics (e.g. caused by the switching of lightly filtered phase-controlled rectifiers) and noise they can severely affect the operation of any analyzer or protection relay. In order to show this condition and the robustness property of the enhanced DFT algorithm Figure 11 is presented. The used conditions are as follows: after t=0.098 s the voltage signal contains a decaying DC component with a time constant of 30 ms and the following harmonics: 5th with 15%, and 7th with 10%, and arbitrary phase; the frequency is 50 Hz and the S/N ratio is 20 dB. This is a severe condition; all the parameters related to power quality are affected: the positive and negative sequences vary during almost two grid cycles; the estimated frequency suffers a strong transient. Comparing with Figure 10 it can be concluded that these perturbations are mainly caused by the DC component; low-frequency harmonics and a certain noise level are quite well tolerated by the DFT algorithm.

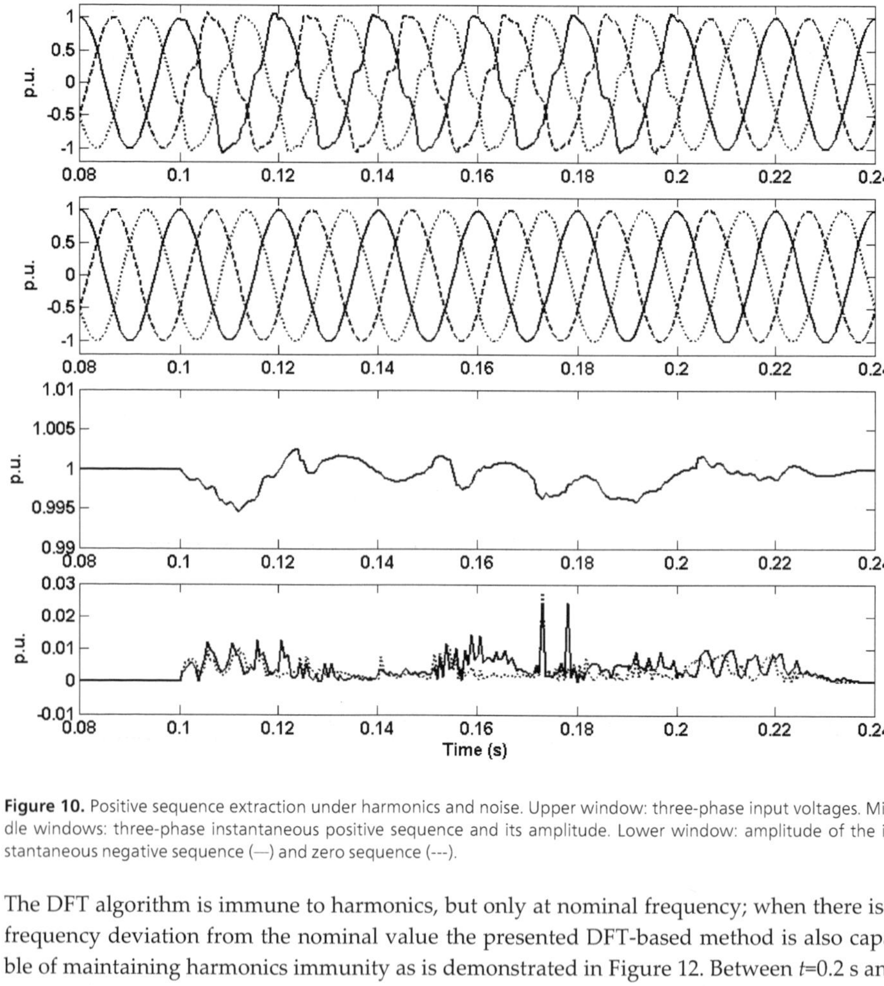

Figure 10. Positive sequence extraction under harmonics and noise. Upper window: three-phase input voltages. Middle windows: three-phase instantaneous positive sequence and its amplitude. Lower window: amplitude of the instantaneous negative sequence (—) and zero sequence (---).

The DFT algorithm is immune to harmonics, but only at nominal frequency; when there is a frequency deviation from the nominal value the presented DFT-based method is also capable of maintaining harmonics immunity as is demonstrated in Figure 12. Between t=0.2 s and t=0.3 s, the frequency goes to 48 Hz and the signal contains the following harmonics: 2nd with 5%, 3rd with 20%, 5th with 20%, and 7th, with 10%, and arbitrary phase.

Different noise types generated by electromagnetic interference, digital circuits or power electronics converters are always present in the acquired signals. Also, low frequency harmonics due to nonlinear loads or saturated magnetic circuits are common in the grid voltage. Figure 13 shows the noise rejection capability of the presented DFT-based method when the three-phase voltages contain the same harmonic level as in Figure 12 and are corrupted by non-correlated random noise with a signal to noise ratio as low as 20 dB between t=0.1 s and t=0.5 s, and 30 dB between t=0.6 s and t=1.0 s.

http://dx.doi.org/10.5772/53138

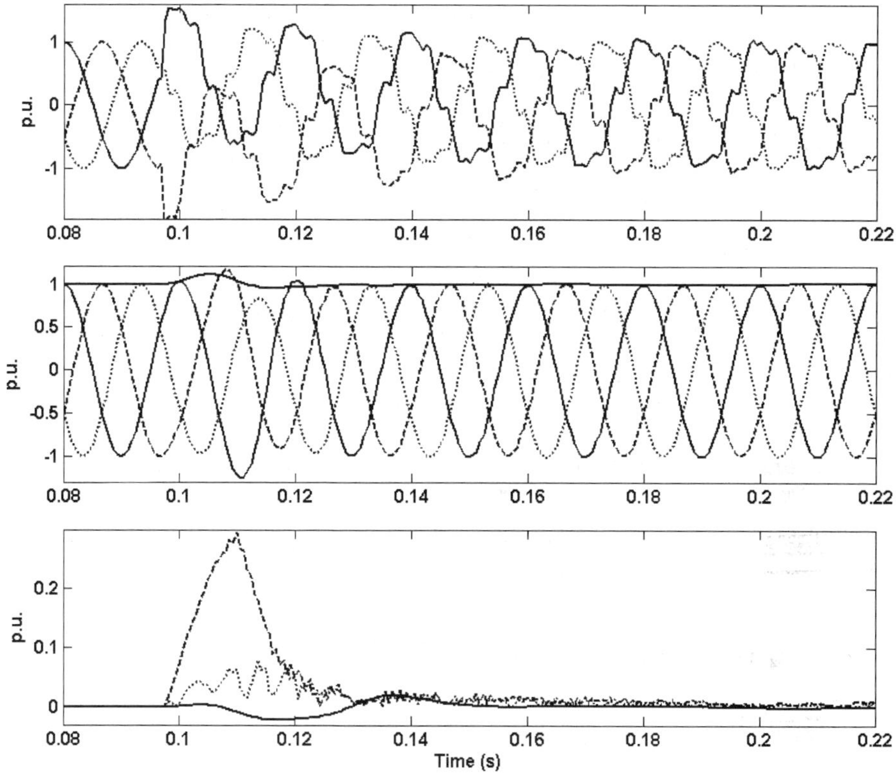

Figure 11. Exponentially decaying DC component, with harmonics and noise from *t*=0.098 s. Upper window: three-phase input voltages. Middle window: three-phase instantaneous positive sequence and its amplitude. Lower window: amplitude of the instantaneous negative sequence (---), zero sequence (····) and frequency deviation (—).

There is no noticeable perturbation in the amplitude detection or in the instantaneous positive sequence component estimation.

All the presented results are dependent on the imposed conditions; some can be managed in real experimental implementations like noise level and low frequency harmonic distortion. The others are uncontrollable: sags, swells, AC fluctuation, DC decaying components, frequency deviations and phase steps will occur in an unpredictable way and level. Any phasor estimation method should be prepared to deal with them guaranteeing appropriate dynamics, stability, precision and robustness; the presented method does.

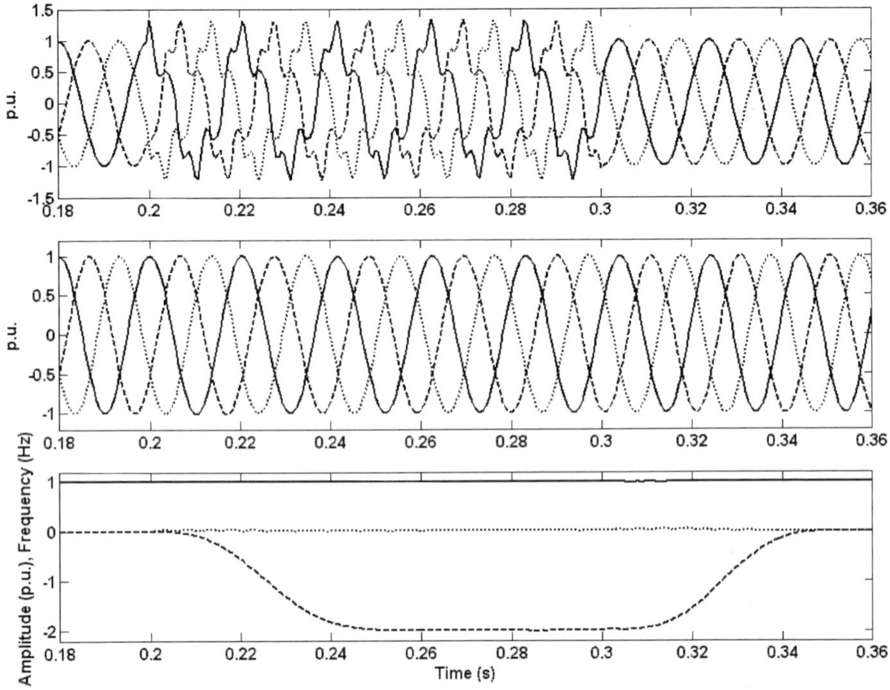

Figure 12. Harmonics immunity under frequency deviation. Upper window: three-phase input voltages. Middle window: three-phase instantaneous positive sequence voltages. Lower window: positive (—) and negative sequence amplitude (·····); and frequency deviation estimation (----).

5. Applications

As discussed in Section III.C and according to the diagram in Figure 1 and the flow diagram in Figure 2, the application of the DFT-based algorithm in the power quality domain can be divided into two categories: real-time operation and off-line processing. Hard real-time is used in the synchronization of power electronics converters like STATCOMS or FACTS, in phasor estimation or control and in protection functions. Nearly real-time processing (or off-line) is used in quasi-steady-state conditions to evaluate power quality parameters like symmetrical components estimation, voltage sags, swells and harmonics, [1, 26-27].

The extraction of harmonics is made according to (21). Like the fundamental component, the *h* order harmonic can be estimated using the absolute or recursive version of (21).

http://dx.doi.org/10.5772/53138

$$X_h(k) = \frac{2}{N} \sum_{i=0}^{N-1} x(k+i-N) e^{-\frac{2\pi}{N} ih} \tag{21}$$

The algorithm is based on intensive data processing and uses trigonometric functions and nonlinear functions to estimate the power quality parameters. A fundamental issue arising in the case of harmonics detection is the number of samples needed (*N*); it must satisfy the Nyquist criterion and highly increases the number of operations (and the required time) needed to estimate a range of harmonics. However, the use of fast A/D converters in conjunction with FPGAs or DSPs allows an efficient solution to be used in spectrum and power quality analyzers, [1].

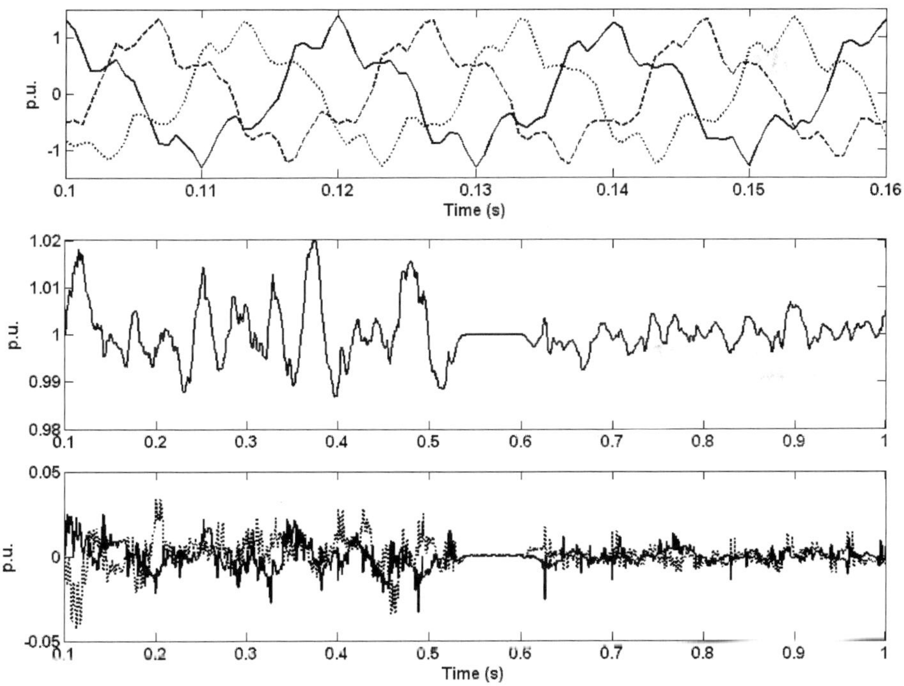

Figure 13. Symmetrical components estimation precision with low frequency harmonics and noise presence. From *t*=0.1 s to *t*=0.5 s, *S*/*N*=20 dB; from *t*=0.6 s to *t*=1.0 s, *S*/*N*=30 dB. Upper window: three-phase input voltages. Middle window: amplitude of the instantaneous positive sequence. Lower window: amplitude of the instantaneous negative sequence (—) and zero sequence (----).

In case of variable frequency conditions (or frequency deviation) the power quality parameters are estimated in quasi-steady-state. Instead of using a constant and fixed sampling frequency a variable one is preferred and there are no errors due to a non-matched window.

As referred in [1], despite some issues (e.g. computational complexity, memory requirements, and data synchronization) it is predicted that new designs of PQ instruments will use the FFT algorithm.

6. Conclusions

Power quality monitoring and power systems control and protection need fast and accurate frequency and amplitude estimation. Also, instantaneous symmetrical components estimation with amplitude and phase detection is needed for power systems stability analysis. Phenomena like high amplitude voltage sags and swells, decaying DC components, phase and frequency deviations, harmonics and noise are becoming more frequent and more intense, especially in weak grids. The corrupted voltage is difficult to manage in all conditions.

In this chapter, the recognized robustness of the DFT algorithm is extended to handle this new and more demanding grid voltage behaviour. The main errors caused by large frequency deviations and DC decaying components occurring in the DFT algorithm are acknowledged and analyzed, and the associated corrections to deal with the referred parameters are presented. The results, obtained in very unfavourable conditions, shown that it is needed a careful signal conditioning and a computationally powerful control platform in order to obtain fast dynamics and high accuracy.

In terms of power quality monitoring, the enhanced DFT method is capable of detecting all related parameters: symmetrical components, voltage sags and swells, frequency deviations and a range of harmonics. Its use is a requirement imposed by some Standards but its specific implementation in each PQ instrument or protection device has different possibilities. The presented DFT-based method has improved capabilities namely substantially improves the robustness to decaying DC components and frequency deviations.

Nomenclature

Δf frequency deviation

ϕ phase reference

τ time constant of exponential component

ω angular frequency

A magnitude of exponential component

a complex operator: a=exp($j2\pi/3$)

b operator: b=exp(-1($f_o N\tau$))

http://dx.doi.org/10.5772/53138

e error

f_o nominal frequency

f_s sampling frequency

N number of samples per period

PS power sum

S_{120}, S_{240} time delay operators

S/N signal to noise ratio

X magnitude of input signal

x(t) continuous input signal

$\bar{x}$, $\bar{x}^*$ phasor, complex conjugate of $\bar{x}$

$x_c(k)$ corrected samples

$v_a(t)$, $v_b(t)$, $v_c(t)$ phase-neutral voltage, phases a, b, c

$v_a^p(t)$, $v_b^p(t)$, $v_c^p(t)$ positive sequence, phases a, b, c

$v_a^n(t)$, $v_b^n(t)$, $v_c^n(t)$ negative sequence, phases a, b, c

$v_a^0(t)$, $v_b^0(t)$, $v_c^0(t)$ negative sequence, phases a, b, c

Acknowledgments

The Author wishes to thank José Miguel Ferreira for helpful discussions related to this work.

Author details

António Pina Martins

Address all correspondence to: ajm@fe.up.pt

Department of Electrical and Computer Engineering, Faculty of Engineering, University of Porto, Rua Roberto Frias, s/n, Porto, Portugal

References

[1] Tarasiuk T. Comparative Study of Various Methods of DFT Calculation in the Wake of IEC Standard 61000-4-7. IEEE Transactions on Instrumentation and Measurement, October 2009; 58(10) 3666-3677.

[2] Warichet J, Sezi T, Maun J-C. Considerations about Synchrophasors Measurement in Dynamic System Conditions. Electrical Power and Energy Systems, 2009; 31, 452–464.

[3] International Electrotechnical Comission. IEC Standard 61000-4-30: Testing and Measurement Techniques – Power Quality Measurement Methods. 2003.

[4] Phadke AG, Kasztenny B. Synchronized Phasor and Frequency Measurement under Transient Conditions. IEEE Transactions on Power Delivery, January 2009; 24(1) 89-96.

[5] McGrath BP, Holmes DG, Galloway J. Improved Power Converter Line Synchronisation using an Adaptive Discrete Fourier Transform (DFT). Proceedings of the IEEE Power Electronics Specialists Conference, Cairns, Queensland, Australia, June 2002, vol. 2, 821-826.

[6] Yang J-Z, Liu C-W. A Precise Calculation of Power System Frequency and Phasor. IEEE Transactions on Power Delivery, April 2000; 15(2) 494-499.

[7] Yang J-Z, Liu C-W. A Precise Calculation of Power System Frequency. IEEE Transactions on Power Delivery, July 2001; 16(3) 361-366.

[8] Jauch C, Sorensen P, Bak-Jensen B. International Review of Grid Connection Requirements for Wind Turbines. Proceedings of the Nordic Wind Power Conference, Chalmers University of Technology, Sweden, March 2004.

[9] Hart D, Novosel D, Hu Y, Smith B, Egolf M. A New Frequency Tracking and Phasor Estimation Algorithm for Generator Protection. IEEE Transactions on Power Delivery, July 1997; 12(3) 1064-1073.

[10] Funaki T, Matsuura K, Tanaka S. Error Correction for Phase Detection by Recursive Algorithm Real Time DFT. Electrical Engineering in Japan, 2002; 141(1) 8-17.

[11] Wang M, Sun Y. A Practical, Precise Method for Frequency Tracking and Phasor Estimation. IEEE Transactions on Power Delivery, October 2004; 19(4) 1547-1552.

[12] Nakano K, Ota Y, Ukai H, Nakamura K, Fujita H. Frequency Detection Method based on Recursive DFT algorithm. Proceedings of the 14th Power Systems Computation Conference (PSCC), Sevilla, Spain, Session 1, Paper 5, June 2002.

[13] Benmouyal G. Removal of DC-offset in Current Waveforms Using Digital Mimic Filtering. IEEE Transactions on Power Delivery, April 1995; 10(2) 621-630.

[14] Gu J-C. Yu S-L. Removal of DC Offset in Current and Voltage Signals Using a Novel Fourier Filter Algorithm. IEEE Transactions on Power Delivery, January 2000; 15(1) 73-79.

[15] Guo Y, Kezunovic M, Chen D. Simplified Algorithms for Removal of the Effect of Exponentially Decaying DC-Offset on the Fourier Algorithm. IEEE Transactions on Power Delivery, July 2003; 18(3) 711-717.

[16] Chen C-S, Liu C-W, Yang J-Z. A DC Offset Removal Scheme with a Variable Data Window for Digital Relaying. Proceedings of the Power Systems and Communications Infrastructures for the Future Conference, Beijing, September 2002.

[17] Yang J-Z, Liu C-W. Complete Elimination of DC Offset in Current Signals for Relaying Applications. Proceedings of the IEEE Power Engineering Society Winter Meeting, vol. 3, pp. 1933-1038, Singapore, January 2000.

[18] Stevenson WD. Elements of Power System Analysis. New York: McGraw-Hill, 1995.

[19] Chen C-C, Zhu Y-Y. A Novel Approach to the Design of a Shunt Active Filter for an Unbalanced Three-Phase Four-Wire System under Nonsinusoidal Conditions. IEEE Transactions on Power Delivery, October 2000; 15(4) 1258-1264.

[20] Hsu J-S. Instantaneous Phasor Method for Obtaining Instantaneous Balanced Fundamental Components for Power Quality Control and Continuous Diagnostics. IEEE Transactions on Power Delivery, October 1998; 13(4) 1494-1500.

[21] Stankovic AM, Aydin T. Analysis of Asymmetrical Faults in Power Systems Using Dynamic Phasors. IEEE Transactions on Power Delivery, August 2000; 15(3) 1062-1068.

[22] Lobos T. Fast Estimation of Symmetrical Components in Real Time. IEE Proceedings-C, January 1992; 139(1) 27-30.

[23] Phadke AG, Thorp JS, Adamiak MG. A New Measurement Technique for Tracking Voltage Phasors, Local System Frequency, and Rate of Change of Frequency. IEEE Transactions on Power Apparatus and Systems, May 1983; 102(5) 1025-1038.

[24] Andria G, Salvatore L. Inverter Drive Signal Processing via DFT and EKF. IEE Proceedings, March 1990; 137, Pt. B, (2) 111-119.

[25] IEEE. IEEE Std 1159: Recommended Practice for Monitoring Electric Power Quality. 2009.

[26] Caciotta M, Giarnetti S, Leccese F, Leonowicz Z. Comparison between DFT, Adaptive Window DFT and EDFT for Power Quality Frequency Spectrum Analysis. Proceedings of the Modern Electric Power Systems Conference 2010, September 20-22, 2010, Wroclaw, Poland.

[27] Gallo D, Langella R, Testa A. Desynchronized Processing Technique for Harmonic and Interharmonic Analysis. IEEE Transactions on Power Delivery, July 2004; 19(3) 993-1001.

Chapter 4

Monitoring Power Quality in Small Scale Renewable Energy Sources Supplying Distribution Systems

Nicolae Golovanov, George Cristian Lazaroiu,
Mariacristina Roscia and Dario Zaninelli

Additional information is available at the end of the chapter

http://dx.doi.org/10.5772/53464

1. Introduction

The integration of renewable sources within the existing power system affects its traditional principles of operation. The renewable energy sources (RES) can be used in small, decentralized power plants or in large ones, they can be built in small capacities and can be used in different locations [1]. In isolated areas where the cost of the extension of the power systems (from utilities point of view) or the cost for interconnection with the grid (from customer's point of view) are very high with respect to the cost of the RES system, these renewable sources are suitable. The RES systems are appropriate for a large series of applications, such as stand-alone systems for isolated buildings or large interconnected networks. The modularity of these systems makes possible the extension in the case of a load growth.

The increasing penetration rate of RES in the power systems is raising technical problems, as voltage regulation, network protection coordination, loss of mains detection, and RES operation following disturbances on the distribution network [2]. The utilization of these alternative sources presents advantages and disadvantages. The impact of the wind turbines and photovoltaic systems on network operation and power quality (harmonics, and voltage fluctuations) is highly important. The capability of the power system to absorb the power quality disturbances is depending on the fault level at the point of common coupling. [3] In weak networks or in power systems with a high wind generation penetration, the integration of these sources can be limited by the flicker level that must not exceed the standardized limits. The wind generators and PV systems interconnected to the main grid with the help of power electronics converters can cause important current harmonics.

INTECH
open science | open minds

© 2013 Golovanov et al.; licensee InTech. This is a paper distributed under the terms of the Creative Commons Attribution License (http://creativecommons.org/licenses/by/3.0), which permits unrestricted use, distribution, and reproduction in any medium, provided the original work is properly cited.

2. Power quality and renewable energy sources

Nowadays, the renewable sources generation is rapidly developing in Europe. In the last 17th years, the average growth rate is of wind generation is 15.6% annually [4]. As these renewable sources are increasingly penetrating the power systems, the impact of the RES on network operation and power quality is becoming important. The intermittent character of the wind and solar irradiation constrains the power system to have an available power reserve. Due to the output power variations of the wind turbines, voltage fluctuations are produced. In weak networks or in power systems with a high wind generation penetration, the integration of these sources can be limited by the flicker level that must not exceed the standardized limits.

The photovoltaic (PV) installations, interconnected to the mains supply, can be single-phase connected (photovoltaic installations with capacity less than 5 kW) or three-phase connected (photovoltaic installations with capacity greater than 5 kW). The direct-coupled PV systems, without electrical energy storage, inject in the power system a generated power that follows the intermittency of the primary energy source. In this case, important voltage variations can occur at the PCC. The connection of PV systems to the low voltage grid can determine voltage variations and harmonic currents [5].

2.1. Voltage fluctuations

Determination of voltage fluctuations (flicker effect) due to output power variations of renewable sources is difficult, because depend of the source's type, of generator's characteristics and network impedance. For the case of wind turbines, the long term flicker coefficient P_{lt} due to commutations, computed over a 120 min interval and for step variations, becomes [6]:

$$P_{lt} = \frac{8}{S_{sc}} \cdot N_{120}^{0.31} \cdot k_f\left(\psi_{sc}\right) \cdot S_r \tag{1}$$

where N_{120} is the number of possible commutations in a 120 min interval, $k_f(\psi_{sc})$ is the flicker factor defined for angle $\psi_{sc} = \arctan(X_{sc}/R_{sc})$, S_r – rated power of the installation, and S_{sc} – fault level at point of common coupling (PCC).

For a 10 minutes interval, the short-term flicker P_{st} is defined [6]:

$$P_{lt} = \frac{18}{S_{sc}} \cdot N_{10}^{0.31} \cdot k_f\left(\psi_{sc}\right) \cdot S_r \tag{2}$$

where N_{10} is the number of possible commutations in a 10 min interval.

The values of flicker indicator for wind turbines, due to normal operation, can be evaluated using flicker coefficient $c(\psi_{sc}, \upsilon_a)$, dependent on average annual wind speed, υ_a, in the point where the wind turbine is installed, and the phase angle of short circuit impedance, ψ_{sc}:

$$P_{st} = P_{lt} = c(\psi_{sc}, \upsilon_a)\frac{S_r}{S_{sc}}. \tag{3}$$

The flicker coefficient $c(\psi_{sc}, \upsilon_a)$ for a specified value of the angle ψ_{sc}, for a specified value of the wind speed υ_a and for a certain installation is given by the installation manufacturer, or can be experimentally determined based on standard procedures. Depending on the voltage level where the wind generator (wind farms) is connected, the angle ψ_{sc} can take values between 30° (for the medium voltage network) and 85° (for the high voltage network). Flicker evaluation is based on the IEC standard 61000-3-7 [7] which gives guidelines for emission limits for fluctuating loads in medium voltage and high voltage networks. Table 1 reports the recommended values.

Flicker severity factor	**Planning levels**	
	MV	HV
P_{st}	0.9	0.8
P_{lt}	0.7	0.6

Table 1. Flicker planning levels for medium voltage (MV) and high voltage (HV) networks

The flicker evaluation determined by a wind turbine of 650kW is analyzed. The wind turbine has a tower height of 80 meters, the rotor diameter is 47 m, and the swept area is 1735 m^2. The electrical energy production during two winter months is 127095 kWh, respectively 192782 kWh. The average wind speeds, measured at 60m height, during the first monitoring month was 6.37m/s while during march was 7.32m/s. The variation of turbine output power is shown in Fig. 1. The intermittent character of the produced power is clearly highlighted. The tower

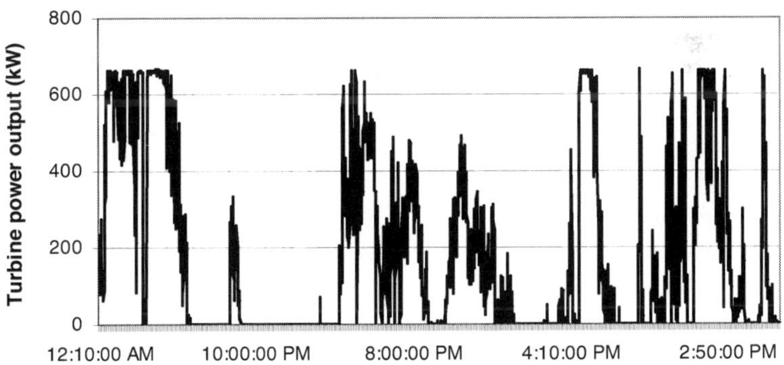

Figure 1. Turbine output power variation during the 1 month monitoring period

shadow effect for the wind generator determines a variation of the absorbed energy, which is measured as a power variation at generator terminals. Fig. 2(a) shows the wind generator, while Fig. 2(b) illustrates the tower shadow effect corresponding variation of the generator output power.

The measured values of the flicker coefficient $c(\psi_{sc}, \upsilon_a)$ for different values of the annual average wind speed υ_a and for different network impedance angle ψ_{sc} are reported in Table 2. Table 3 reports the flicker coefficient k_f values for voltage step variations, for the same wind generator.

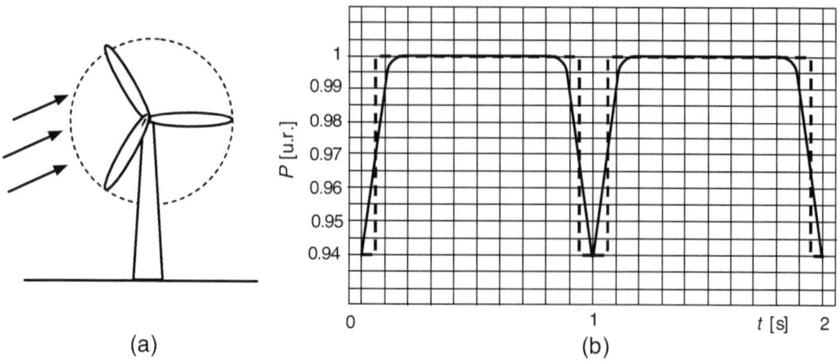

Figure 2. Variation of the wind generator output power due to the tower shadow effect.

Annual wind speed v_a [m/s]	Network impedance angle ψ_{sc} [°]			
30°	50°	70°	85°	
6	3.1	2.9	3.6	4.0
7.5	3.1	3.0	3.8	4.2
8.5	3.1	3.0	3.8	4.2
10	3.1	3.1	3.8	4.2

Table 2. Values of the flicker factor for various values of the wind speed v_a and for various angles ψ_{sc}

		Network impedance angle ψ_{sc} [°]			
30°	50°	70°	85°		
Flicker factor k_f for voltage step variations	With start at minimum speed	0.02	0.02	0.01	0.01
	With start at rated speed	0.12	0.09	0.06	0.06
	Installation is sized for $N_{10} = 3$; N_{120}=35				

Table 3. Values of the flicker factor k_f

The computations based on the values reported in Table 2 and Table 3 lead to the flicker indicator values:

i. continuous operation, annual average wind speed υ_a=7.5, interconnection with the medium voltage network (ψ_{sc}=50°, S_{sc} = 300 MVA, S_r = 0.65MVA)

$$P_{st}=P_{lt}=\frac{S_r}{S_{sc}}\cdot c(\psi_{sc}, v_a)=\frac{0.65}{300}\cdot 3=0.0065$$

2. generator interconnection at minimum speed of the wind turbine

$$P_{st}=18\cdot N_{10}^{0,31}\cdot k_f(\psi_{sc})\cdot\frac{S_r}{S_{sc}}=18\cdot 3^{0,31}\cdot 0,\ 02\cdot\frac{0.65}{300}=0,\ 00109;$$

$$P_{lt}=8\cdot N_{120}^{0,31}\cdot k_f(\psi_{sc})\cdot\frac{S_r}{S_{sc}}=8\cdot 35^{0,31}\cdot 0,\ 02\cdot\frac{0.65}{300}=0,\ 00104.$$

2. generator interconnection at rated speed of the wind turbine

$$P_{st}=18\cdot N_{10}^{0,31}\cdot k_f(\psi_{sc})\cdot\frac{S_r}{S_{sc}}=18\cdot 3^{0,31}\cdot 0,\ 09\cdot\frac{0.65}{300}=0,\ 0049;$$

$$P_{lt}=8\cdot N_{120}^{0,31}\cdot k_f(\psi_{sc})\cdot\frac{S_r}{S_{sc}}=8\cdot 35^{0,31}\cdot 0,\ 09\cdot\frac{0.65}{300}=0,\ 0047.$$

Due to the output power variations of the wind turbines, voltage fluctuations are produced. Voltage fluctuations are produced due to the wind turbine switching operations (start or stop), and due to the continuous operation. The presented voltage fluctuations study, made for one turbine, becomes necessary in large wind farms as the wind power penetration level increases quickly.

2.2. PV impact on steady state voltage variations

The variability nature of solar radiation, the weather changes or passing clouds can cause important variation of PV output power [8]. The variation of the power produced by a 30kW PV system is illustrated in Fig. 3. Fig. 4 (a) shows the generation power in a sunny day, while Fig. 4 (b) illustrates the generation power in a cloudy day.

The connection of these variable renewable sources can determine a voltage rise at PCC and in the grid. The utility has the general obligation to ensure that customer voltages are kept within prescribed limits. A voltage variation ΔV between V_{max} and V_{min}, can appear on short periods. This voltage variation can highly stress the electrical devices supplied by the power system, and in particular the owner of the photovoltaic facility (as shown in Fig. 5). Fig. 5 illustrates the possible case of a summer mid-day, when the load downstream PCC is relatively small and the PV output power exceeds the demand.

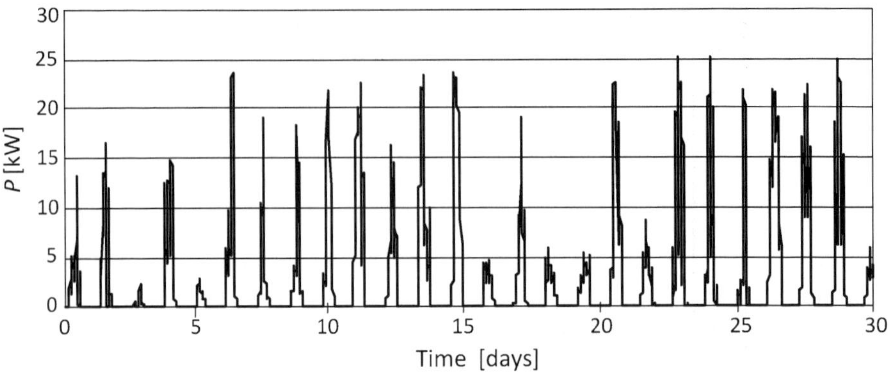

Figure 3. Variation of PV system power output during a month

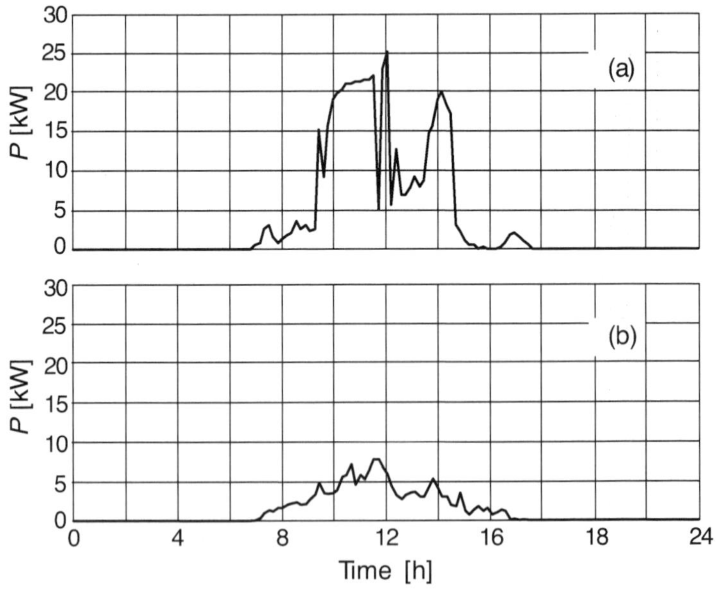

Figure 4. Variation of PV system power output during: (a) sunny day, (b) cloudy day

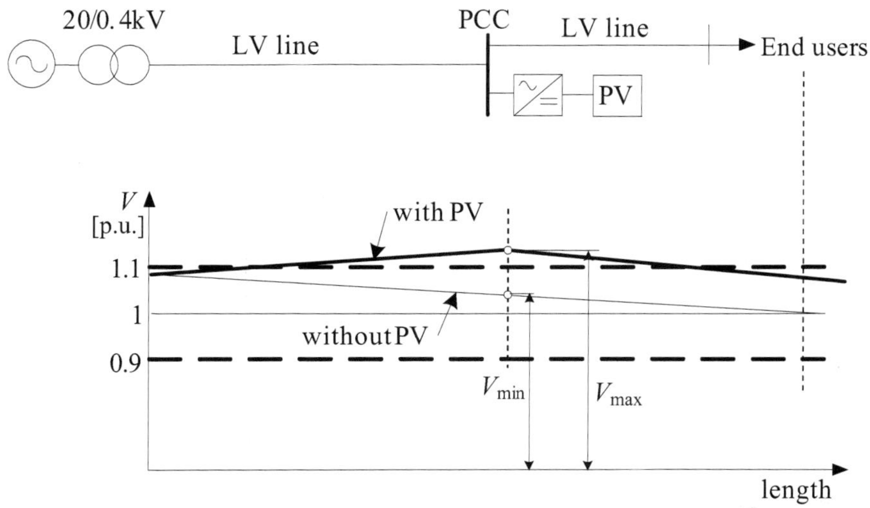

Figure 5. Influence of PV on voltage level

Voltage variations can influence the characteristics of the electrical equipment and household appliances (loss of the guaranteed performances, modifications of the efficiency) leading in some cases even to the interruption of operation. The voltage variation at PCC can be expressed as:

$$\Delta V = \frac{S_{PV}}{S_{sc}} \cdot \cos(\psi_{sc} - \varphi) \tag{4}$$

where S_{PV} is the power produced by PV, S_{sc} is the short circuit power at PCC, ψ_{sc} = arctan(X/R) is the angle of the network short circuit impedance, φ is the phase angle of the PV output current (we consider that the electric quantities are sinusoidal). In existing power systems, there are measures such that the line voltage to be sinusoidal. In (4), the system harmonics are not considered.

At low and medium voltage levels, the utilities have established limits for the amplitude of the voltage variations, which must not be exceeded during normal operation. Due to the statistical nature of the steady state voltage variations, the standard EN 50160 stipulates statistical limits [9]. In some Countries the limit ±10% established in EN 50160 is applied, while other present guidelines, elaborated in different Countries, impose more restrictive limits for the voltage variations. The relevant variations of the voltage will overlap the voltage's variations caused by load modification and can lead to the widening of the voltage limit bands.

2.3. Current harmonic perturbations

Measurement results of a 200 MW wind farm reveals the harmonic current and voltage spectra, active and reactive power variations, and the relationship between wind farm harmonic emission level and output power. It is considered that the harmonics are determined by the converter at the interconnection point with the main power system. In order to connect the PV power systems with the grid, an inverter that transforms the dc output power of the PV to the 50Hz ac power is required. The small capacity PV systems are interconnected to the main grid with the help of simple single-phase inverters, which can cause important current harmonics. General requirements can be found in standards, especially those for the interconnection of distributed generation systems to the grid and for photovoltaic systems [1, 10]. In the standard IEEE 1547, the harmonic current injection of RES at the PCC must not exceed the limits stated in table 4.

Individual harmonic order	$h<11$	$11\le h<17$	$17\le h<23$	$23\le h<35$	$35\le h$	TDD
Percent (%)	4	2	1.5	0.6	0.3	5

Table 4. Maximum harmonic current distortion in percent of current *I*, where *I* is the fundamental frequency current at full system output. Even harmonics in these ranges shall be <25% of the odd harmonic limits listed [1].

The current variation on phase *a*, during one week monitoring period of the 200 MW wind farm, is illustrated in Fig. 6. The high current variability function of wind speed can be clearly observed. The variation of total current harmonic distortion (*THDI*) during the monitoring period is illustrated in Fig. 7. The inverse relationship between the RMS electrical current and *THDI* can be seen in Fig. 6 and Fig. 7, at the same time instants. When the generated power is high (large value of RMS electrical current), the fundamental current is high and the *THDI* is small. For small generated powers, the fundamental current is small and the *THDI* is high. From practical point of view, this fact is not highly important as the small current values do not influence the voltage quality at point of common coupling.

Fig. 8 illustrates the output power variation of a PV system and the total current harmonic distortion variation during a day [11]. When the PV system has an output power close to the rated power, the *THDI* is relatively low. During the shadowing period of the PV system, the *THDI* is taking high values.

The analysis of the total harmonic distortion factor has to consider that, for high variability of the primary energy source, the large *THDI* values can lead to inappropriate conclusions. For the periods with small primary energy source, the electric current injected into the grid presents a reduced fundamental component, resulting in a high distortion factor. As the electrical current has small values, the voltage distortion and the voltage drop in the power system are negligible, and thus the voltage waveform at the point of common coupling is not affected.

http://dx.doi.org/10.5772/53464

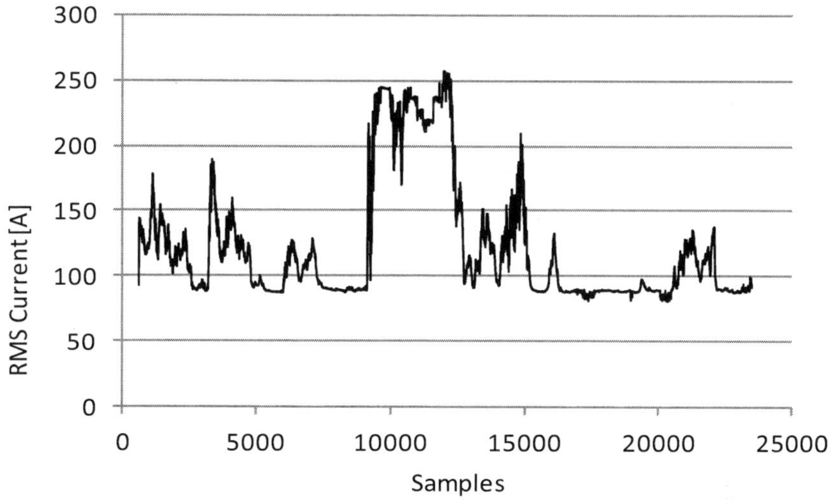

Figure 6. Variation of RMS current, phase a, during the monitoring period

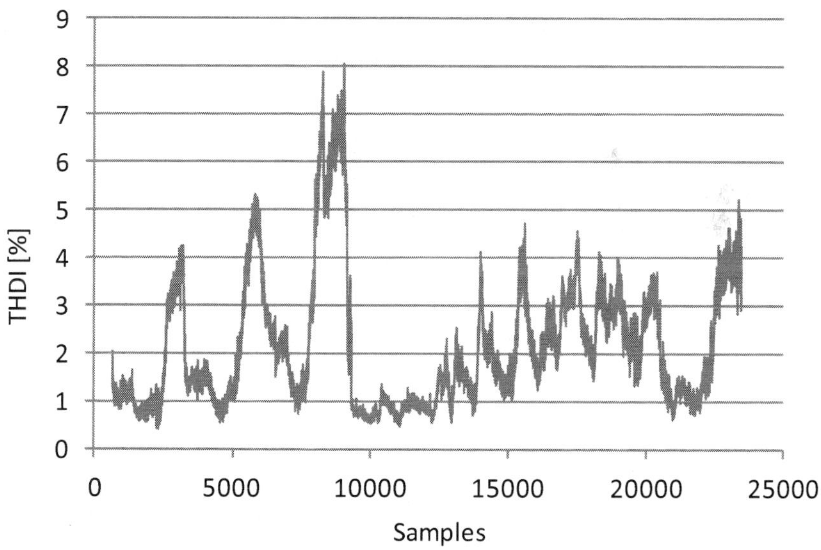

Figure 7. Variation of *THDI* during the monitoring period

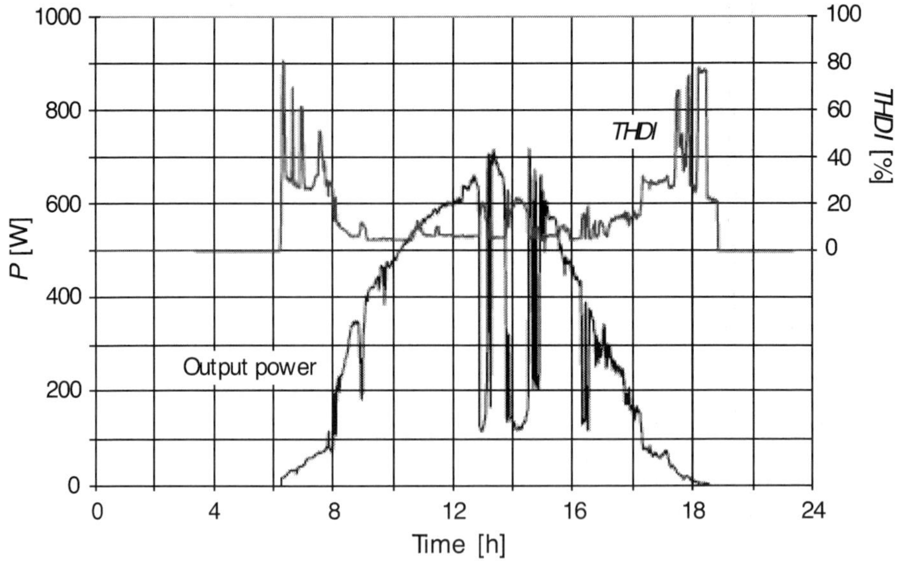

Figure 8. Variation of PV system output power and variation of THDI, during a day.

3. Conclusions

The renewable sources interconnected with the main supply can influence the power quality at the point of common coupling and can pollute the electrical network with harmonic components that must not exceed the stipulated limits. The existing trend of installing more and more small capacity sources implies the establishment as accurate as possible of their impact on power system operation.

The voltage fluctuations determined wind power variations are analyzed, both for the wind turbine switching operations (start or stop), as well as for the continuous operation. The voltage flicker study becomes necessary as the wind power penetration level increases quickly. The connection of variable renewable sources, like photovoltaic systems, can determine a voltage rise at PCC and in the grid which can affect the electrical characteristics of the equipments.

The wind generators and photovoltaic sources, connected to the power system through power electronic converters, can pollute the electrical network with harmonic components that must not exceed the stipulated limits. A better characterization, from the practical point of view, of the total current harmonic distortion determined by renewable energy sources interconnected to the mains supply through power electronic converters is necessary.

Author details

Nicolae Golovanov[1], George Cristian Lazaroiu[1], Mariacristina Roscia[2] and Dario Zaninelli[3]

1 Department of Power Systems, University Politehnica of Bucharest, Romania

2 Department of Design and Technology, University of Bergamo, Italy

3 Dipartimento di Energia, Politecnico di Milano, Italia

References

[1] IEEE Standard for interconnecting distributed resources with electric power systems, IEEE Standard 1547, 2003.

[2] N. Jenkins and R. Allan, Embedded generation, The Institution of Electrical Engineers, London, 2000, pp. 49-85.

[3] R. C. Dugan, H. W. Beaty, and M. F. McGranagham, Electrical power systems quality, New York: Mc.Graw-Hill, 1996, pp. 1-8.

[4] European Wind Energy Association [Online]. Available: http://www.ewea.org/fileadmin/files/library/publications/statistics/Wind_in_power_2011_European_statistics.pdf

[5] F. Gagliardi, F. Iannone, G.C. Lazaroiu, M. Roscia, D. Zaninelli, Sustainable building for the quality of power supply, in Proc. IEEE PowerTech Conference 2005, Sankt Petersburg, Russia, June 27-30, 2005, pp. 6

[6] IEC 61400-21, Wind turbine generator systems - Measurement and assessment of power quality characteristics of grid connected wind turbines, 2001

[7] IEC 61000-3-7, Assessment of emission limits for the connection of fluctuating load installation to MV, HV and EHV power systems, 2008.

[8] M. Simonov, M. Mussetta, F. Grimaccia, S. Leva, R. E. Zich, "Artificial Intelligence forecast of PV plant production for integration in smart energy systems," International Review of Electrical Engineering, vol. 7, no. 1, pp. 3454-3460

[9] EN 50160/2010, Voltage characteristics of electricity supplied by public distribution systems.

[10] IEEE Recommended Practice for Utility Interface of Photovoltaic (PV) Systems, IEEE Standard 929-2000, April 2000

[11] G. Chicco, J. Schlabbach, F. Spertino, Characterisation and Assessment of the Harmonic Emission of Grid-Connected Photovoltaic Systems, in Proc. IEEE Power Tech 2005, St. Petersburg, Russia, June 27-30, 2005, pp.7

Chapter 5

Active Power Conditioners to Mitigate Power Quality Problems in Industrial Facilities

João L. Afonso, J. G. Pinto and Henrique Gonçalves

Additional information is available at the end of the chapter

http://dx.doi.org/10.5772/53189

1. Introduction

Non-linear loads are commonly present in industrial facilities, service facilities, office buildings, and even in our homes. They are the source of several Power Quality problems such as harmonics, reactive power, flicker and resonance [1-3]. Therefore, it can be observed an increasing deterioration of the electrical power grid voltage and current waveforms, mainly due to the contamination of the system currents with harmonics of various orders, including inter-harmonics. Harmonic currents circulating through the line impedance produces distortion in the system voltages (see Figure 1). Moreover, since many of the loads connected to the electrical systems are single-phase ones, voltage unbalance is also very common in three-phase power systems [2]. The distortion and unbalance of the system voltages causes several power quality problems, including the incorrect operation of some sensitive loads [4,5]. Figure 1 presents a power system with sinusoidal source voltage (v_S) operating with a linear and a non-linear load. The current of the non-linear load (i_{L1}) contains harmonics. The harmonics in the line-current (i_S) produce a non-linear voltage drop (Δv) in the line impedance, which distorts the load voltage (v_L). Since the load voltage is distorted, even the current at the linear load (i_{L2}) becomes non-sinusoidal.

The problems caused by the presence of harmonics in the power lines can be classified into two kinds: instantaneous effects and long-term effects. The instantaneous effects problems are associated with interference problems in communication systems, malfunction or performance degradation of more sensitive equipment and devices. Long-term effects are of

INTECH
open science | open minds

© 2013 Afonso et al.; licensee InTech. This is a paper distributed under the terms of the Creative Commons Attribution License (http://creativecommons.org/licenses/by/3.0), which permits unrestricted use, distribution, and reproduction in any medium, provided the original work is properly cited.

thermal nature and are related to additional losses in distribution and overheating, causing a reduction of the mean lifetime of capacitors, rotating machines and transformers. Because of these problems, the issue of the power quality delivered to the end consumers is, more than ever, an object of great concern. International standards concerning electrical power quality (IEEE-519, IEC 61000, EN 50160) impose that electrical equipments and facilities should not produce harmonic contents greater than specified values, and also indicate distortion limits to the supply voltage. According to the European COPPER Institute – Leonard Energy Initiative, costs related to power quality problems in Europe are estimated in more than €150.000.000 per year. Therefore, it is evident the necessity to develop solutions that are able to mitigate such disturbances in the electrical systems, improving their power quality.

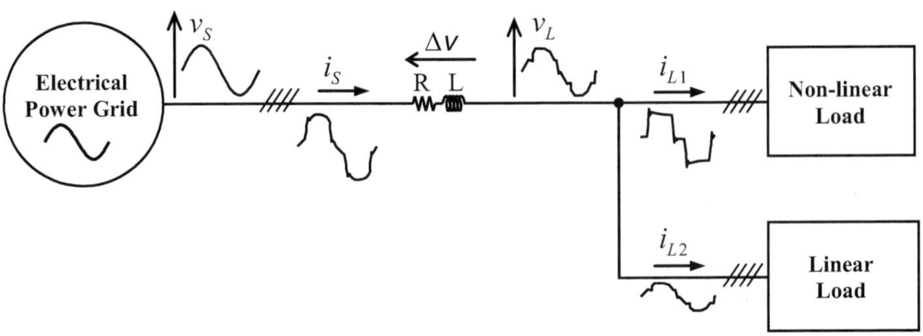

Figure 1. Single line block diagram of a system with non-linear loads.

Passive filters have been used as a solution to solve harmonic current problems, but they present several disadvantages, namely: they only filter the frequencies they were previously tuned for; their operation cannot be limited to a certain load; the interaction between the passive filters and other loads may result in resonances with unpredictable results [6]. To cope with these disadvantages, in the last years, research engineers have presented various solutions based in power electronics to compensate power quality problems [6-12]. These equipments are usually designated as Active Power Conditioners. Examples of such devices are the Shunt Active Power Filter, the Series Active Power Filter, and the Unified Power Quality Conditioner (UPQC).

Active Power Filters are conditioners connected in parallel or in series with the electrical power grid. When connected in parallel is called Shunt Active Power Filter, and when connected in series is named Series Active Power Filter. The Shunt Active Power Filter behaves as a controlled current-source draining the undesired components from the load currents, such that the currents in the electrical power grid become sinusoidal, balanced, and in phase with fundamental positive sequence component of the system voltages. On the other hand, the Series Active Power Filter works as a voltage-source connected in series with the electrical power grid, compensating voltage harmonics, sags (sudden reduction of the voltage followed by its recovery after a brief interval), swells (sudden increase of the voltage followed by its recovery after a brief interval) and flicker (cyclic variation of light intensity of lamps caused by fluctuation of the supply voltage). Three-phase Series Active Power Filters can also compensate unbalances in the phase voltages [12]. If the DC link of the Series Active Power Filter inverter is connected to a power supply, its compensation capabilities increases, allowing also the compensation of long term undervoltages and overvoltages [10,11].

The Unified Power Quality Conditioner (UPQC) is composed by two power converters sharing the DC Link. One of these power converters is connected in series with the electrical power grid and the other is connected in parallel with the electrical power grid. This conditioner offers many compensation options in function of the type of control used. Some of the power quality problems that the UPQC can compensate are: voltage harmonics, voltage unbalance, voltage sags, voltage swells, current harmonics, current unbalance, undervoltages, overvoltages, reactive power, and neutral wire current in three-phase four wire systems.

The utilization of equipment like Shunt Active Power Filters, Series Active Power Filters and UPQCs presents significant advantages to the power system. In this way, the research of new topologies and new control algorithms to improve the performance and capabilities of these equipments is object of great interest [13,14].

The aforementioned Active Power Conditioners are the most suited for industrial applications, so they will be presented with more detail in the following topics.

2. Shunt Active Power Filter

The Shunt Active Power Filter is a device which is able to compensate for both current harmonics and power factor. Furthermore, in three-phase four wire systems it allows to balance the currents in the three phases, and to eliminate the current in the neutral wire [15-17]. Figure 2 presents the electrical scheme of a Shunt Active Power Filter for a three-phase power system with neutral wire.

The power stage is, basically, a voltage-source inverter with a capacitor in the DC side (the Shunt Active Filter does not require any internal power supply), controlled in a way that it acts like a current-source. From the measured values of the phase voltages (v_a, v_b, v_c) and load currents (i_a, i_b, i_c), the controller calculates the reference currents (i_{ca}^*, i_{cb}^*, i_{cc}^*, i_{cn}^*) used by the inverter to produce the compensation currents (i_{ca}, i_{cb}, i_{cc}, i_{cn}). This solution requires 6 current sensors: 3 to measure the load currents (i_a, i_b, i_c) for the control system and 3 for the closed-loop current control of the inverter (in both cases the fourth current, the neutral wire currents, i_n and i_{cn}, are calculated by adding the three measured currents of phases *a*, *b*, *c*). It also requires 4 voltage sensors: 3 to measure the phase voltages (v_a, v_b, v_c) and another for the closed-loop control of the DC link voltage (V_{dc}). For three-phase balanced loads (three-phase motors, three-phase adjustable speed drives, three-phase controlled or non-controlled rectifiers, etc) there is no need to compensate for the current in neutral wire, so the forth wire of the inverter is not required, simplifying the Shunt Active Power Filter hardware. Since they compensate the power quality problems upstream to its coupling point they should be installed as near as possible of the non-liner loads, avoiding the circulation of current harmonics, reactive currents and neutral wire currents through the facility power lines. Therefore it is advantageous to use various small units, spread along the electrical installation, instead of using a single high power Shunt Active Power Filter at the input of the industry, at the PCC (Point of Common Coupling – where the electrical installation of the industry is connected to the electrical power distribution system).

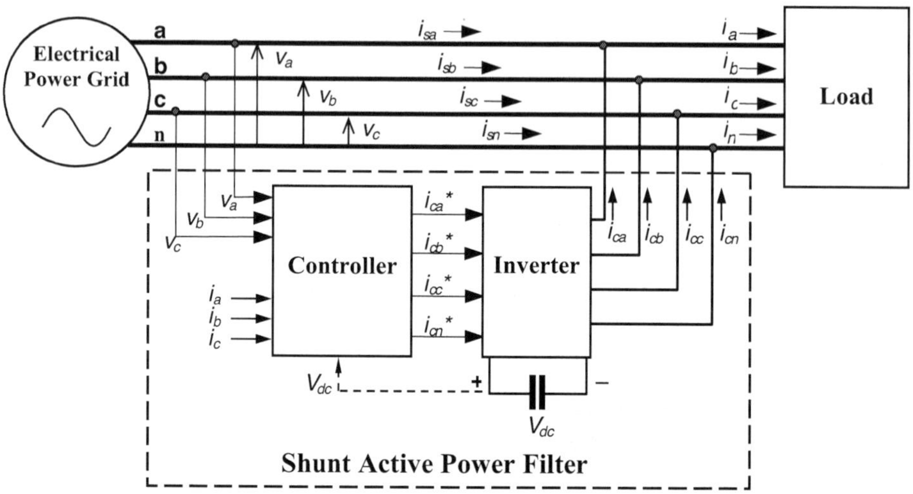

Figure 2. Shunt Active Power Filter for a three-phase power system with neutral wire.

2.1. Typical Waveforms

Typical waveforms of an electrical installation equipped with a Shunt Active Power Filter are presented in Figure 3. It can be seen that the currents in the load present high harmonic content (THD% of 58%, in average, see Figure 4), and are also unbalanced, which results in a considerable neutral wire current (Figure 3 (d)). The Shunt Active Power Filter makes the currents in the source sinusoidal and balanced (see Figure 3 (b)). The THD% of the source currents is only of about 1% (Figure 4).

In Figure 4 is presented the THD% of the different currents in the system (at the load, source and active filter). The THD% was, in all the cases, calculated in relation to the fundamental frequency of the power grid source (50 Hz). That is why the values of THD% presented for the compensation currents injected by the active filter are so high.

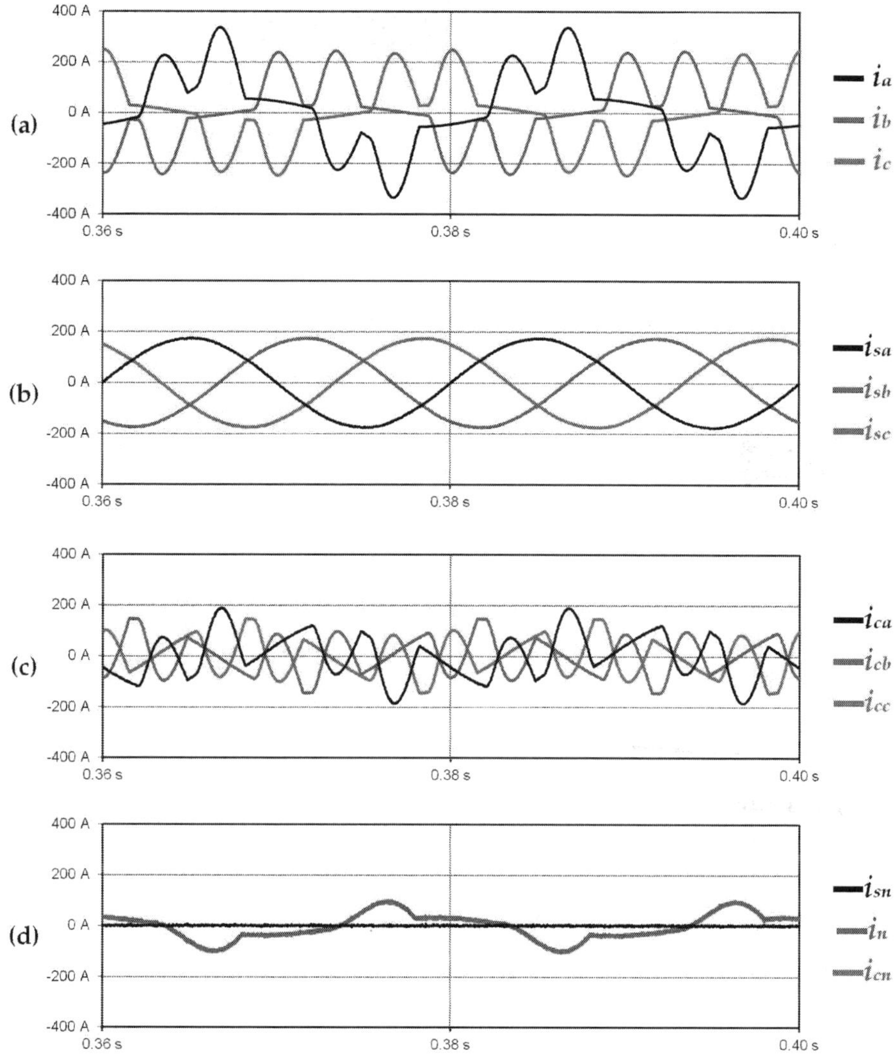

Figure 3. Typical waveforms of an installation with a Shunt Active Power Filter: (a) Load currents; (b) Source currents; (c) Active filter compensation currents; (d) Neutral wire currents.

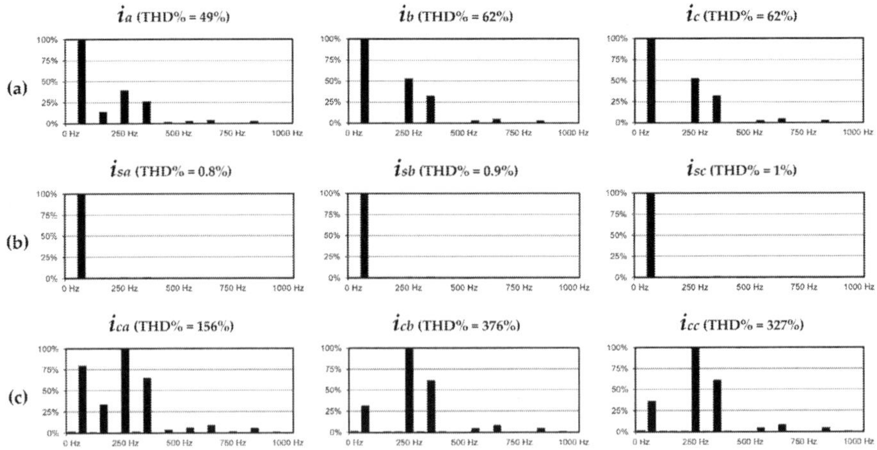

Figure 4. Harmonic spectrum and THD% of the currents in an installation with a Shunt Active Power Filter: (a) Load currents; (b) Source currents; (c) Compensation currents.

3. Series Active Power Filter

The Series Active Power Filter is the dual of the Shunt Active Power Filter, and is able to compensate for voltage harmonics, voltage sags, voltage swells and flicker, making the voltages applied to the load almost sinusoidal (compensating for voltage harmonics) [18,19]. The three-phase Series Active Filter can also balance the load voltages [20]. Figure 5 shows the electrical scheme of a Series Active Power Filter for a three-phase power system.

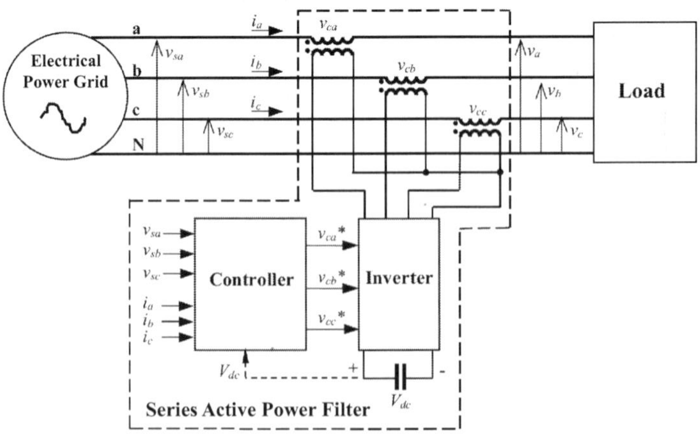

Figure 5. Series Active Power Filter for a three-phase power system.

The Series Active Power Filter consists of a voltage-source inverter (behaving as a controlled voltage-source) and requires 3 single-phase transformers to interface with the power system. However, some authors have presented research results of Series Active Power Filter topologies without the use of line transformers [21,22]. From the measured values of the phase voltages at the source side (v_{sa}, v_{sb}, v_{sc}) and of the load currents (i_a, i_b, i_c), the controller calculates the reference compensation voltages (v_{ca}^*, v_{cb}^*, v_{cc}^*), used by the inverter to produce the compensation voltages (v_{ca}, v_{cb}, v_{cc}). The Series Active Power Filter does not compensate for load current harmonics but it acts as high-impedance to the current harmonics coming from the electrical power grid side. Therefore, it guarantees that passive filters eventually placed at the load side will work appropriately and not drain harmonic currents from the rest of the power system.

3.1. Typical Waveforms

Typical waveforms of an installation equipped with a Series Active Power Filter are presented in Figure 6.

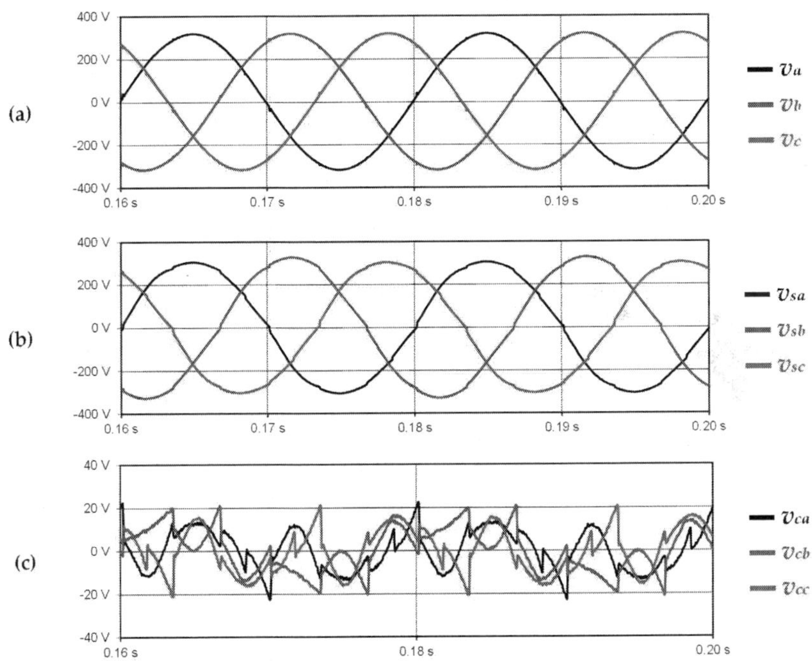

Figure 6. Typical waveforms of an installation with a Series Active Power Filter: (a) Load voltages; (b) Source voltages; (c) Active filter compensation voltages.

In Figure 7 is presented the THD% of the different voltages in the system (load, source and series active filter). It can be seen that the voltages in the source present some harmonic content (THD% between 2.5% and 4%). The Series Active Power Filter makes the voltages in the load practically sinusoidal, with almost none distortion (see Figure 6 (a)). The THD% of the load voltages is below or equal to 0.4%. The THD% was, in all the cases, calculated in relation to the fundamental frequency of the power grid source (50 Hz). That is the motive way the values presented for the compensation voltages produced by the active filter are so high.

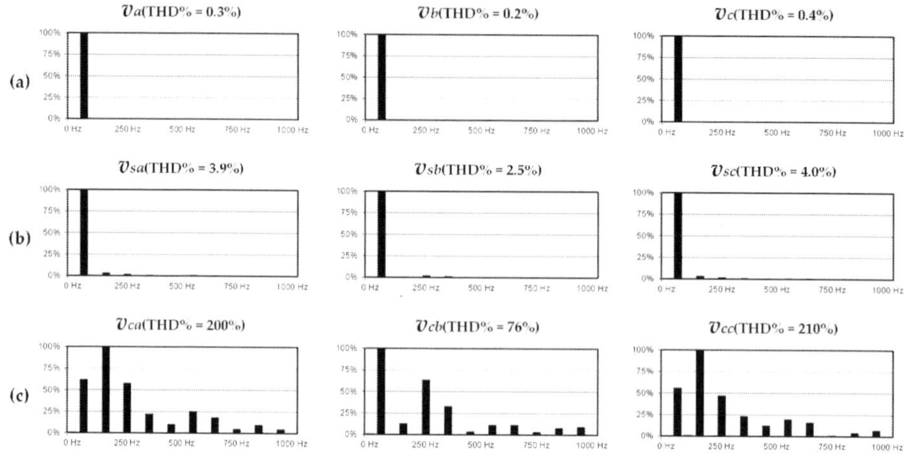

Figure 7. Harmonic spectrum and THD% of the voltages in an installation with a Series Active Power Filter: (a) Load voltages; (b) Source voltages; (c) Compensation voltages.

4. Unified Power Quality Conditioner

The Unified Power Quality Conditioner (UPQC) combines the Shunt Active Power Filter with the Series Active Power Filter, sharing the same DC Link, in order to compensate both voltages and currents, so that the load voltages become sinusoidal and at nominal value, and the source currents become sinusoidal and in phase with the source voltages [23,24]. In the case of three-phase systems, a three-phase UPQC can also balance the load voltages and the source currents, and eliminate the source neutral current. Figure 8 shows the electrical scheme of a Unified Power Quality Conditioner for a three-phase power system.

From the measured values of the source phase voltages (v_{sa}, v_{sb}, v_{sc}) and load currents (i_a, i_b, i_c), the controller calculates the reference compensation currents (i_{ca}^*, i_{cb}^*, i_{cc}^*, i_{cn}^*) used by the inverter of the shunt converter to produce the compensation currents (i_{ca}, i_{cb}, i_{cc}, i_{cn}). Using the measured values of the source phase voltages, and source currents (i_{sa}, i_{sb}, i_{sc}), the controller calculates the reference compensation voltages (v_{ca}^*, v_{cb}^*, v_{cc}^*) used by the inverter of the series converter to produce the compensation voltages (v_{ca}, v_{cb}, v_{cc}).

http://dx.doi.org/10.5772/53189

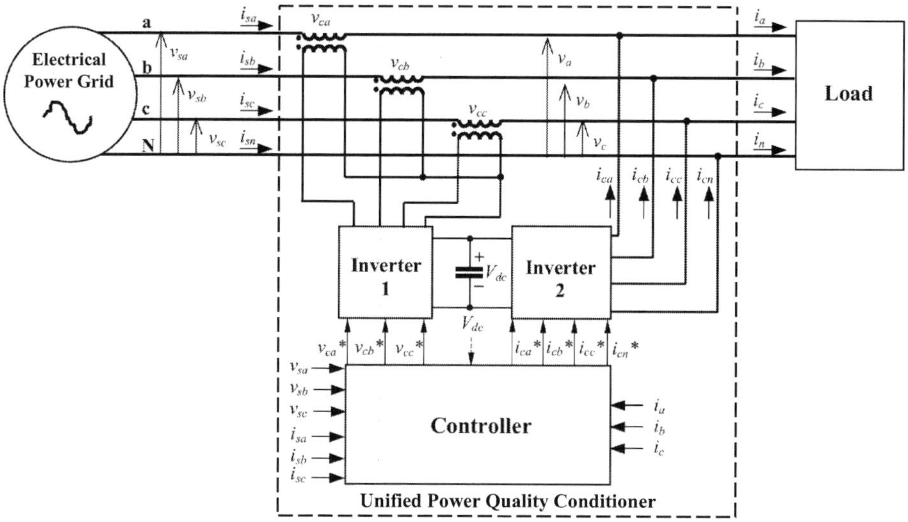

Figure 8. Unified Power Quality Conditioner for a three-phase power system.

4.1. Typical Waveforms

Typical waveforms of an installation equipped with a Unified Power Quality Conditioner are presented next. In Figure 9 are shown the load currents, source currents, compensation currents, neutral wire currents, load voltages, and source voltages. It can be seen that the currents in the load present a high harmonic content (THD% between 32% and 41%, see Figure 10 (a)), and are also unbalanced, which results in a considerable neutral wire current (i_n in Figure 9 (d)). The THD% of the source voltages is also high (about 6%, as shown in Figure 10 (e)).

By the action of the Unified Power Quality Conditioner the currents in the source become sinusoidal, in phase with the voltages, and balanced (Figure 6 (b)). The THD% of the source currents is reduced to about 1% (Figure 10 (b)). Also, the load voltages become sinusoidal with almost none distortion (Figure 6 (e)). The THD% of the load voltages is reduced to only 0.4% (Figure 10 (d)).

In Figure 10 is presented the THD% of the different currents and voltages in the electrical system (at the load, source and UPQC). The THD% was, in all the cases, calculated in relation to the fundamental frequency of the power grid source (50 Hz). That is way the values presented for the compensation currents injected by the UPQC are so high.

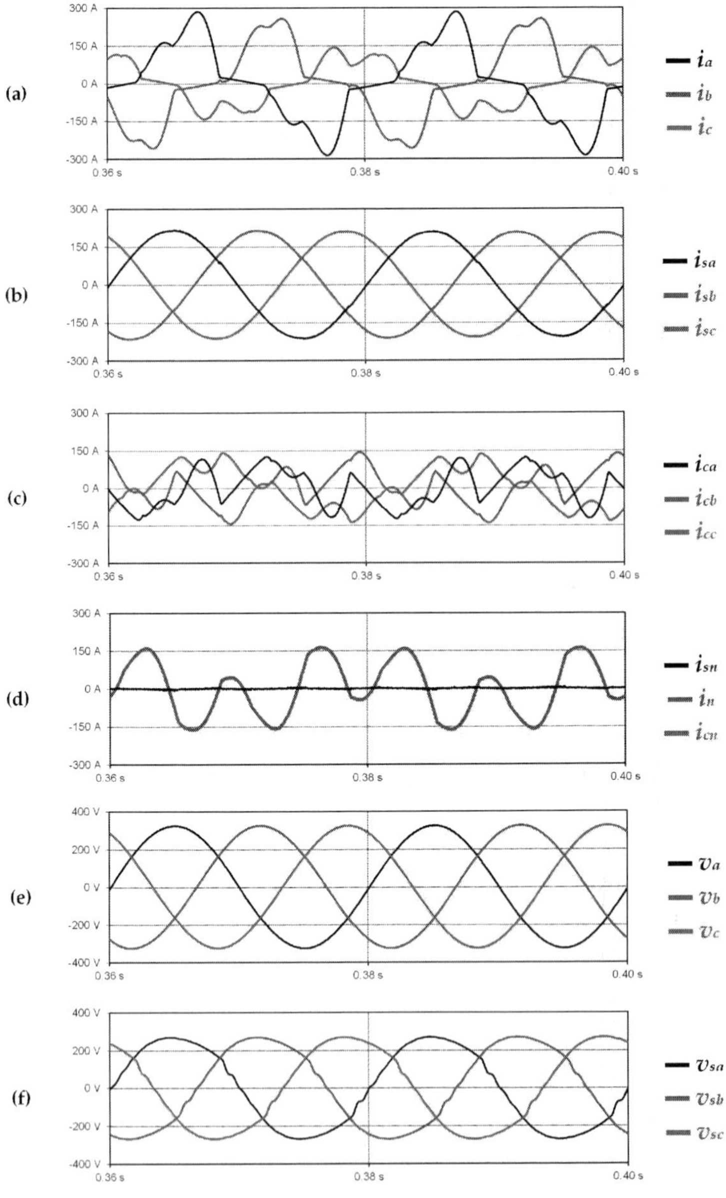

Figure 9. Typical waveforms of an installation with a UPQC: (a) Load currents; (b) Source currents; (c) Compensation currents; (d) Neutral wire currents; (e) Load voltages; (f) Source voltages.

http://dx.doi.org/10.5772/53189

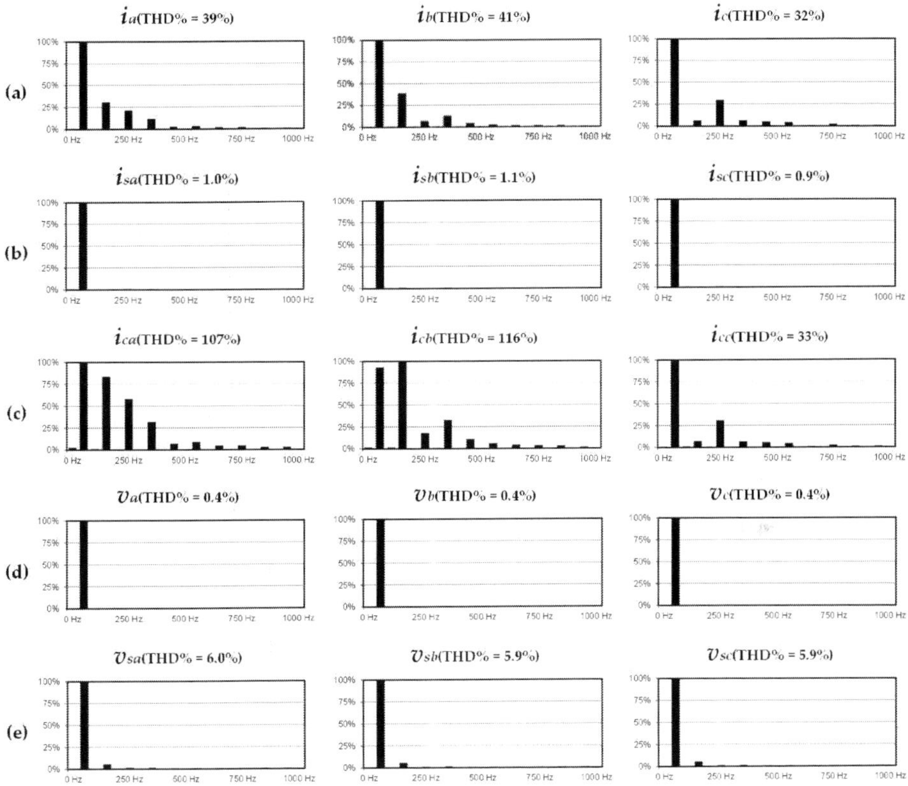

Figure 10. Harmonic spectrum and THD% of the currents and voltages in an installation with an UPQC: (a) Load currents; (b) Source currents; (c) Compensation currents; (d) Load voltages; (e) Source voltages.

5. Control Methods for Active Power Filters

The control methods applied to Active Power Filters and Unified Power Quality Conditioners are decisive in achieving the goals of compensation, in the determination of the conditioner power rate, and in theirs dynamic and steady-state performances. Basically, the different approaches regarding the calculation of the compensation currents and voltages from the measured distorted quantities can be grouped into two classes: frequency-domain and time-domain.

The frequency-domain approach implies the analysis of the Fourier transform, which leads to a huge amount of calculations, making the control method very heavy in terms of processing time and required computational capacity. The time-domain approach uses the traditional concepts of circuit analysis and algebraic transformations associated with changes of reference frames, which greatly simplify the control task. In general, power definitions in the time domain offer a more appropriate basis for the design of controllers for power electronic devices, because they are also valid during transients. This is especially true for applications in three-phase electrical power systems if the definitions are done already considering a three-phase circuit, instead of considering a single-phase circuit and then summing up to have a three-phase system [14].

In a three-phase electrical power system the three-phase power delivered to a load by the source has the well-known expression:

$$p_3(t) = v_a(t) i_a(t) + v_b(t) i_b(t) + v_c(t) i_c(t) \tag{1}$$

where $v_a(t)$, $v_b(t)$, and $v_c(t)$ represents the instantaneous load voltages referred to the neutral point, and $i_a(t)$, $i_b(t)$, and $i_c(t)$ are the load instantaneous currents. However, for the given voltages, there is more than one set of currents producing the same instantaneous power. So, what is the optimal set of currents for a given power? One possible answer is the set of currents that minimizes power loss in the lines. On the other hand, it is known that for a balanced sinusoidal system, in voltage and current, the instantaneous power is constant, and equal to the active power, since this value corresponds to the average value of the instantaneous power. Therefore, the best set of currents can be the one that leads to a constant instantaneous power.

Different time-domain power definitions can be found in the literature. The most important are: the p-q Theory (Instantaneous Power Theory) proposed by Akagi et al. [25,26]; FBD (Fryze - Buchholz - Depenbrock) proposed by Depenbrock [27]; the CPT (Conservative Power Theory) proposed by Tenti [28]; and the CPC (Current's Physical Components) proposed by Czarnecki [29,30]. It can be also found in the literature p-q Theory inspired control algorithms for switching compensators, as for example, the p-q-r Theory [31-33]. A comparison involving the p-q-r and the p-q theories is provided in [33]. The control algorithm denominated as Synchronous Reference Frame (SRF) [34] also presents similar aspects related with the p-q-r and the p-q theories. The SRF control algorithm is defined in the *d-q-0* reference frame. All of these control algorithms can be applied to control switching compensators connected in three-phase systems, with or without neutral wire.

5.1. The p-q Theory Fundamentals

In 1983, Akagi et al. [25,26] have proposed "The Generalized Theory of the Instantaneous Reactive Power in Three-Phase Circuits", also known as Instantaneous Power Theory, or p-q Theory, for the control of Active Power Filters. The fact of being a time-domain theory makes it viable for operation in steady state or transient state, as well as for generic voltage and current waveforms, allowing a real time control of the Active Power Filters. Another advantage of the p-q Theory is the simplicity of its calculations, which consists only in algebraic operations, being the only exception the extraction of the average and alternating components of the calculated powers.

5.1.1. Clarke for Three-Phase Four-Wire Electrical Power Systems

The p-q Theory was initially developed for three-phase electrical power systems without neutral wire, with a short reference to three-phase systems with neutral wire. Later, Watanabe et al. [35] and Aredes et al. [36] extended it to three-phase electrical power systems with neutral wire.

This theory consists in an algebraic transformation (the Clarke transformation) of the three-phase voltages and currents in the *a-b-c* coordinates to the *α-β-0* coordinates, where *α-β* are orthogonal, and the *0* coordinate corresponds to the zero-sequence component. The p-q Theory transformation applied to the electrical power grid voltages and load currents is given by:

$$\begin{bmatrix} v_0 \\ v_\alpha \\ v_\beta \end{bmatrix} = \sqrt{2/3} \begin{bmatrix} 1/\sqrt{2} & 1/\sqrt{2} & 1/\sqrt{2} \\ 1 & -1/2 & -1/2 \\ 0 & \sqrt{3}/2 & -\sqrt{3}/2 \end{bmatrix} \begin{bmatrix} v_a \\ v_b \\ v_c \end{bmatrix}$$
$$\begin{bmatrix} i_0 \\ i_\alpha \\ i_\beta \end{bmatrix} = \sqrt{2/3} \begin{bmatrix} 1/\sqrt{2} & 1/\sqrt{2} & 1/\sqrt{2} \\ 1 & -1/2 & -1/2 \\ 0 & \sqrt{3}/2 & -\sqrt{3}/2 \end{bmatrix} \begin{bmatrix} i_a \\ i_b \\ i_c \end{bmatrix} \tag{2}$$

The instantaneous three-phase electrical power, in the *a-b-c* coordinates is defined as:

$$p_3 = p_a + p_b + p_c = v_a i_a + v_b i_b + v_c i_c \tag{3}$$

In the *α-β-0* coordinates the instantaneous three-phase electrical power is defined as:

$$p_3 = p + p_0 = v_\alpha i_\alpha + v_\beta i_\beta + v_0 i_0 \tag{4}$$

The two components of p_3 are defined as follow:

$$p = v_\alpha i_\alpha + v_\beta i_\beta \qquad \text{Instantaneous real power} \tag{5}$$

$$p_0 = v_0 i_0 \qquad \text{Instantaneous zero-sequence power} \tag{6}$$

The instantaneous imaginary power is defined as:

$$q = v_\beta i_\alpha - v_\alpha i_\beta \tag{7}$$

The $q$power difers form the conventional reactive three-phase electrical power, since it also takes into consideration all the voltage and current harmonics.

Since the $p$and $q$powers do not depend on the zero-sequence components of the voltages and currents, but only on the same α-β components, they can be written together:

$$\begin{bmatrix} p \\ q \end{bmatrix} = \begin{bmatrix} v_\alpha & v_\beta \\ v_\beta & -v_\alpha \end{bmatrix} \begin{bmatrix} i_\alpha \\ i_\beta \end{bmatrix} \tag{8}$$

5.1.2. Physical Meaning of the p-q Theory Electrical Powers

The different p-q Theory electrical powers are illustrated in Figure 11, for an electrical power system represented in *α-β-0* and in Figure 12 for an electrical power system represented in *a-b-c* coordinates, and have the following physical meaning:

$\bar{p}$: mean value of the instantaneous real power – corresponds to the energy per time unity that is transferred from the power supply to the load, through the α-β coordinates, or through the *a-b-c* coordinates, in a balanced way (it is the desired power component).

$\tilde{p}$: alternated value of the instantaneous real power – It is the energy per time unity that is exchanged between the power supply and the load, through the α-β coordinates, or through the *a-b-c* coordinates.

q: instantaneous imaginary power – corresponds to the power that is exchanged between the α-β coordinates, or between the *a-b-c* coordinates. This power does not imply any transference or exchange of energy between the power supply and the load, but is responsible for the existence of undesirable currents, which circulate between the system phases. In the case of a balanced sinusoidal voltage supply and a balanced load, with or without harmonics, $\bar{q}$ (the mean value of the instantaneous imaginary power) is equal to the conventional reactive power ($\bar{q}=3VI_1\sin\phi_1$).

$\bar{p}_0$: mean value of the instantaneous zero-sequence power – corresponds to the energy per time unity which is transferred from the power supply to the load through the zero-sequence components of voltage and current.

$\tilde{p}_0$:alternated value of the instantaneous zero-sequence power – it means the energy per time unity that is exchanged between the power supply and the load through the zero-sequence components.

The zero-sequence power, p_0, only exists in three-phase systems with neutral wire. Furthermore, the systems must have unbalanced voltages and currents and/or 3rd harmonics in both voltage and current of at least one phase.

http://dx.doi.org/10.5772/53189

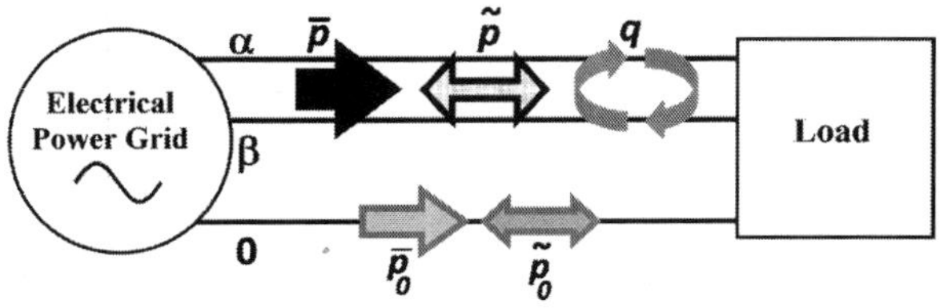

Figure 11. Power components of the p-q Theory in *α-β-0* coordinates.

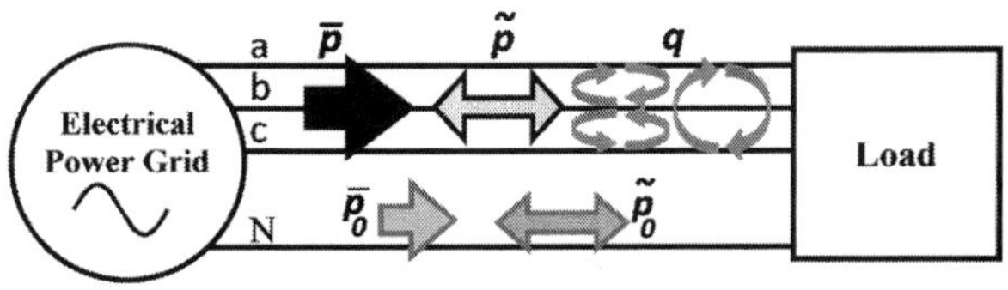

Figure 12. Power components of the p-q Theory in *a-b-c* coordinates.

5.1.3. The p-q Theory Powers Compensation

From the concepts seen before, $\bar{p}$and $\bar{p}_0$ are usually the only desirable p-q Theory power components that the source must supply. The other power components can be compensated using a Shunt Active Power Filter. Figure 13 shows the Shunt Active Power Filter for an electrical power system represented in *a-b-c* coordinates, and Figure 14 shows the Shunt Active Power Filter for an electrical power system represented in *α-β-0* coordinates.

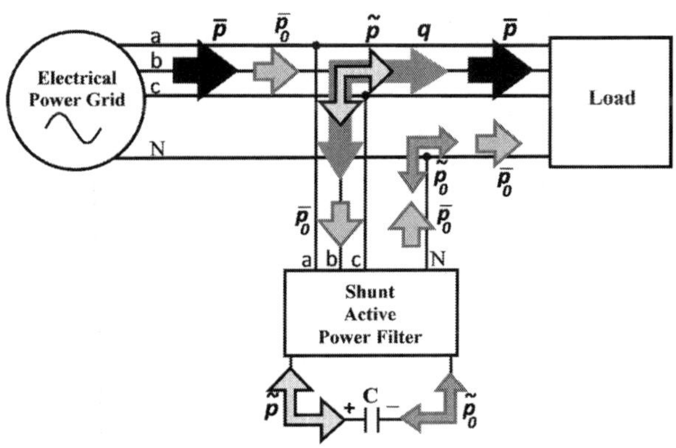

Figure 13. Compensation of power components $\tilde{p}$, q, $\bar{p}_0$, and $\tilde{p}_0$ in *a-b-c* coordinates.

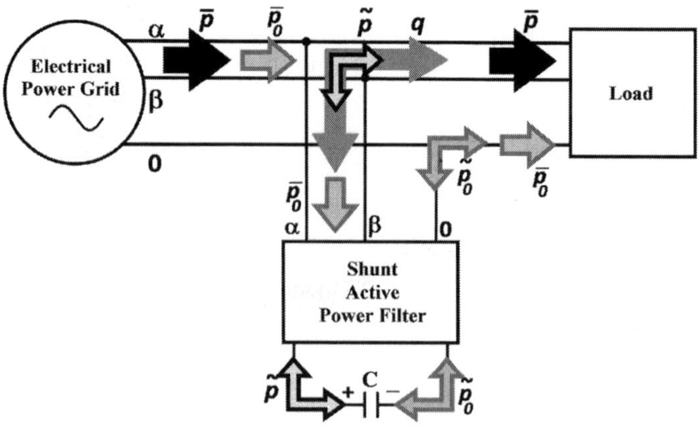

Figure 14. Compensation of power components $\tilde{p}$, q, $\tilde{p}_0$, and $\bar{p}_0$ in *α-β-0* coordinates.

With the Shunt Active Power Filter in operation, the $\tilde{p}$ and $\tilde{p}_0$ power components cease to be exchanged between the load and the electrical power source and start to be exchanged between the load and the Shunt Active Power Filter DC link capacitor, which continuously stores and delivers energy, to compensate these pulsating electrical powers.

The power component q is not associated with any energy transference, so the currents associated with this electrical power component start to circulate only between the Shunt Active Power Filter and the load, and not anymore through the electrical power grid.

The power component p_0 only can exist in three-phase four-wire systems with voltage and current distortions or/and unbalances (when simultaneously $i_0 \neq 0$ and $v_0 \neq 0$, at the same frequencies), in these conditions it is necessary to compensate the electrical power component p_0 to allow the balancing of the currents, and to make the current in the neutral wire assume a null value upstream of the Shunt Active Power Filter, or in other words, to make that the zero-sequence component of the current between the electrical power source and the Shunt Active Power Filter is eliminated.

The compensation of $\bar{p}_0$, requires that the Shunt Active Power Filter delivers energy to the load. To do this there are two possibilities:

- Include a power supply on the Shunt Active Power Filter inverter DC link to deliver this energy.
- Drain the energy required for the $\bar{p}_0$ compensation from the electrical power grid itself, in a balanced way by the three-phases.

The second possibility, was proposed by Aredes et al. [36], and is implicit in Figure 13. It is also possible to conclude that the Shunt Active Power Filter DC link capacitor is only necessary to compensate $\tilde{p}$ and $\tilde{p}_0$, since these quantities must be stored in this component at one moment to be later delivered back to the load. The instantaneous imaginary power (q), which includes the conventional reactive power, is compensated without the contribution of this capacitor. This means that, the size of the DC link capacitor does not depend on the amount of reactive power to be compensated.

5.2. Calculations for theShunt Active Power Filter Control

The p-q Theory presents some interesting features when applied to the control of Active Power Filters for three-phase power systems, namely:

- It is inherently a three-phase system theory;
- It can be applied to any three-phase system (balanced or unbalanced, with or without harmonics, for compensation of both voltages and/or currents);
- It is based in instantaneous values, allowing excellent dynamic response;
- Its calculations are relatively simple (it only includes algebraic expressions that can be implemented using a simple controller);
- It allows two control strategies: "constant instantaneous real power at source"and "sinusoidal current at source".

As can be seen in Figure 15, the inputs of the control system are the instantaneous values of the voltages and currents in the phases that feed the load to be compensated (v_a, v_b, v_c and i_a, i_b, i_c).

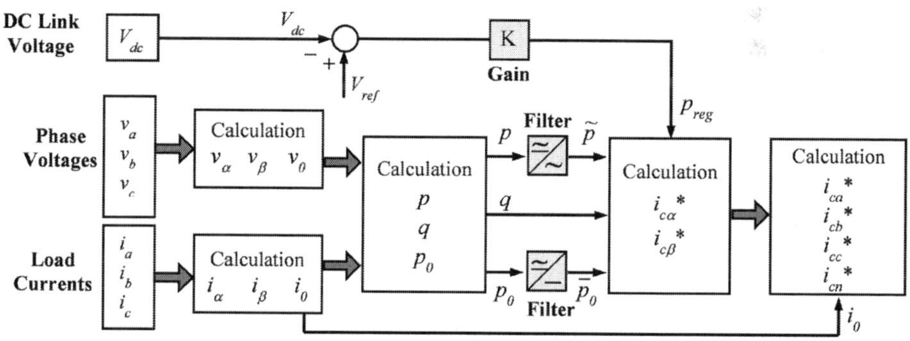

Figure 15. Control system structure for the control strategy "constant instantaneous real power at source".

These currents and voltages are calculated in the *α-β-0* coordinates through the equations given in (2). Using the equations (5), (6) and (7) are calculated the instantaneous powers p, p_0 and q, respectively. The separation of the p-q Theory power components in their average and alternating values can be obtained using analog or digital filters, according to the type of control system.

To calculate the reference compensation currents in the *α-β-0* coordinates is used the equation (9):

$$\begin{bmatrix} i_{c\alpha}^* \\ i_{c\beta}^* \end{bmatrix} = \frac{1}{v_\alpha^2 + v_\beta^2} \begin{bmatrix} v_\alpha & -v_\beta \\ v_\beta & v_\alpha \end{bmatrix} \begin{bmatrix} p_x \\ q_x \end{bmatrix} \tag{9}$$

In the previous equation, p_x and $q_x$are the values of the power components to be provided by the Shunt Active Power Filter. Always that the Shunt Active Power Filter compensates the zero-sequence power (p_0), $p_x$must be subtracted of the average value of the zero-sequence power,$\bar{p}_0$, as presented in equation (10). In this way the energy per time unit, which $\bar{p}_0$ represents, can be delivered to the load by the electrical power grid. The $q_x$power component usually assumes the value of$q$, as shown in equation (11).

$$p_x = \tilde{p} - \bar{p}_0 \tag{10}$$

$$q_x = q \tag{11}$$

Since the zero-sequence current must be compensated, the reference compensation current in the *0* coordinate is i_0 itself:

$$i_{c0}^* = i_0 \tag{12}$$

For a proper operation of the inverter of the Shunt Active Power Filter, the DC link voltage (V_{dc}), which corresponds to the capacitor voltage, should be regulated to be kept within appropriate levels. The p-q Theory calculations allow a simple method to regulate that voltage: if the Shunt Active Power Filter receives energy from the electrical power grid, it is stored in the capacitor and its voltage (V_{dc}) will increase, otherwise, V_{dc}will decrease. It is set a regulation power (p_{reg}), that is included in the value ofp_x:

$$p_x = \tilde{p} - \bar{p}_0 - p_{reg} \tag{13}$$

And the regulation power,p_{reg}, can be calculated according to:

$$p_{reg} = K(V_{ref} - V_{dc}) \tag{14}$$

where K is a proportional gain[1], V_{ref} is the reference of the desired voltage in the DC link, and V_{dc} is the average voltage in the DC link.

So:

- If $V_{dc}>V_{ref}$ – the Shunt Active Power Filter delivers energy to the electrical power grid and V_{dc} decreases.
- If $V_{dc}<V_{ref}$ – the Shunt Active Power Filter absorbs energy from the electrical power grid and V_{dc} increases.

The reference compensation currents in the *a-b-c* coordinates can be obtained by the transformation given in equation (15) and equation (16):

$$\begin{bmatrix} i_{ca}^* \\ i_{cb}^* \\ i_{cc}^* \end{bmatrix} = \sqrt{2/3} \begin{bmatrix} 1/\sqrt{2} & 1 & 0 \\ 1/\sqrt{2} & -1/2 & \sqrt{3}/2 \\ 1/\sqrt{2} & -1/2 & -\sqrt{3}/2 \end{bmatrix} \begin{bmatrix} i_{c0}^* \\ i_{c\alpha}^* \\ i_{c\beta}^* \end{bmatrix} \tag{15}$$

$$i_{cn}^* = -\left(i_{ca}^* + i_{cb}^* + i_{cc}^*\right) \tag{16}$$

The calculations presented so far are synthesized in Figure 15, and correspond to a Shunt Active Power Filter control strategy for "constant instantaneous real power at source". This approach, when applied to a three-phase system with balanced sinusoidal voltages, produces the following results:

- The phase supply currents become sinusoidal, balanced, and in phase with the voltages (in other words, the power supply "sees" the load as a purely resistive symmetrical load);
- The neutral current is made equal to zero (even 3rd order current harmonics are compensated);
- The three-phase instantaneous power supplied, equation (17), is made constant.

$$p_{3s} = v_a i_{sa} + v_b i_{sb} + v_c i_{sc} \tag{17}$$

The p-q Theory is also a valid control strategy for the Shunt Active Power Filter when the voltages are distorted and/or unbalanced, and sinusoidal supply currents are desired. However, with this strategy the total instantaneous power supplied will not be constant, since it is not physically possible to achieve both sinusoidal currents and constant power in systems with unbalanced and/or distorted voltages.

---

1 Is also possible to use a PI controller to regulate de DC link voltage. Whit the PI controller it is possible to eliminate the steady-state error.

In the case of a non-sinusoidal or unbalanced supply voltage, with the control strategy "constant instantaneous real power at source", the compensated supply currents will include harmonics, but in practical cases, when the voltage distortion and the voltage unbalance are within the limits established by the standards for the supply voltage at industries, the distortion in the source currents will be negligible after the compensation made by the Shunt Active Power Filter.

With the control strategy "sinusoidal current at source", even with highly distorted and/or unbalanced source voltages, are obtained sinusoidal supply currents with the compensation made by the Shunt Active Power Filter. When this approach is used the results are:

- The phase supply currents become sinusoidal, balanced, and in phase with the fundamental voltages;
- The neutral current is made equal to zero (even 3rd order current harmonics are compensated);
- The total instantaneous power supplied (p_{3s}) is not made constant, but in real cases when voltages and unbalance are within normal limits, it will present a small ripple (much smaller than before the compensation).

The only difference of the control strategy "sinusoidal current at source" in relation to the control strategy "constant instantaneous real power at source" is that its control system uses the fundamental positive sequence component of the system voltages, instead of using the real measured system voltages. It is usually accomplished using a PLL (Phase Locked Loop) algorithm, as described in [37-39].

6. Shunt Active Power Filter Implementation and Field Results

The Shunt Active Power Filter previously described in this chapter was implemented in the form of prototypes in order to validate the topology and control algorithms. To strength this validation it is advisable to test the active filter in different operation conditions, so it were developed four prototypes to be tested in real operation conditions in four different electrical installations, with different load profiles.

The target installations were previously monitorized, and simulation models of each installation were developed using a simulation tool. The simulation models were used to foresee the Shunt Active Power Filter behavior and to help sizing the hardware components and the protection systems.

According to the performed measurements and studies, the four Shunt Active Power Filters were constructed within three different compensation ranges: two 20 kVA prototypes to be used in a computation center and in an hospital, a 35 kVA prototype to be used in a textile industry installation, and a 55 kVA prototype to be applied in a medical drugs distribution warehouse.

In terms of hardware the main components that were used are:

- A DSP (Digital Signal Processor) from Texas Instruments (the control system was implemented using only fixed point calculations in order to enhance performance in terms of execution time);
- Hall effect sensors (used to measure the voltages and currents);
- Semikron IGBTs (the inverter stage was implemented using 4 Semikron IGBT modules - one for each leg of the inverter).

Two of the most important aspects when an equipment prototype is installed in field environment are security and reliability. The security of the human operators, the security of the industry plant, and the integrity of the equipment are factors that must be evaluated carefully. Therefore, it is very important to protect the Shunt Active Power Filter prototype against phenomena that usually do not exist in a laboratory environment, but that may occur in real industry installations. To accomplish these constraints, the laboratory prototypes were designed to be assembled in an electric switchboard (Figure 16). To prevent that anomalous operations could damage the Shunt Active Power Filter components, or other equipment connected to the electrical installation, various protections schemes were implemented.

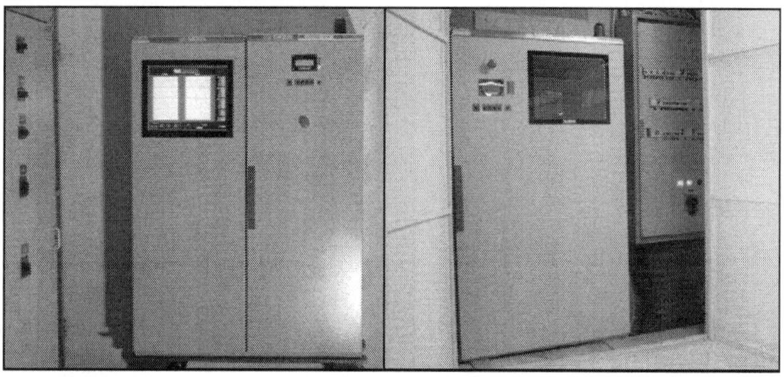

Figure 16. Two of the four final prototypes of Shunt Active Power Filters.

A supervision and protection system was developed to permanently monitor the Shunt Active Power Filter operation parameters, and to disconnect the device if any anomalous values are detected. Some of the implemented protections have two levels of actuation, in a first level the problem can be detected through software algorithms, and the Shunt Active Power Filter is softly turned off if the problem persists. More extreme malfunctions will activate implemented hardware protections that instantaneously disconnect the Shunt Active Power Filter from the electrical power grid and also discharge the DC link capacitors. The supervision and protection system also has the responsibility to correctly operate the Shunt Active Power Filter. It is responsible for the soft connection of the Shunt Active Power Filter to the electrical power grid, performing the pre-charge of the DC capacitor. Some of the implemented protections are:

- Protection against abnormal system voltages (protections for different values of transitory and RMS values are implemented).
- Protection against overcurrents produced by the Shunt Active Power Filter (the maximum compensation currents are limited by software, but several malfunctions can origin a current higher than the parameterized limit, triggering the protection).
- Protections against high temperature are also implemented through temperature sensors assembled in various representative points. Temperature sensors also allow the ON/OFF control of the electric board ventilation fans, which are responsible for cooling the heatsinks of the IGBTs modules, and the inductors (that connect the inverter to the electrical-power grid).

In Figure 17 is presented the generic electrical diagram of the case studies installations with the Shunt Active Power Filter. It shows the main electrical signals that were measured to validate the installation's power quality improvement achieved with the active filter. In blue are the source currents that are expected to become sinusoidal and balanced by the action of the active filter. In red are the non-sinusoidal currents of the load. In green are represented the compensation currents produced by the Shunt Active Power Filter.

The experimental results achieved in the four demonstration installations are presented in the following topics.

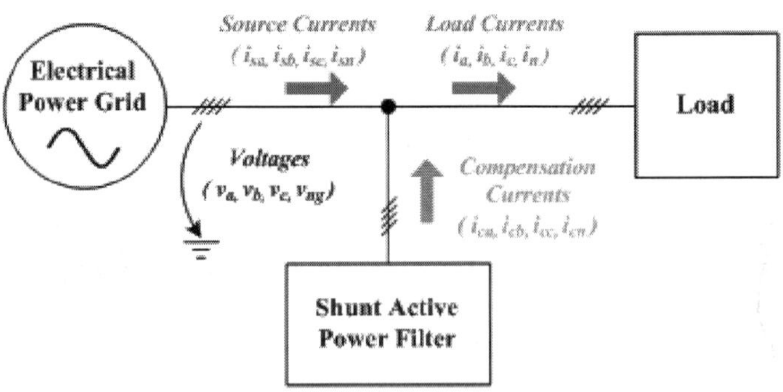

Figure 17. Generic electrical diagram of the case studies installations with the Shunt Active Power Filter.

6.1. Results at the Textile Industry

The first place selected to test the Shunt Active Power Filters consisted in an electrical switchboard that feeds a cloth whitening machine, in a large textile industry. In this place, the load is composed by eight variable speed drives with different power rates. Figure 18 shows the voltage and current waveforms and RMS values measured with the Shunt Active Power Filter in operation. In this figure it is possible to see that at the load side (waveforms

http://dx.doi.org/10.5772/53189

of current in red color) the three phase currents are distorted, and at the source side (blue waveforms) the three phase currents become almost sinusoidal, and in phase with the system voltages (black waveforms). The total power factor increased from 0.82 to 1.

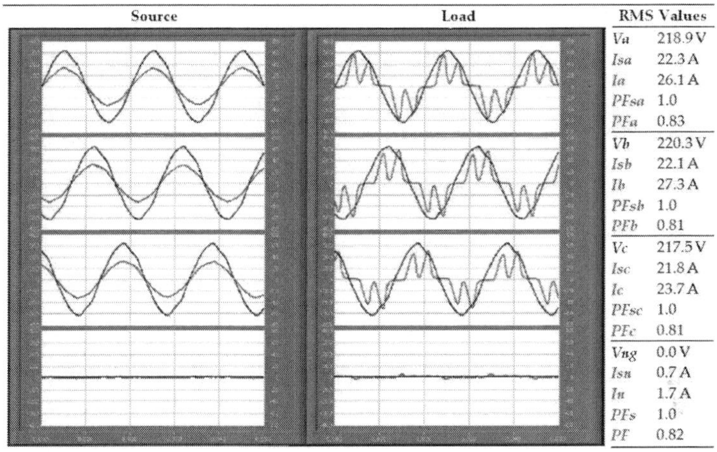

Figure 18. System voltages (black) and currents waveforms at Load (red) and Source (blue) sides of the Shunt Active Power Filter, registered in installation 1 (Textile Industry).

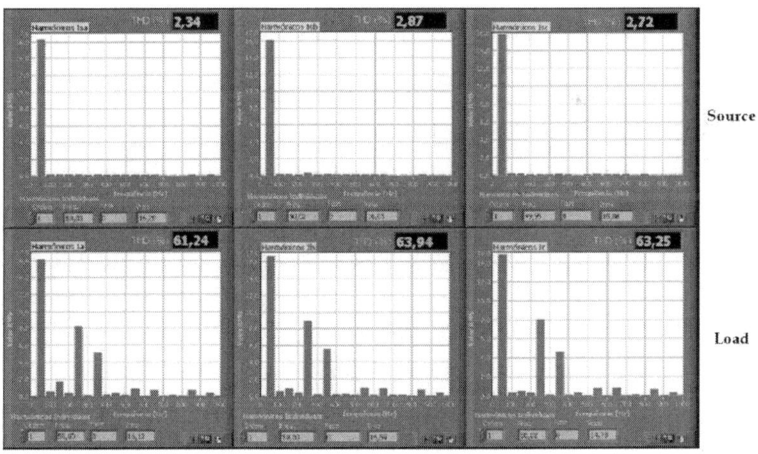

Figure 19. Current harmonics and THD% at Load and Source sides of the Shunt Active Power Filter, registered in installation 1 (Textile Industry).

The load currents presented a Total Harmonic Distortion (THD%) greater than 60% in all the three phases, the fifth and seventh harmonics are the highest ones, but other harmonics are also present (see Figure 19 - Load). In this load the neutral wire current was nearly zero. According to the measurements presented in Figure 19, the source current THD% of all the three phases decreased to values smaller than 3%.

6.2. Results at the Computational Center

The second test installation consisted in the main electrical switchboard of a computational center, at the University of Minho, where the main loads are computers, deskJet and laser printers, lighting and air-conditioning circuits.

At this electrical installation the load current presented a THD% near to 50%, the third harmonic was especially high, although other harmonics were present (Figure 21). The load presented significant unbalances at certain periods of the day, and the neutral current was high, not only due to the unbalance, but specially due to the third order harmonics at the phase currents, resulting in a neutral wire current with a higher value at the frequency of 150 Hz. As result of the Shunt Active Power Filter operation, the three phase currents were enhanced, the waveforms became approximately sinusoidal (Figure 20), with a THD% around 6%. At the source side the three phase currents became balanced, the neutral wire current was reduced from 16.5 A to 1 A, and the total power factor was increased from 0.88 to 0.99.

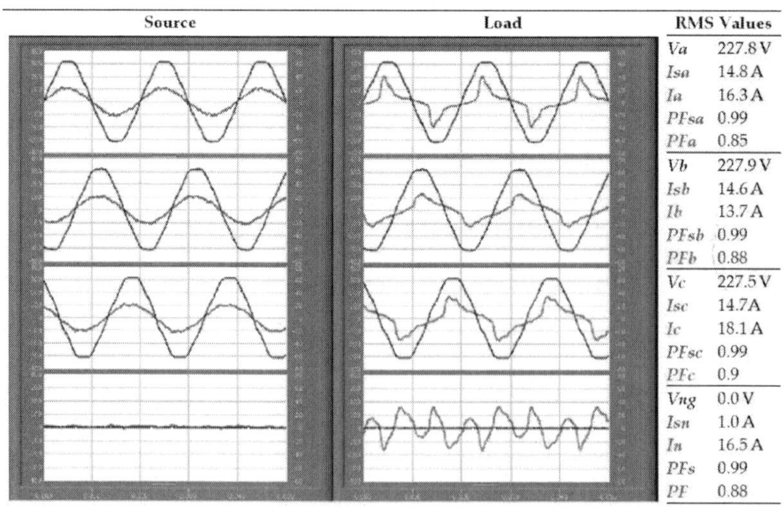

Figure 20. System voltages (black) and currents waveforms at Load (red) and Source (blue) sides of the Shunt Active Power Filter, registered in installation 2 (Computational Center)

http://dx.doi.org/10.5772/53189

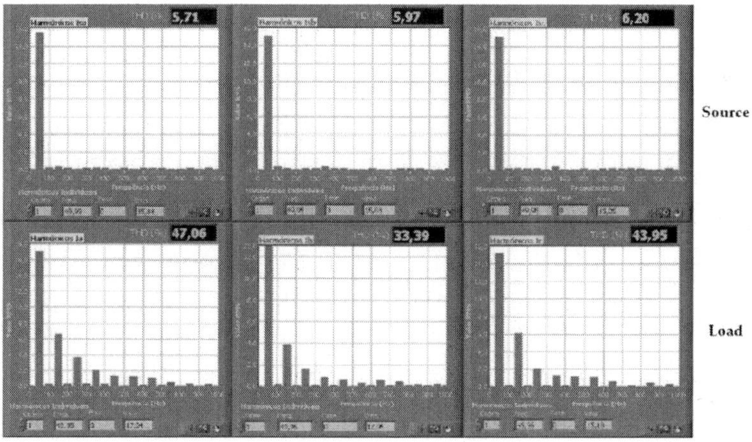

Figure 21. Current harmonics and THD% at Load and Source sides of the Shunt Active Power Filter, registered in installation 2 (Computational Center).

6.3. Results at the Clinical Analysis Laboratory of an Hospital

The third test site was the electrical switchboard of the clinical analyses laboratory of a hospital. Here, the loads were composed by some computers, diverse medical equipments, and lighting and air-conditioning circuits.

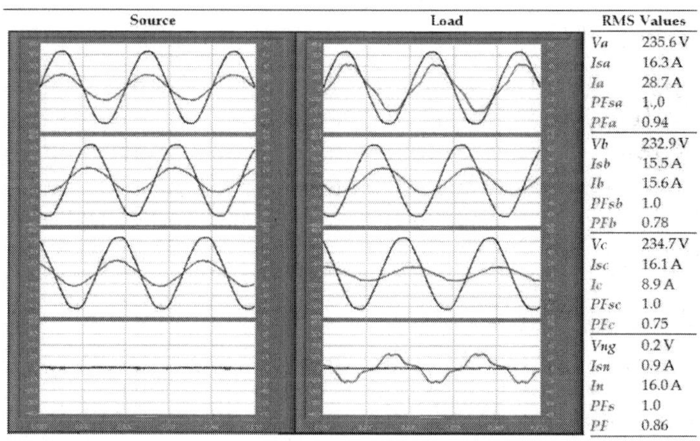

Figure 22. System voltages (black) and currents waveforms at Load (red) and Source (blue) sides of the Shunt Active Power Filter, registered in installation 3 (Clinical Analysis Laboratory).

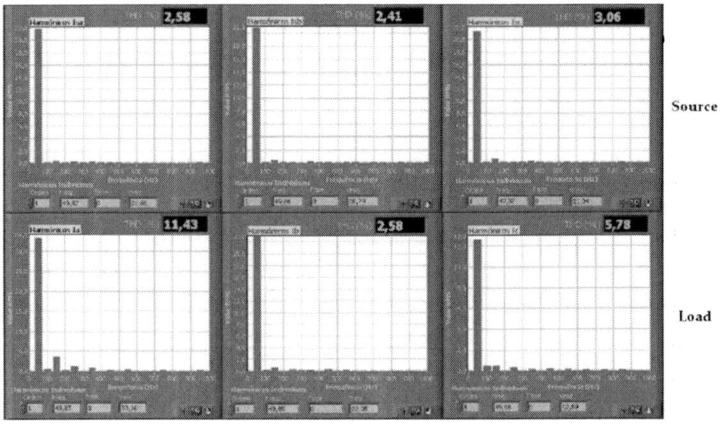

Figure 23. Current harmonics and THD% at Load and Source sides of the Shunt Active Power Filter, registered in installation 3 (Clinical Analysis Laboratory).

The load currents presented low harmonic distortion (the worst case was phase *A* with a THD% near to 11%), but the unbalance was very significant during certain periods of the day (see Figure 21 and Figure 23). In Figure 22 it is possible to see that the phase *A* current was almost 29 A, while the phase *B* current was smaller than 9 A. The phases *B* and *C* also presented low power factor (less than 0.78). When the Shunt Active Power Filter was operating, the current THD% at the source side decreased in all the three phases, reaching values near to 3%, and became balanced with unitary power factor. The current in the neutral wire decreased from 16 A to approximately 1 A.

6.4. Results at the Medical Drugs Distribution Warehouse

The fourth test site consisted in a medical drugs distribution warehouse. Here, the Shunt Active Power Filter was installed at the main switchboard of the warehouse. The principal loads of this installation were illumination circuits (composed by a large number of fluorescent tube lamps with magnetic ballasts), chest refrigerators, conveyor belt systems, and a central air-conditioning unit. The load current presented low distortion (the worst case was in phase *A* with a THD% near to 5%), as it can be seen in Figure 25. The current unbalance was also small, resulting in a neutral wire current of around only 8 A (also there were not large values of third order harmonics). The major problem of this installation was the power factor. According to the Portuguese legislation, if an installation presents a $tan\ \varphi$ higher than 0.4 (equivalent to a $cos\ \varphi$ lower than 0.93), the Reactive Energy is taxed. It is possible to see in Figure 24 that the total power factor of the installation was lower than 0.7. When the Shunt Active Power Filter was connected, the current THD% at the electrical power grid side decreased in all the three phases, reaching values of less than 2%. The three phase currents became sinusoidal, in phase with the system voltages, and perfectly balanced. The power factor increased from 0.69 to 1, and the current in the neutral wire decreased from 8 A to 3 A.

http://dx.doi.org/10.5772/53189

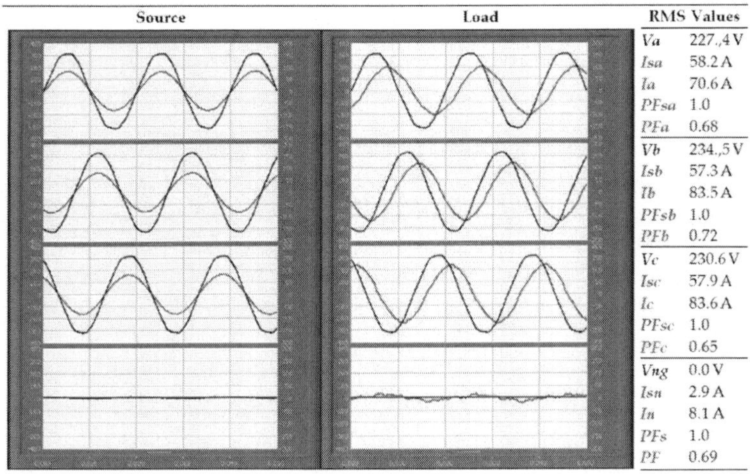

Figure 24. System voltages (black) and currents waveforms at Load (red) and Source (blue) sides of the Shunt Active Power Filter, registered in installation 4 (Medical Drugs Distribution Warehouse).

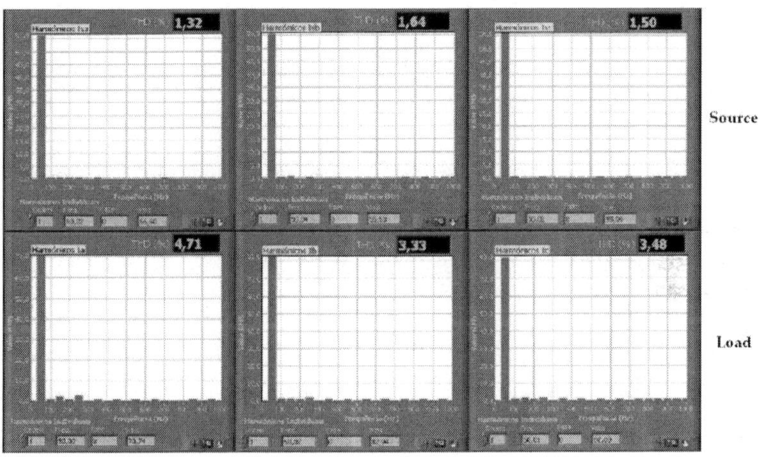

Figure 25. Current harmonics and THD% at Load and Source sides of the Shunt Active Power Filter, registered in installation 4 (Medical Drugs Distribution Warehouse).

The presented results confirm the ability of the Shunt Active Power Filters to compensate problems like current harmonics, current unbalance and power factor. The developed prototypes presented a good performance in all the four demonstration installations.

7. Conclusions

The growing use of non-linear loads in industrial facilities is the source of several power quality problems, such as harmonics, reactive power, flicker and resonance. These problems affect not only the facility but also the electrical power system by distorting the voltage and current waveforms with harmonics of various orders, including inter-harmonics. Active Power Conditioners are an up-to-date solution to mitigate these power quality problems. It can be found conditioners to mitigate current problems, others to mitigate voltage problems, and others that mitigate both current and voltage problems, both in power systems and in industrial facilities. In this chapter were presented the Active Power Conditioners more suitable for use in industrial facilities, explaining in detail their concepts, presenting their power electronics topologies and typical waveforms.

Shunt Active Power Filters allow the compensation of problems related to the consumed currents, like current harmonics and current unbalance, together with power factor correction, and can be a much better solution than the conventional approach (capacitors for power factor correction and passive filters to compensate for current harmonics). They are most suitable for facilities with a high level of distortion and/or unbalance of the consumed currents. There are some situations in which the use of Shunt Active Power Filters to compensate the current problems also improves the power grid voltage waveforms due to the reduction of the current harmonics flowing through the line impedances.

Series Active Power Filters permit the compensation of problems related to the supplied voltages, like voltage harmonics, voltage unbalance, sags, swells and flicker. They are most suitable for facilities with loads sensitive to voltage problems. In installations that use shunt passive filters to mitigate current harmonics they also improve the behavior of those passive filters and the overall installation power quality.

Unified Power Quality Conditioners (UPQCs) can compensate both and simultaneously problems related to the consumed currents and to the supplied voltages. So, it is suitable for facilities that have problems in the consumed currents and that also have loads which are sensitive to voltage problems. The UPQC topology allows the power flow between the shunt and series conditioners, so it is able to compensate undervoltages and overvoltages in steady-state. This is a great advantage comparing to the use of shunt and series active power filters operating independently.

The control of the conditioners is also a matter of great importance, and different control theories can be found. The p-q Theory is a suitable tool to the analysis of three-phase electrical systems with non-linear loads and for the control of Active Power Conditioners. Based on this theory, two control strategies for Shunt Active Power Filters were described in this chapter, one leading to constant instantaneous real power at source and the other leading to sinusoidal currents at source.

The experimental results obtained in four different test facilities, and presented in this chapter, show that the developed Shunt Active Power Filters have a good performance.They dynamically compensate for harmonic currents, and correct power factor. They also compensate for load current unbalance, and almost eliminate the current in the neutral wire at the source side. Therefore, the Shunt Active Power Filters allow the power source to see an unbalanced and non-linear load, with reactive power consumption, as if the load is a symmetrical linear resistive load. By the action of the Shunt Active Power Filters, the currents at the three phases of the source side become almost sinusoidal and in phase with the voltages, and the neutral wire current become almost null. Since all the source currents are reduced in relation to the load currents, the electrical installation losses also decrease.

Acknowledgements

This work is financed by FEDER Funds, through the Operational Program for Competitiveness Factors – COMPETE, and by National Funds through FCT – Foundation for Science and Technology, under the projects: DEMTEC/020/1/03, PTDC/EEA-EEL/104569/2008, and FCOMP-01-0124-FEDER-022674.

Author details

João L. Afonso*, J. G. Pinto and Henrique Gonçalves

*Address all correspondence to: jla@dei.uminho.pt

Centro Algoritmi, Universityof Minho, Guimarães, Portugal

References

[1] IEEE Working Group on Power System Harmonics, "The Effects of Power System Harmonics on Power System Equipment and Loads," Power Apparatus and Systems, IEEE Transactions on, vol. PAS-104, 1985, pp. 2555-2563.

[2] A. Bachry; Z. A. Styczynski, "An Analysisof Distribution System Power Quality Problems Resulting from Load Unbalance and Harmonics," IEEE PES – Transm. and Dist. Conf. and Exp., Vol 2, 7-12 Sept. 2003 pp.763–766.

[3] V.E. Wagner, J.C. Balda, D.C. Griffith, A. McEachern, T.M. Barnes, D.P. Hartmann, D.J. Phileggi, A.E. Emannuel, W.F. Horton, W.E. Reid, R.J. Ferraro, and W.T. Jewell, "Effects of harmonics on equipment," Power Delivery, IEEE Transactions on, vol.8, 1993, pp. 672-680.

[4] E.F. Fuchs, D.J. Roesler, and F.S. Alashhab, "Sensitivity of Electrical Appliances to Harmonics and Fractional Harmonics of the Power SYSTEM's Voltage. Part I: Transformers and Induction Machines," Power Delivery, IEEE Transactions on, vol. 2, 1987, pp. 437-444.

[5] E.F. Fuchs, D.J. Roesler, and K.P. Kovacs, "Sensitivity of Electrical Appliances to Harmonics and Fractional Harmonics of the Power System's Voltage. Part II: Television Sets, Induction Watthour Meters and Universal Machines," Power Delivery, IEEE Transactions on, vol. 2, 1987, pp. 445-453.

[6] João L. Afonso, Carlos Couto, Júlio Martins, "Active Filters with Control Based on the p-q Theory," IEEE Industrial Electronics Society Newsletter, vol. 47, nº 3, Sept. 2000, pp. 5-10.

[7] J. G. Pinto, Pedro Neves, D. Gonçalves, João L. Afonso, "Field Results on Developed Three-Phase Four-Wire Shunt Active Power Filters," IECON 2009 - The 35th Annual Conference of the IEEE Industrial Electronics Society, 3 5 November, Porto, Portugal.

[8] Pedro Neves, Gabriel Pinto, Ricardo Pregitzer, Luís Monteiro, João L. Afonso, "Experimental Results of a Single-Phase Shunt Active Filter Prototype with Different Switching Techniques," Proceedings of ISIE 2007- 2007 IEEE International Symposium on Industrial Electronics, 4-7 June, 2007, Vigo, Spain.

[9] J. G. Pinto, Pedro Neves, Ricardo Pregitzer, Luís F. C. Monteiro, João L. Afonso, "Single-Phase Shunt Active Filter with Digital Control," Proceedings of ICREPQ'07- International Conference on Renewable Energies and Power Quality, 28-30 March Seville, Spain.

[10] M. Aredes, J. Häffner and K. Heumman, "A combined series and shunt active power filters," IEEE/KTH– Stockholm Power Tech. Conf., SPT PE 07-05-0643, vol. Power Elect., Sweden, June 1995, pp. 237–242.

[11] H. Fujita, H. Akagi,"The Unified Power Quality Conditioner:The Integration of Series and Shunt Active Filters," IEEE Trans. On Power Electronics, vol.13, No.2, March 1998, pp. 315-322.

[12] J. G. Pinto, R. Pregitzer, Luís. F. C. Monteiro, Carlos Couto, João. L. Afonso, "A Combined Series Active Filter and Passive Filters for Harmonics, Unbalances and Flicker Compensation," Proceedings of POWERENG - First International Conference on Power Engineering, Energy and Electrical Drives, 12 14 April, 2007, Setubal, Portugal, pp 54-59 ISBN: 1-4244-0895-4.

[13] L. F. C. Monteiro, João L. Afonso, J. G.Pinto, E. H. Watanabe, M. Aredes, H. Akagi, "CompensationAlgorithmsbasedonthe p-q andCPCTheories for SwitchingCompensators in Micro-Grids," Revista Eletrônica de Potência – Associação Brasileira de Eletrônica de Potência – SOBRAEP, ISSN 1414-8862, Vol. 14, no. 4, Novembro de 2009, pp. 259-268.

[14] E. H. Watanabe, J. L. Afonso, J. G. Pinto, L. F. C. Monteiro, M. Aredes, H. Akagi, "Instantaneous p-q Power Theory for Control of Compensators in Micro-Grids," IEEE ISNCC - International School on Nonsinusoidal Currents and Compensation, 15-18 June 2010, àagów, Poland.

[15] R. Pregitzer, J. G. Pinto, Luís F.C. Monteiro, João L. Afonso," Shunt Active Power Filter with Dynamic Output Current Limitation" , Proceedings of ISIE 2007- 2007 IEEE International Symposium on Industrial Electronics, 4-7 June, 2007, Vigo, Spain.

[16] Pedro Neves, Gabriel Pinto, Ricardo Pregitzer, Luís Monteiro, João L. Afonso," Experimental Results of a Single-Phase Shunt Active Filter Prototype with Different Switching Techniques", Proceedings of ISIE 2007- 2007 IEEE International Symposium on Industrial Electronics, 4-7 June, 2007, Vigo, Spain.

[17] Filipe Ferreira, Luís Monteiro, João L. Afonso, Carlos Couto, "A Control Strategy for a Three-Phase Four-Wire Shunt Active Filter", IECON'08 - The 34th Annual Conference of the IEEE Industrial Electronics Society, Orlando, Florida, USA, 10-13 Nov. 2008, pp.411-416.

[18] J. G. Pinto, Pedro Neves, Ricardo Pregitzer, Luís F. C. Monteiro, João L. Afonso, "Single-Phase Shunt Active Filter with Digital Control", Proceedings of ICREPQ'07- International Conference on Renewable Energies and Power Quality, 28-30 March Seville, Spain.

[19] H. Carneiro, J. G. Pinto, J. L. Afonso, "Single-Phase Series Active Conditioner for the Compensation of Voltage Harmonics, Sags, Swell and Flicker", ISIE 2011 - 20th IEEE International Symposium on Industrial Electronics, pp. 384-389, 27-30 June 2011, Gdansk, Poland.

[20] J. G. Pinto, R. Pregitzer, Luís. F. C. Monteiro, Carlos Couto, João. L. Afonso, "A Combined Series Active Filter and Passive Filters for Harmonics, Unbalances and Flicker Compensation", Proceedings of POWERENG - First International Conference on Power Engineering, Energy and Electrical Drives, 12 14 April, 2007, Setubal, Portugal.

[21] J. G. Pinto; Helder Carneiro, Bruno Exposto, Carlos Couto, João L. Afonso, "Transformerless Series Active Power Filter to Compensate Voltage Disturbances", Proceedings of the 14th European Conference on Power Electronics and Applications (EPE 2011), pp. 1-6, Birmingham, United Kingdom, Aug. 30 - Sept. 1 2011.

[22] A. J. Visser, J. H. R. Enslin, H. du T. Mouton, "Transformerless Series Sag Compensation With a Cascaded Multilevel Inverter," IEEE Transactions on Industrial Electronics, vol. 49, nº 4, August 2002, pp. 824-831.

[23] Carneiro, H.; Exposto, B.; Gonçalves, H.; Pinto, J.G.; Afonso, J.L. , "Single-phase Series Active Conditioner Active Power Flow in a Harmonic Free Electrical System During Sag and Swell Events", Proceedings of the 14th European Conference on Power Electronics and Applications (EPE 2011), pp. 1-10, Aug. 30 - Sept. 2011, Birmingham, United Kingdom.

[24] Luís F. C. Monteiro, José C. C. Costa, Maurício Aredes, João L. Afonso, "A Control-Strategy for a Three-LevelUnifiedPowerQualityConditioner," Proceedings (CD-ROM) 8th COBEP - Congresso Brasileiro de Eletrônica de Potência, Recife, PE, Brasil, 14 17 Julho 2005.

[25] H. Akagi, Y. Kanazawa, A. Nabae, "Generalized Theory of the Instantaneous Reactive Power in Three-Phase Circuits," IPEC'83 - Int. Power Electronics Conf., Tokyo, Japan, 1983, pp. 1375-1386.

[26] H. Akagi, Y. Kanazawa, A. Nabae, "Instanataneous Reactive Power Compensator Comprising Switching Devices without Energy Storage Compenents," IEEE Trans. Industry Applic., vol. 20, May/June 1984.

[27] M. Depenbrock, "The FBD-Method, a Generally Applicable Tool for Analysing Power Relations," IEEE Transactions on Power Systems, vol. 8, no. 2, pp. 381-387, May 1993.

[28] P. Tenti, E. Tedeschi, P. Mattavelli, "Cooperative Operation of Active Power Filters by Instantaneous Complex Power Control," 7th International Conference on Power Electronics and Drive Systems, 2007 (PEDS '07), pp. 555-561, November 2007.

[29] L. S. Czarnecki, "On some misinterpretations of the instantaneous reactive power p–q theory," IEEE Transactions on Power Electronics, vol. 19, no. 3, pp. 828–836, May 2004.

[30] L. S. Czarnecki, "Instantaneous reactive power p–q theory and power properties of three-phase systems," IEEE Transactions on Power Delivery, vol. 21, no. 1, pp. 362–367, January 2006.

[31] H. S. Kim, H. Akagi, "The instantaneous power theory on the rotating p-q-r reference frames," in Proc. IEEE/PEDS 1999 Conf., Hong Kong, Jul., pp. 422–427.

[32] M. Depenbrock, V. Staudt, H. Wrede, "Concerning instantaneous power compensation in three-phase systems by using p–q–r theory," IEEE Transactions on Power Electronics, vol. 19, no. 4, pp. 1151–1152, Jul. 2004.

[33] M. Aredes, H. Akagi, E. H. Watanabe, E. V. Salgado, L. F. Encarnação, "Comparisons Between the p–q and p–q–r Theories in Three-Phase Four-Wire Systems," IEEE Transactions on Power Electronics, , vol. 24, no. 4, pp. 924-933, April 2009.

[34] R. I. Bojoi, G. Griva, V. Bostan, M. Guerreiro, F. Farina, F. Profumo, "Current Control Strategy for Power Conditioners Using Sinusoidal Signal Integrators in Synchronous Reference Frame," IEEE Transactions on Power Electronics, vol. 20, no. 6, pp. 1402–1412, November 2005.

[35] E. H. Watanabe, R. M. Stephan, M. Aredes, "New Concepts of Instantaneous Active and Reactive Powers in Electrical Systems with Generic Loads," IEEE Trans. Power Delivery, vol. 8, no. 2, April 1993, pp. 697-703.

[36] M. Aredes, E. H. Watanabe, "New Control Algorithms for Series and Shunt Three-Phase Four-Wire Active Power Filters," IEEE Trans. Power Delivery, vol 10, no. 3, July 1995, pp. 1649-1656.

[37] H. Carneiro, L. F. C. Monteiro, João L. Afonso, "Comparisons between Synchronizing Circuits to Control Algorithms for Single-Phase Active Converters", IECON 2009

http://dx.doi.org/10.5772/53189

- The 35th Annual Conference of the IEEE Industrial Electronics Society, 3 5 November 2009, Porto, Portugal.

[38] Helder Carneiro; Bruno Exposto; João L. Afonso, "Evaluation of Two Fundamental Positive-Sequence Detectors for Highly Distorted and Unbalanced Systems", IEEE EPQU 2011 - Electrical Power Quality and Utilization Conference, Lisbon, Portugal, 17-19 Oct. 2011.

[39] L. G. Barbosa Rolim, D. Rodrigues da CostaJr., and M. Aredes, "Analysis and Software Implementation of a Robust Synchronizing PLL Circuit Based on the pq Theory," IEEE Transactions on Industrial Electronics, vol. 53, no. 6, pp. 1919-1926, Dec. 2006.

Chapter 6

Power Quality Management Based Security OPF Considering FACTS Using Metaheuristic Optimization Methods

Belkacem Mahdad

Additional information is available at the end of the chapter

http://dx.doi.org/10.5772/52875

1. Introduction

Power quality management based environmental optimal power flow (OPF) is a vital research area for power system operation and control, this type of problem is complex with non-linear constraints where many conflicting objectives are considered like fuel cost, gaz emission, real power loss, voltage deviation, and voltage stability[1-2]. The solution of this multi objective problem becomes more complex and important when flexible ac transmission systems (FACTS) devices and renewable energy are considered. In the literature a wide variety of optimization techniques have been applied, a number of approaches have been developed for solving the standard optimal power flow problem using mathematical programming, lambda iteration method, gradient method, linear programming, quadratic programming and interior point method [3-4-5]. However all these developed techniques rely on the form of the objective function and fail to find the near global optimal solution, authors in [6-7] provide a valuable introduction and surveys the first optimization category based conventional optimization methods.

To overcome the major problem related specially to restriction on the nature of the objective function, researchers have proposed a second optimization category based evolutionary algorithms for searching near-optimum solutions considering non linear objective function characteristic and practical generating units constraints such as: Genetic algorithms (GA), simulated annealing (SA), tabu search (TS), and evolutionary programming (EP) [8-9-10], which are the forms of probabilistic heuristic algorithm, author in [11] gives a recent and significant review of many global optimization methods applied for solving many problems related to power system operation, planning and control.

INTECH
open science | open minds

© 2013 Mahdad; licensee InTech. This is a paper distributed under the terms of the Creative Commons Attribution License (http://creativecommons.org/licenses/by/3.0), which permits unrestricted use, distribution, and reproduction in any medium, provided the original work is properly cited.

In order to enhance the performance of these standard global optimization methods for solving complex and practical problem, many hybrid variants based evolutionary algorithm have been proposed and applied with success for solving many problems related to power system operation and control like: hybrid EP–SQP[12], Quantum genetic Algorithm (QGA) [13], artificial immune system (AIS) [14], Adaptive particle swarm optimization (APSO) [15], improved PSO (IPSO) [16], improved chaotic particle swarm optimization (ICPSO) [17], adaptive hybrid differential evolution (AHDE) [18], and a hybrid multi agent based particle swarm optimization (HMPSO) [19]. These methods have a better searching ability in finding near global optimal solution compared to mathematical methods and to the standard evolutionary algorithms. Very recently a significant review of recent non-deterministic and hybrid methods applied for solving the optimal power flow problem is proposed in [7].

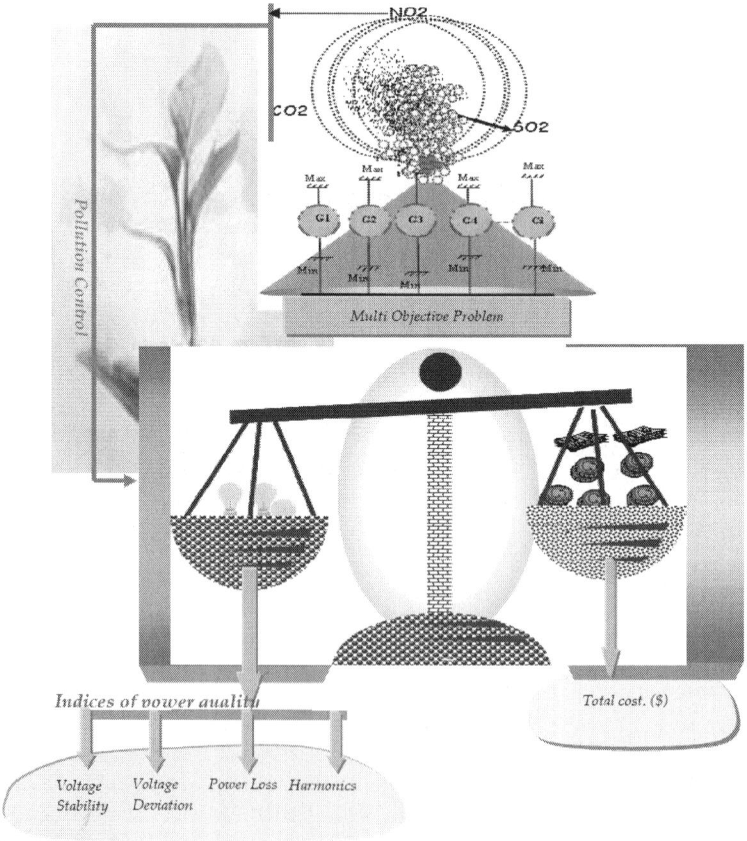

Figure 1. Strategy of the power quality management based SOPF.

Recently a new metaheuristic optimization method called Firefly research method introduced and attracted many researchers due to its capacity and efficiency to find the global optimal solution, this method applied with success to solving many real optimization problems [20-21]. In this chapter firefly algorithm is proposed to solve the combined economic dispatch problem, and a decomposed genetic algorithm considering multi shunt FACTS proposed for solving the multi objective optimal power flow [22-23]. The remainder of this chapter is organized as follows; section 2 provides a formulation of the multi objective optimization problem, Section 3 introduces the mechanism search of the firefly algorithm. Brief description and modelling of shunt FACTS devices is developed in section 4. Simulation results based Matlab program are demonstrated in section 5, finally, the chapter is concluded in section 6.

2. Energy planning strategy: Economic issue and power quality

It is important to underline the importance of energy efficiency planning in power systems, the combined term energy planning is usually associated with power quality; how energy is produced (economic aspect), how energy is consumed at the point of end use (technical aspect), and what is the impact of the total energy produced on the environment (gaz emissions). Figure 1 shows the structure of energy planning strategy considering FACTS technology; the following points should be taken by expert engineers and researchers to assure energy efficiency.

1. *Environmental Issue*: The environment (gaz emissions) can be considered as the first step towards the improvement of energy efficiency by forcing the utilities and expert engineers to introduce the environmental constraints to the standard optimal power flow problem.

2. *Technology Issue:* Energy planning strategy becomes a complex problem and difficult to solving efficiently with the wide integration of two new technologies, flexible ac transmission systems (FACTS) and renewable energy.

3. *Economic Issue*: How to estimate economically the number, the size of FACTS Controllers coordinated with renewable sources to be installed in a practical and large electrical network?

3. Multi objective optimal power flow formulation

The real life problems involve several objectives and the decision maker would like to find solution, which gives compromise between the selected objectives. The multi objective OPF is to optimize the settings of control variables in terms of one or more objective functions while satisfying several equality and inequality constraints.

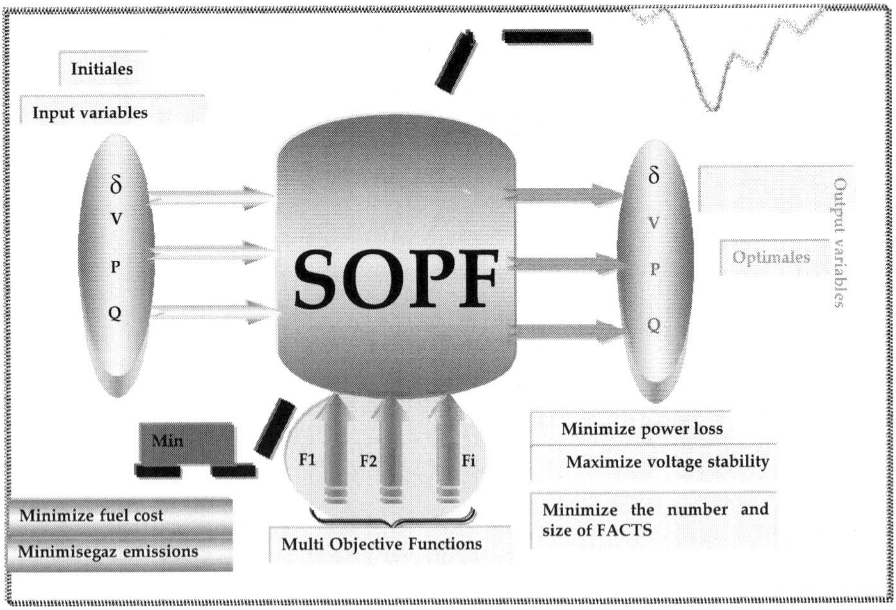

Figure 2. Schematic presentation of security multi objective OPF problem

In multi objective OPF we have to optimize two or more objective functions. Figure 2 shows a general presentation of the multi objective OPF problem. The mathematical problem can be formulated as follows:

Minimize

$$J_i(x,u) \quad i=1,\ldots,\ N_{obj} \tag{1}$$

Subject to:

$$g(x,u)=0 \tag{2}$$

$$h(x,u)\leq 0 \tag{3}$$

Where J_i is the *ith* objective function, and N_{obj} is the number of objectives,$g(x,\ u)$ and $h\,(x,\ u)$ are respectively the set of equality and inequality constraints, x is the state variables and u is the vector of control variables. The control variables are generator active and reactive power outputs, bus voltages, shunt capacitors/reactors and transformers tap-setting. The state variables are voltage and angle of load buses.

3.1. Active power planning with smooth cost function

For optimal active power dispatch, the objective function f is the total generation cost expressed in a simple form as follows:

Min

$$f = \sum_{i=1}^{N_g} \left(a_i + b_i P_{gi} + c_i P_{gi}^2 \right) \tag{4}$$

where N_g is the number of thermal units, P_{gi} is the active power generation at unit i and a_i, $b_i$and c_i are the cost coefficients of the *ith* generator.

The equality constraints $g(x)$ are the power flow equations.

The inequality constraints $h(x)$ reflect the limits on physical devices in the power system as well as the limits created to ensure system security.

3.2. Emission objective function

An alternative dispatch strategy to satisfy the environmental requirement is to minimize operation cost under environmental requirement. Emission control can be included in conventional economic dispatch by adding the environmental cost to the normal dispatch. The objective function that minimizes the total emissions can be expressed as the sum of all the three pollutants (NO_x,CO_2, SO_2) resulting from generator real power [22].

In this study,NO_x emission is taken as the index from the viewpoint of environment conservation. The amount of NO_x emission is given as a function of generator output (in Ton/hr), that is the sum of quadratic and exponential functions [9].

$$f_e = \sum_{i=1}^{Ng} 10^{-2} \times \left(\alpha_i + \beta_i P_{gi} + \gamma_i P_{gi}^2 + \omega_i \exp\left(\mu_i P_{gi} \right) \right) \text{Ton / h} \tag{5}$$

where α_i, β_i, γ_i, ω_i and μ_i are the parameters estimated on the basis of unit emissions test results.

The pollution control can be obtained by assigning a cost factor to the pollution level expressed as:

$$f_{ce} = \omega . f_e \$ / \text{h} \tag{6}$$

where $\omega$is the emission control cost factor in \$/Ton.

Fuel cost and emission are conflicting objective and can not be minimized simultaneously. However, the solutions may be obtained in which fuel cost an emission are combined in a single function with different weighting factor. This objective function is described by:

Minimize

$$F_T = \alpha f + (1-\alpha) f_{ce} \tag{7}$$

where α is a weighting factor that satisfies$0 \le \alpha \le 1$.

In this model, when weighting factor α=1, the objective function becomes a classical economic dispatch, when weighting factor α=0, the problem becomes a pure minimization of the pollution control level.

3.3. Active power planning with valve-point loading effect

The valve-point loading is taken in consideration by adding a sine component to the cost of the generating units [9]. Typically, the fuel cost function of the generating units with valve-point loadings is represented as follows:

$$f_T = \sum_{i=1}^{NG} \left(a_i + b_i P_{gi} + c_i P_{gi}^2 \right) + \left| d_i \sin\left(e_i \left(P_{gi}^{\min} - P_{gi} \right) \right) \right| \tag{8}$$

$d_i$and $e_i$are the cost coefficients of the unit with valve-point effects.

3.4. Reactive power planning

The main role of reactive power planning (RPP) is to adjust dynamically the control variables individually or simultaneously to reduce the total power loss, transit power flow, voltages deviation, and to improve voltage stability, but still satisfying specified constraints (generators constraints and security constraints). The basic vector control structure is well presented in Fig 3.

3.4.1. Power loss objective function

The objective function here is to minimize the active power loss (P_{loss}) in the transmission system. It is given as:

$$P_{loss} = \sum_{k=1}^{N_l} g_k \left[\left(t_k V_i \right)^2 + V_j^2 - 2 t_k V_i V_j \cos \delta_{ij} \right] \tag{9}$$

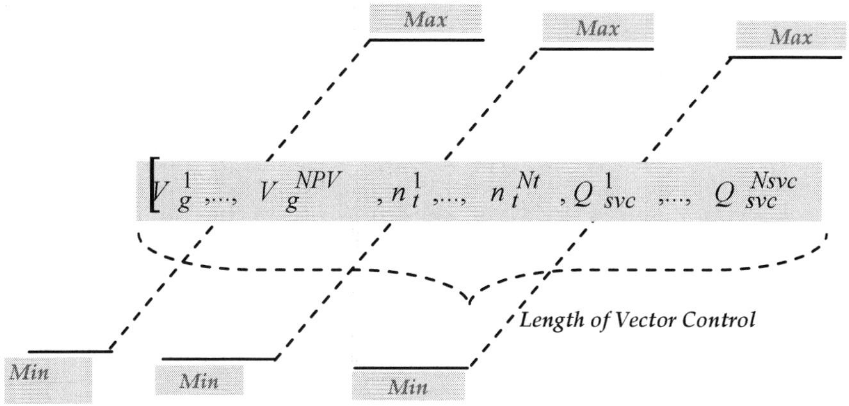

Figure 3. Basic vector control structure for reactive power planning.

Where, N_l is the number of transmission lines; g_k is the conductance of branch k between buses i and j; t_k the tap ration of transformer k; V_i is the voltage magnitude at bus i; δ_{ij} the voltage angle difference between buses *i* and *j*.

3.4.2. Voltage deviation objective function

One of the important indices of power system security is the bus voltage magnitude. The voltage magnitude deviation from the desired value at each load bus must be as small as possible. The deviation of voltage is given as follows:

$$\Delta V = \sum_{k=1}^{N_{PQ}} \left| V_k - V_k^{des} \right| \tag{10}$$

where, $N_{PQ}$is the number of load buses and V_k^{des} is the desired or target value of the voltage magnitude at load bus k.

3.5. Constraints

3.5.1. Equality constraints

The equality constraints $g(x)$ are the real and reactive power balance equations.

$$P_{gi} - P_{di} = V_i \sum_{j=1}^{N} V_j \left(g_{ij} \cos\delta_{ij} + b_{ij} \sin\delta_{ij} \right) \tag{11}$$

$$Q_{gi} - Q_{di} = V_i \sum_{j=1}^{N} V_j \left(g_{ij} \sin \delta_{ij} - b_{ij} \cos \delta_{ij} \right) \qquad (12)$$

Where N is the number of buses, P_{gi}, Q_{gi} are the active and the reactive power generation at bus i;P_{di}, $Q_{di}$are the real and the reactive power demand at bus i, V_i, V_j, the voltage magnitude at bus i, j, respectively; δ_{ij} is the phase angle difference between buses i and j respectively, g_{ij} and b_{ij} are the real and imaginary part of the admittance (Y_{ij}).

3.5.2. Inequality constraints

The inequality constraints h reflect the generators constraints and power system security limits,

a. Generator Constraints

- Upper and lower limits on the generator bus voltage magnitude:

$$V_{gi}^{\min} \le V_{gi} \le V_{gi}^{\max}, \; i = 1,2,\ldots,NPV \qquad (13)$$

b. Security Limits

The inequality constraints on security limits are given by:

- Constraints on transmission lines loading

$$S_{li} \le S_{li}^{\max}, \; i = 1,2,\ldots,NPQ \qquad (14)$$

- Constraints on voltage at loading buses (PQ buses)

$$V_{Li}^{\min} \le V_{Li} \le V_{Li}^{\max}, \; i = 1,2,\ldots,NPQ \qquad (15)$$

- Upper and lower limits on the tap ratio (t) of transformer.

$$t_i^{\min} \le t_i \le t_i^{\max}, i = 1,2,\ldots,NT \qquad (16)$$

- Parameters of shunt FACTS Controllers must be restricted within their upper and lower limits.

$$X^{\min} \le X_{FACTS} \le X^{\max} \qquad (17)$$

4. Metaheuristic optimization methods

The difficulties associated with using mathematical optimization on large-scale engineering problems have contributed to the development of alternative solutions, during the last two decades; the interest in applying new metaheuristic optimization methods in power system field has grown rapidly.

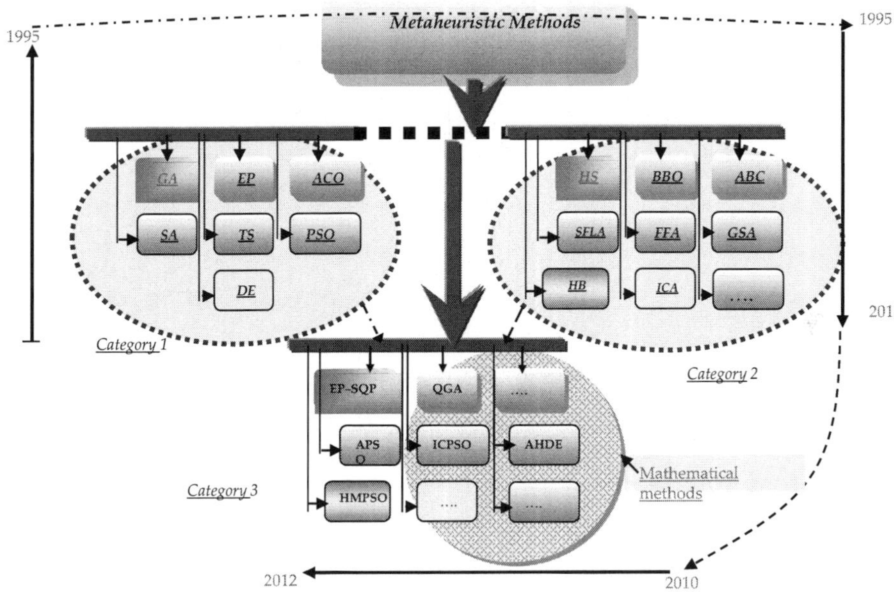

Figure 4. Presentation of metaheuristic optimization methods

In general, metaheuristic optimization methods can be classified into three large categories.

1. The first category such as: GA, TS, SA, ACO, PSO, DE and theirs successive developed variants.

2. The second category includes: Harmony search (HS), Biogeography based optimization (BBO), Artificial bee colony (ABC), Honey bee (HB), Shuffled frog leaping algorithm (SFLA), Firefly algorithm (FFA), Gravitational search algorithm (GSA), Imperialist competition algorithm (ICA) and many other variants.

3. The third category called hybrid optimization methods includes a combination between metaheuristics and conventional methods.

Details about their performances and the application of metaheuristic optimization techniques for solving many practical problem related to power system field particularly multi objective OPF can be found in a recent state of the art of non-deterministic optimization and hybrid methods presented in [11]. Figure 4 shows a simplified presentation about the most popular metaheuristic methods applied by researchers for solving many complex problems.

4.1. Firefly search algorithm

This section introduces and describes a solution to the combined economic dispatch problem using a new metaheuristic nature inspired algorithm. Firefly research algorithm is one of the new Biology inspired metaheuristic algorithms which have recently introduced by Dr. Xin-She Yang at Cambridge University in 2007, as an efficient way to deal with many hard combinatorial optimization problems and non-linear optimization constrained problems [20-21]. In general, the firefly algorithm has three particular idealized rules which are based on some of the major flashing characteristics of real fireflies these are presented on the following schematic diagram (Figure 5), and described in brief as follows:

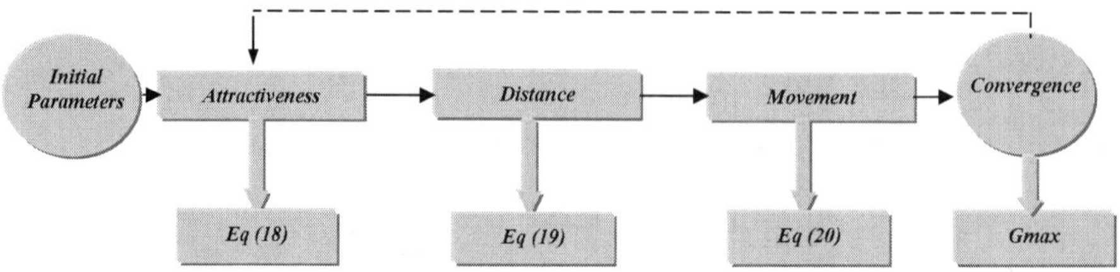

Figure 5. Basic mechanism search of firefly algorithm.

1. All fireflies are unisex, and they will move towards more attractive and brighter ones regardless their sex.

2. The degree of attractiveness of a firefly is proportional to its brightness which decreases as the distance from the other firefly increases due to the fact that the air absorbs light. If there is not a brighter or more attractive firefly than a particular one, it will then move randomly.

3. The brightness or light intensity of a firefly is determined by the value of the objective function of a given problem.

4. Attractiveness

The basic form of attractiveness function of a firefly is the following monotonically decreasing function:

$$\beta(r) = \beta_0 \exp\left(-\gamma r^m\right), \textit{with } m \geq 1, \tag{18}$$

Where r is the distance between any two fireflies, β0 is the initial attractiveness at r=0, and γ is an absorption coefficient which controls the decrease of the light intensity.

- Distance

The distance between two fireflies I and j at positions xi and xj can be defined by the following relation:

$$r_{ij} = \left\| x_i - x_j \right\| = \sqrt{\sum_{k=1}^{d} \left(x_{i,k} - x_{j,k} \right)^2} \tag{19}$$

Where, $x_{i,k}$ is the *kth* component of the spatial coordinate x_i and x_j and d is the number of dimension.

- Movement

The movement of a firefly i which is attracted by a more attractive firefly j is given by the following equation.

$$x_i = x_i + \beta_0 . \exp\left(-\gamma r_{ij}^2\right) . \left(x_j - x_i\right) + \alpha . (rand - 0.5), \tag{20}$$

Where the first term is the current position of a firefly, the second term is used for considering a firefly's attractiveness to light intensity seen by adjacent fireflies, and the third term is for the random movement of a firefly in case there are not any brighter ones.

- Parameters settings

Like many metaheuristic optimization methods, choosing the initial values of parameters is an important task which affects greatly the convergence behaviors of the algorithm. These parameters depend on the nature of the problem to be solved.

$\alpha \in [0,\ 1]$ is a randomization parameter determined based on the complexity of the problem to be solved.

$$\beta_0 = 1 \tag{21}$$

The attractiveness or absorption coefficient $\gamma = 1$.

4.1.1. Solving combined economic emission problem based FFA

The pseudo code depicted in Figure 6 gives a brief description about the adaptation of the firefly algorithm for solving the combined economic emission problem.

At the first stage an initial solution is generated based on the following equation:

$$x_i = rand.\left(x_i^{\max} - x_i^{\min}\right) + x_i^{\min} \tag{22}$$

Where $x_i^{\max}$ and $x_i^{\min}$ are the upper range and lower range of the *ith* firefly (variable), respectively.

```
Begin of algorithm
Generate initial population of fireflies nf
Light Intensity of firefly n is determined by objective function
Specify algorithm's parameters value a, β0 and γ
While t ≤ Max Gen
For i=1: nf %for all fireflies (solutions)
        For j=1: nf %for all fireflies (solutions)
            If (Ii <Ij)
            Then move firefly i towards firefly j (move towards brighter one)
            Attractiveness varies with distance rij via eq (18)
            Generate and evaluate new solutions and update Light Intensity
        End for j loop
End for i loop
Check the ranges of the given solutions and update them as appropriate
Rank the fireflies, find and display the current best % max solution for each iteration.
End of while loop
End of algorithm
```

Figure 6. Pseudo code: Basic firefly algorithm.

After evaluation of this initial solution, the initial solution will be updated based on the mechanism search, final solution will be achieved after a specified number of generations.

Test1: Benchmark function

In this case, the proposed algorithm is tested with many benchmark functions. All these problems are minimization problems. Due to the limited chapter length, only one function are considered for minimization problem, details can be retrieved from [21]. Fig 7 shows the shape of the objective function, Fig 8 shows the path of fireflies during optimization's process, Fig 9 shows the Final stage corresponding to global solution.

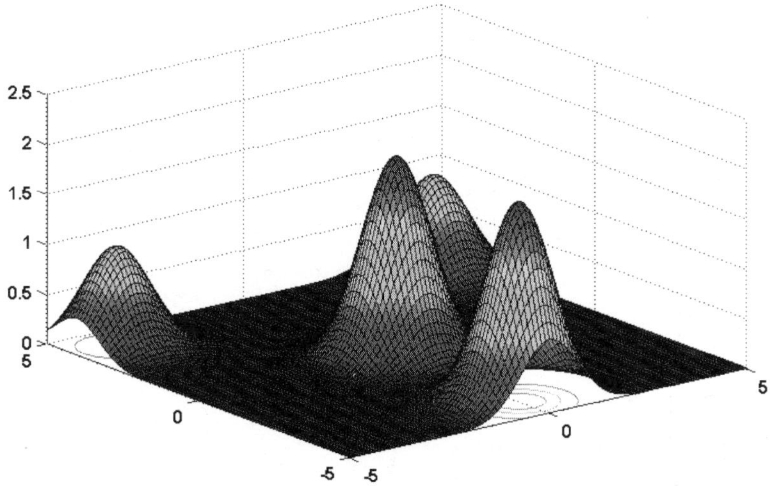

Figure 7. The shape of the objective function

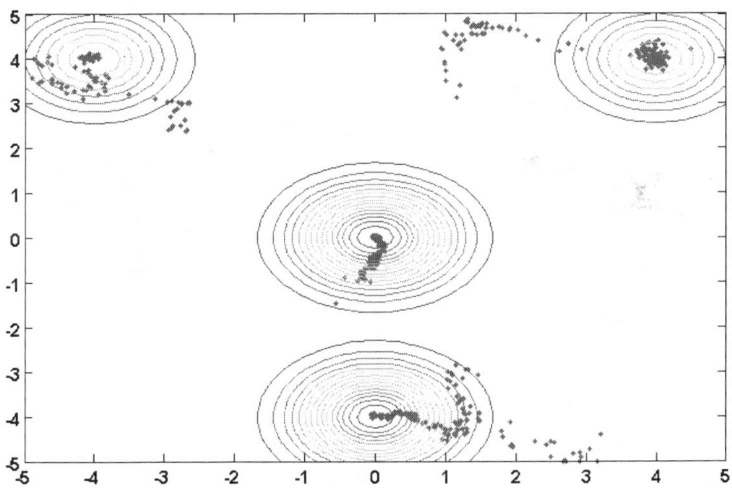

Figure 8. Path of fireflies during optimization's process.

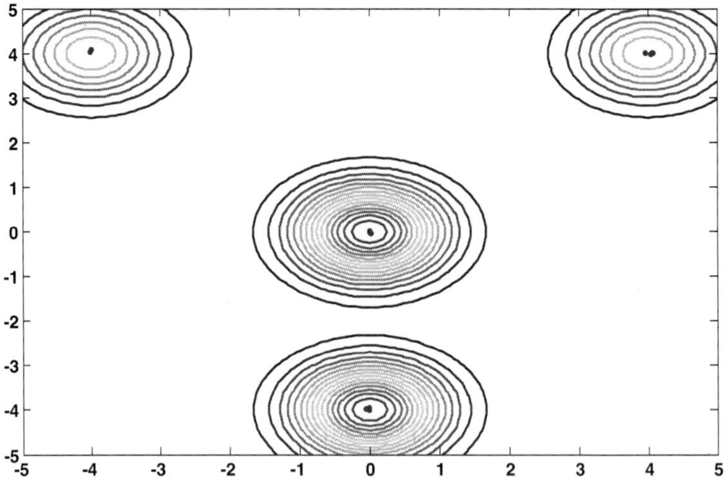

Figure 9. Final stage: global solution

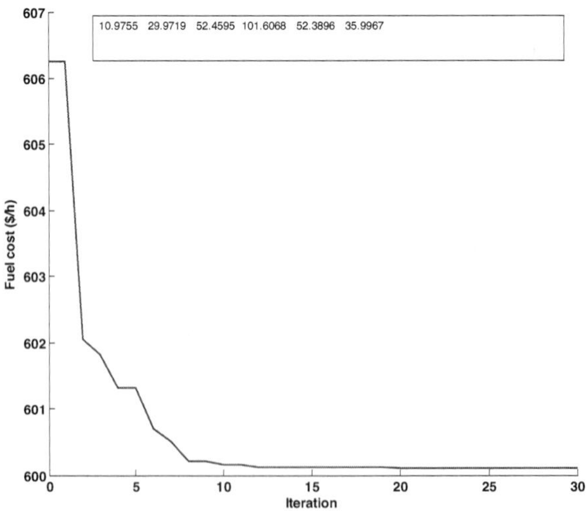

Figure 10. Convergence characteristic for optimal fuel cost.

http://dx.doi.org/10.5772/52875

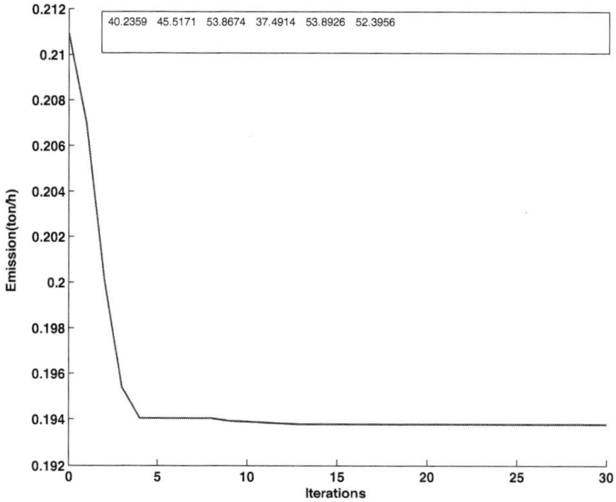

Figure 11. Convergence characteristic for optimal emission.

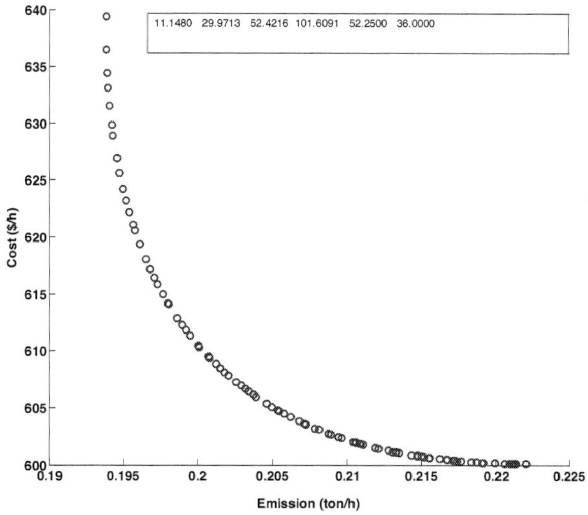

Figure 12. Convergence characteristic for combined fuel cost and emission.

Test 2: Combined economic-emission dispatch with smooth cost function

To demonstrate the performance of the proposed FFA, the algorithm is applied and tested to the six generating units without considering power transmission loss. The problem is solved as single and multi objective optimization. Detail data of this test system is given in [22-23]. For 20 runs, the best cost is 600.1114 ($/h), the algorithm converges at low number of iteration.

Figs 10-11 show the convergence characteristics of the proposed FFA when objective function optimized separately. Fig 12 shows the convergence characteristic for combined fuel cost and emission. Table 1 summarizes the best solutions obtained for three cases.

Control Variables (MW)	Case 1 Fuel cost	Case 2 Emissions	Case3 Combined: Cost-Emissions
P_{G1}	10.9755	40.2359	10.9485
P_{G2}	29.9719	45.5171	29.8834
P_{G5}	52.4595	53.8674	52.1869
P_{G8}	101.6068	37.4914	101.7170
P_{G11}	52.3896	53.8926	52.7216
P_{G13}	35.9967	52.3956	35.9426
FC ($/h)	**600.1114**	638.7807	**612.0227**
Emission (ton/h)	0.2219	**0.1938**	**0.1991**
Loss (MW)	0	**0**	0

Table 1. Control variables optimized using FFA

4.2. Solving security OPF based parallel GA considering SVC controllers

4.2.1. Strategy of the parallel GA

Parallel Genetic Algorithms (PGAs) have been developed to reduce the large execution times that are associated with simple genetic algorithms for finding near-optimal solutions in large search spaces. They have also been used to solve larger problems and to find better solutions. PGAs can easily be implemented on networks of heterogeneous computers or on parallel mainframes. The way in which GAs can be parallelized depends upon the following elements [22-23]:

- How fitness is evaluated.
- How selection is applied locally or globally.
- How genetic operators (crossover and mutation are used and applied)
- If single or multiple subpopulations are used.
- If multiple populations are used how individuals are exchanged.

- How to coordinate between different subpopulations to save the proprieties of the original chromosome.

In the proposed approach the subpopulations created are dependent, efficient load flow used to test the performance of the new subpopulations generated.

The proposed algorithm decomposes the solution of such a modified OPF problem into two coordinated sub problems. The first sub problem is an active power generation planning solved by the proposed algorithm, and the second sub problem is a reactive power planning [14-15] to make fine adjustments on the optimum values obtained from the first stage. This will provide updated voltages, angles and point out generators having exceeded reactive power limits.

4.2.2. Decomposition mechanism

Problem decomposition is an important task for large-scale OPF problem, which needs answers to the following two technical questions.

1. How many efficient partitions needed?
2. Where to practice and generate the efficient inter-independent sub-systems?

The decomposition procedure decomposes a problem into several interacting sub-problems that can be solved with reduced sub-populations, and coordinate the solution processes of these sub-problems to achieve the solution of the whole problem.

4.2.3. Algorithm of the Proposed Approach

1. Initialization based in decomposition mechanism

In the first stage the original network was decomposed in multi sub-systems and the problem transformed to optimize the active power demand associated to each partitioned network. The main idea of the proposed approach is to optimize the active power demand for each partitioned network to minimize the total fuel cost. An initial candidate solution generated for the global N population size.

Suppose the original network under study decomposed into S sub-systems 1,2.....S with Pg1, Pg2,.....PgN, the active power control variables, for this decomposed sub-systems related to S sub-populations, 1,2......S, with population sizes, N1........NS, and (N1+N2+...... NS=N).

- Each sub-population contains NP1 control variables to be optimized.
- Each sub-population updated based on the GA operators.

1-For each decomposition level estimate the initial active power demand:

For NP=2 Do

$$Pd1 = \sum_{i=1}^{M1} P_{Gi} \quad (23)$$

$$Pd2 = \sum_{i=1}^{M2} P_{Gi} = PD - Pd1 \quad (24)$$

Where NP the number of partition

$Pd1$: the active power demand for the first initial partition.

$Pd2$: the active power demand for the second initial partition.

PD: the total active power demand for the original network.

The following equilibrium equation should be verified for each decomposed level:

For level 1:

$$Pd1 + Pd2 = PD + Ploss \quad (25)$$

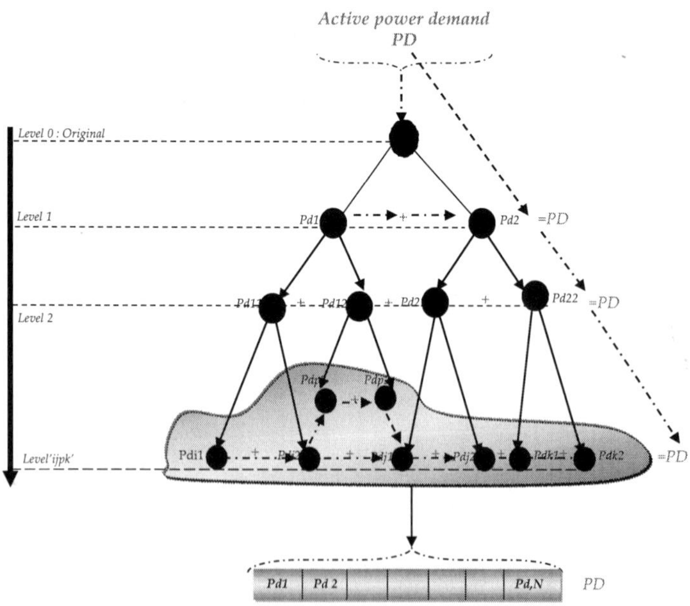

Figure 13. Mechanism of partitioning procedure.

http://dx.doi.org/10.5772/52875

2. Fitness evaluation based power loss

For all sub-systems generated perform a load flow calculation to evaluate the proposed fitness function. A candidate solution formed by all sub-systems is better if its fitness is higher (low total power loss).

3. A global data base generated containing the best technical sub-systems.

4. Consequently under this concept, the final value of active power demand should satisfy the following equations.

$$\sum_{i=1}^{N_g}\left(Pg_i\right)=\sum_{i=1}^{part_i}\left(Pd_i\right)+ploss \tag{26}$$

$$Pg_i^{\min}\le Pg_i\le Pg_i^{\max} \tag{27}$$

4.2.4. Reactive power dispatch

All the sub-systems are collected to form the original network, global database generated based on the best results found from all sub-populations.

The final solution (Global) is found out after reactive power planning procedure to adjust the reactive power generation limits, and voltage deviation, the final optimal cost is modified to compensate the reactive constraints violations. Fig. 13 illustrates the mechanism search partitioning for active power planning, Fig. 14 shows an example of tree network decomposition.

5. FACTS technology

FACTS philosophy was first introduced by Hingorani [24] from the Electric power research institute (EPRI) in the USA. The objective of FACTS devices is to bring a system under control and to transmit power as ordered by the control centers, it also allows increasing the usable transmission capacity to its thermal limits. With FACTS devices we can control the phase angle, the voltage magnitude at chosen buses and/or line impedances. In general these FACTS devices are classified in three large categories as follows:

5.1. Shunts FACTS controllers (SVC, STATCOM)

Principally designed and integrated to adjust dynamically the voltage at specified buses. Fig 15 shows the basic principle of dynamic shunt FATCS controllers (SVC, and STATCOM).

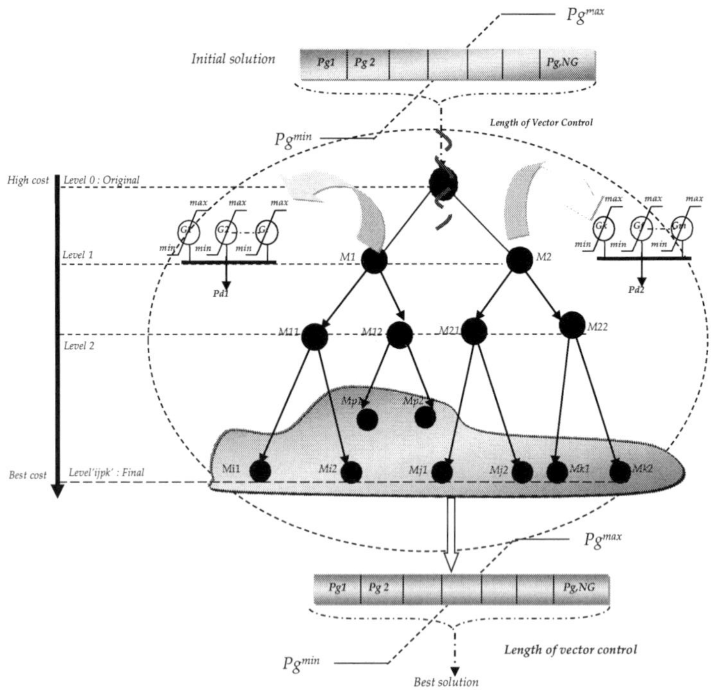

Figure 14. Sample of network in tree decomposition

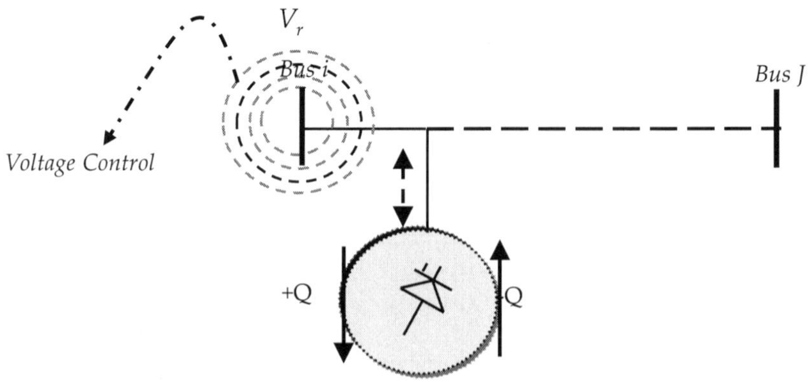

Figure 15. Shunt FACTS Controller.

5.2. Series FACTS controllers (TCSC, SSSC)

Principally designed and integrated to adjust dynamically the transit power at specified lines. Fig 16 shows the basic principle of dynamic series FATCS controllers (TCSC and SSSC).

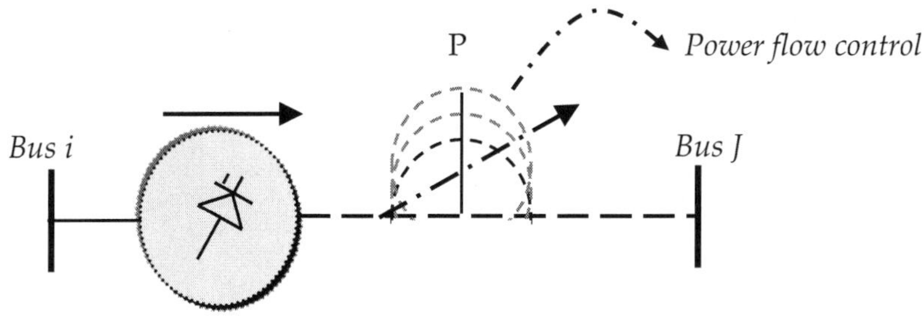

Figure 16. Series FACTS Controllers.

5.3. Hybrid FACTS controllers (UPFC)

Principally designed and integrated to adjust dynamically and simultaneously the voltage, the active power, and the reactive power at specified buses and lines. The basic one line diagram of hybrid controller (UPFC) is well presented in Fig 17.

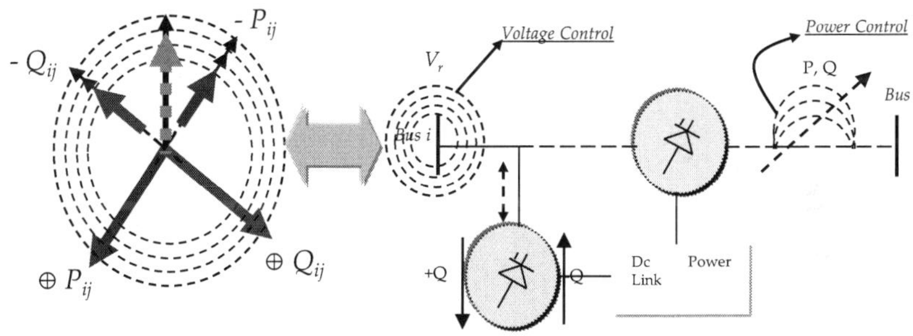

Figure 17. Hybrid FACTS Controller.

5.4. Shunt FACTS modelling

5.4.1. Static VAR Compensator (SVC)

The Static VAr Compensator (SVC) [25] is a shunt connected VAr generator or absorber whose output is adjusted dynamically to exchange capacitive or inductive current so as to maintain or control specific parameters of the electric power system, typically bus voltages. It includes separate equipment for leading and lagging VArs. The steady state model shown in Fig. 18 is used to incorporate the SVC on power flow problems. This model is based on representing the controller as variable susceptance or firing angle control.

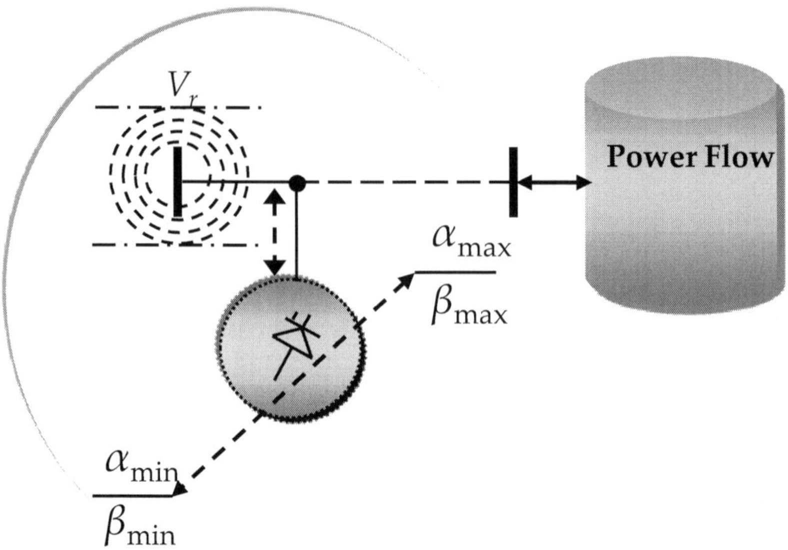

Figure 18. SVC Steady-state model

$$I_{SVC} = jB_{SVC}V \tag{28}$$

$$\left.\begin{aligned} B_{SVC} &= B_C - B_{TCR} = \frac{1}{X_C X_L}\left\{X_L - \frac{X_C}{\pi}\left[2(\pi-\alpha)+\sin(2\alpha)\right]\right\}, \\ X_L &= \omega L, X_C = \frac{1}{\omega C}, \end{aligned}\right\} \tag{29}$$

Where, B_{SVC}, α, X_L , X_C, V are the shunt susceptance, firing angle, inductive reactance, capacitive reactance of the SVC controller, and the bus voltage magnitude to which the SVC is connected, respectively. The reactive power Q_i^{SVC} exchange with the bus i can be expressed as,

$$Q_i^{SVC} = B_i^{SVC}.V_i^2 \tag{30}$$

6. Application

6.1. Test system 1: Algerian electrical network planning

6.1.1. Active Power dispatch without SVC compensators

The proposed algorithm is developed in the Matlab programming language using 6.5 version. The proposed approach has been tested on the Algerian network (1976). It consists of 59 buses, 83 branches (lines and transformers) and 10 generators. Table I shows the technical and economic parameters of the ten generators, knowing that the generator of the bus N°=13 is not in service. Table II shows the generators emission coefficients. The values of the generator emission coefficients are given in Tables 2-3, the generators data and cost coefficients are taken from [23-28], Figs. 19-20 show the topology of the Algerian electrical network test with 59-Bus based one line diagram and in schematic representation. For the purpose of verifying the efficiency of the proposed approach, we made a comparison of our algorithm with others competing OPF algorithm. In [30], they presented a fuzzy controlled genetic algorithm, in [28], authors proposed fast successive linear programming algorithm applied to the Algerian electrical network.

To demonstrate the effectiveness and the robustness of the proposed approach, two cases have been considered with and without consideration of SVC Controllers installation:

Case 1: Minimum total operating cost (α =1).

Case 2: Minimum total emission (α =0).

Bus Number	Pmin [MW]	Pmax [MW]	Qmin [Mvar]	Qmax [Mvar]	a [$/hr]	b [$/MWhr]	c [$/MW2hr]
1	8	72	-10	15	0	1.50	0.0085
2	10	70	-35	45	0	2.50	0.0170
3	30	510	-35	55	0	1.50	0.0085
4	20	400	-60	90	0	1.50	0.0085
13	15	150	-35	48	0	2.50	0.0170
27	10	100	-20	35	0	2.50	0.0170
37	10	100	-20	35	0	2.00	0.0030
41	15	140	-35	45	0	2.00	0.0030
42	18	175	-35	55	0	2.00	0.0030
53	30	450	-100	160	0	1.50	0.0085

Table 2. Technical Admissible paramaters of Generators and the Fuel Cost Coefficients

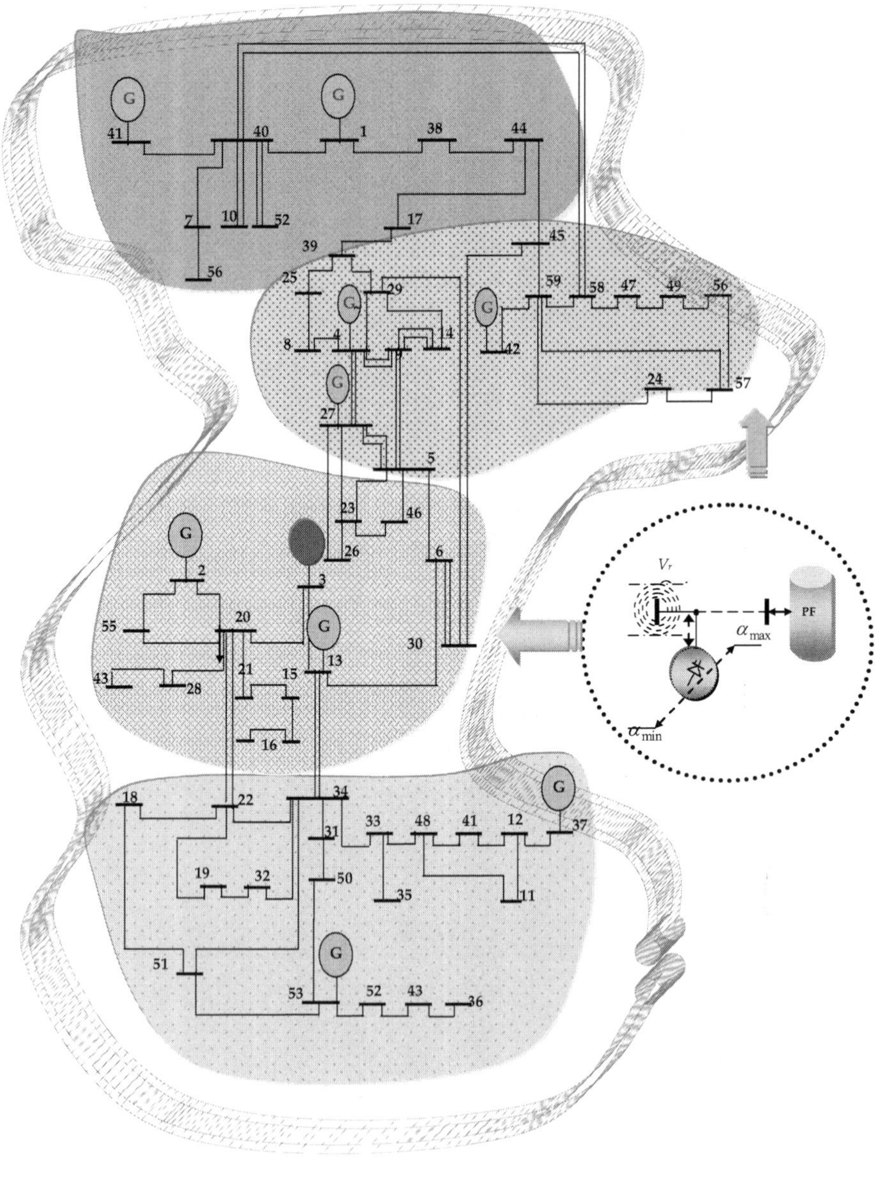

Figure 19. One line representation of the Algerian electrical production and transmission network (Sonelgaz).

http://dx.doi.org/10.5772/52875

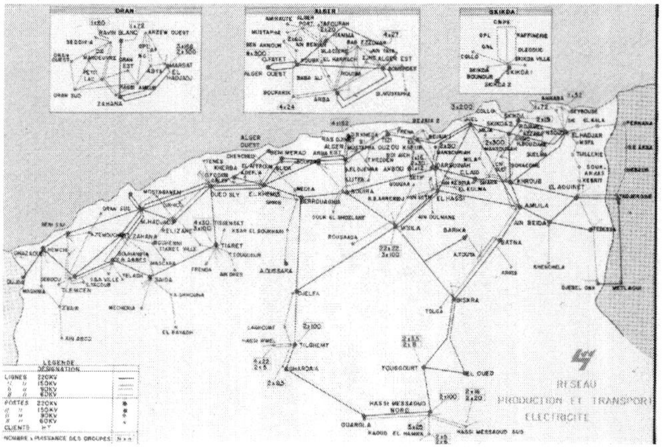

Figure 20. Topology of the Algerian production and transmission network before 1997 (Sonelgaz).

Bus Number	Generator	*a*	*bx1e-2*	*γx1e-4*	*ω*	*μx1e-2*
1	1	4.091	−5.554	6.490	2.00e-04	2.857
2	2	2.543	−6.047	5.638	5.00e-04	3.333
3	3	4.258	−5.094	4.586	1.00e-06	8.000
4	4	5.326	−3.550	3.380	2.00e-03	2.000
13	5	4.258	−5.094	4.586	1.00e-06	8.000
27	6	6.131	−5.555	5.151	1.00e-05	6.667
37	7	4.091	−5.554	6.490	2.00e-04	2.857
41	8	2.543	−6.047	5.638	5.00e-04	3.333
42	9	4.258	−5.094	4.586	1.00e-06	8.000
53	10	5.326	−3.550	3.380	2.00e-03	2.000

Table 3. Generator Emission Coefficients

	Case1: α =1	Case 2: α =0
P_{gi}(MW)	Minimum Cost	Minimum emission
P_{g1}	39.5528	28.2958
P_{g2}	37.3190	70.0000
P_{g3}	133.830	109.400
P_{g4}	142.320	79.8000
P_{g5}	**0.00**	**0.00000**
P_{g6}	24.8000	80.5800
P_{g7}	39.7000	34.8600
P_{g8}	39.5400	70.0400
P_{g9}	119.7800	100.6200
P_{g10}	123.4600	128.0200
Cost ($/h)	**1765.9377**	**1850.155**
Emission (ton/h)	**0.530700**	**0.42170**
Power loss (MW)	**16.20180**	**17.51580**

Table 4. Simulation Results for Three Cases with two optimized shunt Compensators

Generators N°	FGA [29]	GA [28]	ACO [28]	FSLP [28]	Our Approach
P_{g1} (MW)	11.193	70.573	64.01	46.579	39.5528
P_{g2} (MW)	24.000	56.57	22.75	37.431	37.3190
P_{g3} (MW)	101.70	89.27	82.37	134.230	133.830
P_{g4} (MW)	84.160	78.22	46.21	137.730	142.320
P_{g5} (MW)	**0.000**	**0.00**	**0.00**	**0.000**	**0.00**
P_{g6} (MW)	35.22	57.93	47.05	23.029	24.8000
P_{g7} (MW)	56.80	39.55	65.56	35.238	39.7000
P_{g8} (MW)	121.38	46.40	39.55	39.972	39.5400
P_{g9} (MW)	165.520	63.58	154.23	117.890	119.7800
P_{g10} (MW)	117.32	211.58	202.36	131.650	123.4600
PD(MW)	684.10	684.10	684.10	684.10	684.1
Ploss (MW)	33.1930	29.580	39.980	19.65	**16.20180**
Cost[$/hr]	1768.50	1937.10	1815.7	1775.856	**1765.9377**

Table 5. Comparison of the Results Obtained with Conventional and Global Methods: Case Minimum Cost

	FSLP [28]		Our Approach	
	Case1	Case 2	Case1	Case 2
P_{gi}(MW)	α =1	α =0	α =1	α =0
P_{g1}	46.579	28.558	39.5528	28.2958
P_{g2}	37.431	70.000	37.319	70.000
P_{g3}	134.230	114.200	133.83	109.40
P_{g4}	137.730	77.056	142.32	79.800
P_{g5}	**0.000**	**0.000**	**0.00**	**0.00**
P_{g6}	23.029	87.575	24.80	80.580
P_{g7}	35.238	32.278	39.70	34.860
P_{g8}	39.972	63.176	39.54	70.040
P_{g9}	117.890	95.645	119.78	100.62
P_{g10}	131.650	135.540	123.46	128.02
Cost ($/h)	1775.856	1889.805	**1765.9377**	**1850.155**
Emission (ton/h)	0.5328	0.4329	**0.5307**	**0.4217**
Power loss (MW)	19.65	19.93	**16.2018**	**17.5158**

Table 6. Comparison of the results obtained with conventional methods

Table 4 shows simulation results obtained by the proposed approach for the two cases (α =1, α =0), the comparison of the results obtained by the application of the decomposed parallel GA proposed with those found by global optimization (GA, FGA, and ACO) and conventional methods are reported in the Table 5 and Table 6 The proposed approach gives better results for all cases. For example at the case corresponding to the minimum total operating cost (α =1), the fuel cost is 1765.9377 $/h, and power losses 16.2018 MW which are better compared with the results found by the global and conventional methods.

It is important to note that all results obtained by the proposed approach do not violate the physical generation capacity constraints (reactive power). The security constraints are also satisfied for voltage magnitudes (0.9<V<1.1 p.u). Fig 18 shows distribution voltage profile with and without two optimized shunt compensators installed at bus 27 and bus 36. Fig 19 shows the voltage phase profiles, it is clearly identified that all voltage phase profiles are within the constraint limit.

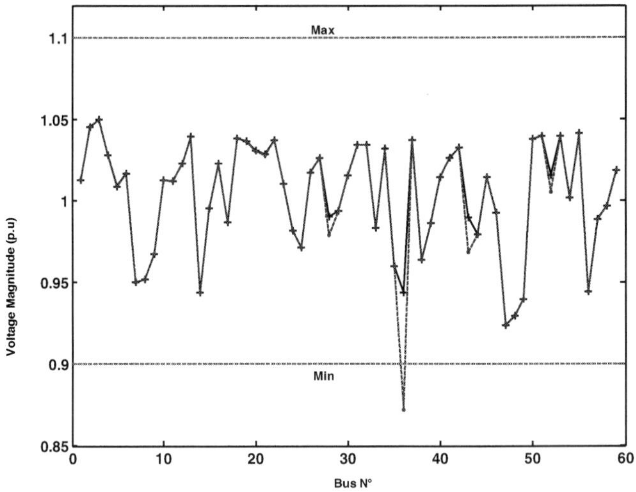

Figure 21. Voltage magnitude profiles with and without compensation: case: minimum cost

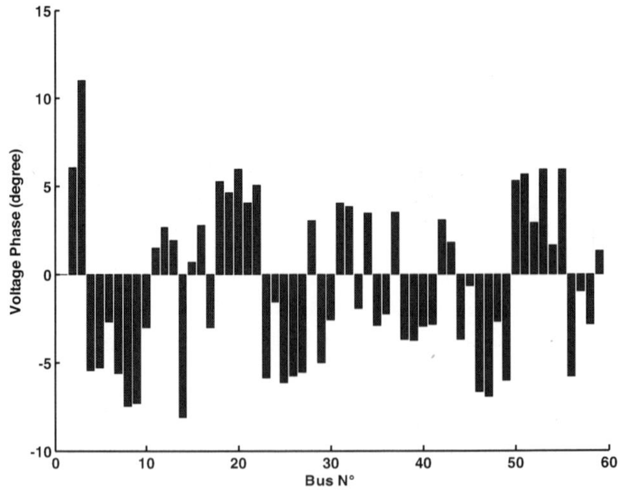

Figure 22. Voltage phase profiles of the Algerian network 59-bus: case 1: minimum fuel cost: with two optimized shunt compensators.

	Case 1	Case 2	Qmax	Qmin
Bus	Qg (Mvar)	Qg (p.u)	Q max	Q min
1	3.6462	7.62240	15.00	-10
2	36.9832	31.8128	45.00	-35
3	12.6903	14.1747	55.00	-35
4	57.312	80.745	90.00	-60
13	1.5281	2.6443	48.00	-35
27	11.7001	-9.7079	35.00	-20
37	21.2983	25.378	35.00	-20
41	37.8163	28.8672	45.00	-35
42	20.3931	23.2128	55.00	-35
53	0.59601	-1.80110	160.00	-100

Table 7. Simulation Results for two Cases: Reactive Power Generation with optimized two shunt compensators

6.1.2. Reactive power dispatch based SVC controllers

For a secure operation of the power system, it is important to maintain required level of security margin, system loadability, voltage magnitude and power loss are three important indices of power quality. In this stage, dynamic shunt compensation based SVC Controllers taken in consideration. The control variables selected for reactive power dispatch (RPD) are the generator voltages, and reactive power of the SVC compensators installed at critical buses.

6.1.3. Optimal location of shunt FACTS

Before the insertion of SVC devices, the system was pushed to its collapsing point by increasing both active and reactive load discretely using continuation load flow [5]. In this test system according to results obtained from the continuation load flow, buses 7, 14, 17, 35, 36, 39, 44, 47, 56 are the best location points for installation and coordination between SVC Compensators and the network. Table 8 gives details of the SVC Data. Table 8, shows results of reactive power generation with SVC Compensators for the minimum cost case, the active power loss and the total cost are reduced to **15.1105 MW, 1763.5771 ($/h)** respectively compared to base case considering two optimized shunt compensators. It is important to note that the system loadability improved to **1.9109 p.u** compared to the base case (**1.8607 p.u**) with two shunt compensators, reactive power of generating units are within their admissible limits.). Fig 20 shows distribution voltage profile considering two fixed shunt compensators and multi SVC compensators. Fig 21 shows the voltage phase profiles, it is clearly identified that all voltage phase profiles are within the constraint limit.

	B_{min} (p.u)	B_{max} (p.u)	B_{init} (p.u)
Susceptance SVC Model	-0.5	0.5	0.025

Table 8. SVCs data.

Bus	B_{SVC}(p.u)	V (p.u) with SVC	Fixed Shunt Capacitors
7	0.04243	1.00	
14	0.23974	1.00	
17	0.01790	1.00	
35	0.01790	1.00	
36		1.00	
39	-0.01116	1.00	
44	0.05846	1.00	
47	0.12876	1.00	
56	0.08352	1.00	
27			
Ploss MW		15.1105	
Cost ($/hr)		1763.5771	
Voltage deviation		0.95<Vi<1.05	
Loading Factor (p.u): without SVC		1.8607	
Loading Factor (p.u): with 8 SVC Controllers		1.9109	

Table 9. Results of the Reactive power dispatch based two static shunt compensators and multi -SVC Controllers: Case 1: Minimum Cost

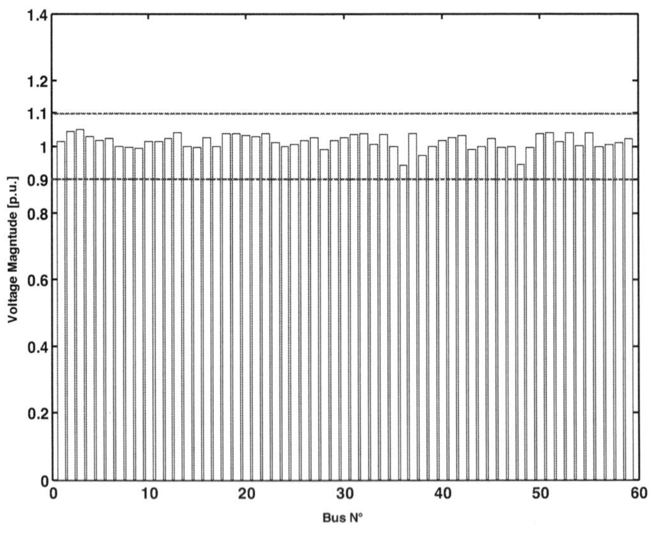

Figure 23. Voltage magnitude profiles with optimized SVC compensation: case: minimum fuel cost

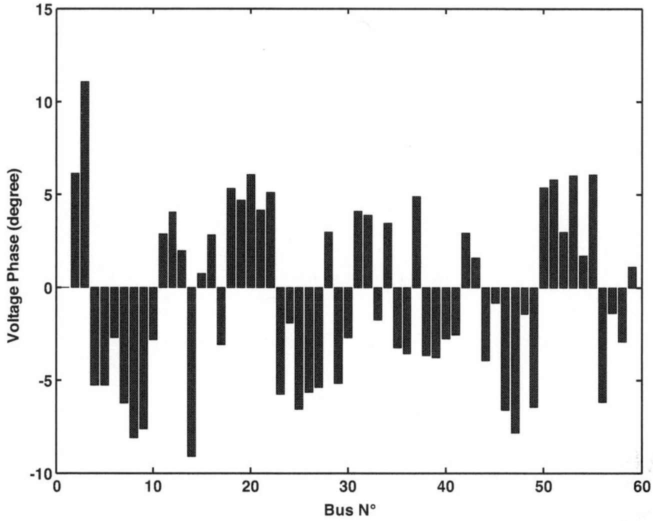

Figure 24. Voltage phase profiles of the Algerian network 59-bus with optimized SVC compensation: case 1: minimum fuel cost:

7. Conclusion

The multi objective power quality management consists in optimizing indices of power quality in coordination with FACTS devices considering fuel cost. In this chapter two techniques based metaheuristic methods have been proposed to solving the multi objective power management problem, many objective functions related to power quality have been proposed like power loss, voltage deviation, system loadability, and pollution constraint. The first proposed method called Firefly research algorithm is used to solve the combined economic emission problem, the algorithm applied to 6 generating units. The second method is a new variant based GA named parallel GA (PGA) proposed as an alternative to improve the performances of the standard GA to solving the security combined economic dispatch considering multi shunt FACTS devices. The proposed PGA applied to solving the security OPF of the Algerian power system (59-Bus) considering three indices of power quality; power loss, voltage deviation and gaz emissions. From simulation results, it is found that the two proposed approaches, FFA and PGA can converge to the near global solution and obtain competitive results in term of solution quality and convergence characteristic compared to the others recent optimization methods.

Due to these efficient properties, in the future work, author will still to apply these two methods to solving the multi objective optimal power flow based multi FACTS devices considering practical generator units (valve point effect and prohibited zones).

Author details

Belkacem Mahdad

Address all correspondence to: bemahdad@mselab.org

Department of Electrical Engineering, Biskra University, Algeria

References

[1] A. A, El-Keib, H. Ma, and J. L. Hart, "Economic dispatch in view of clean air act of 1990," IEEE Trans. Power Systems, vol. 9, no. 2, pp. 972-978, May 1994.

[2] B. Mahdad "Contribution to the improvement of power quality using multi hybrid model based Wind-Shunt FACTS," 10th EEEIC International Conference on Environment and Electrical Engineering, Italy, 2011.

[3] B. Sttot and J. L. Marinho, "Linear programming for power system network security applications," IEEE Trans. Power Apparat. Syst., vol. PAS-98, pp. 837-848, May/June 1979.

[4] M. Huneault, and F. D. Galiana, "A survey of the optimal power flow literature," IEEE Trans. Power Systems, vol. 6, no. 2, pp. 762-770, May 1991.

[5] J. A. Momoh and J. Z. Zhu, "Improved interior point method for OPF problems, " IEEE Trans. Power Syst., vol. 14, pp. 1114-1120, Aug. 1999.

[6] P. Pezzinia, O. Gomis-Bellmunt, and A. Sudrià-Andreua, "Optimization techniques to improve energy efficiency in power systems, " Renewable and Sustainable Energy Reviews, pp.2029-2041, 2011.

[7] S. Frank, I. Steponavice, and S. Rebennak, "Optimal power flow: a bibliographic survey I, formulations and deterministic methods," Int. J. Energy.System (Springer-Verlag), 2012.

[8] A. G. Bakistzis, P. N. Biskas, C. E. Zoumas, and V. Petridis, "Optimal power flow by enhanced genetic algorithm, "IEEE Trans. Power Systems, vol. 17, no. 2, pp. 229-236, May 2002.

[9] J. G. Vlachogiannis, and K. Y. Lee, "Economic dispatch-A comparative study on heuristic optimization techniques with an improved coordinated aggregation-based PSO," IEEE Trans. Power Systems, vol. 24, no. 2, pp. 991-10001, 2009.

[10] C. A. Roa-Sepulveda, B. J. Pavez-Lazo, " A solution to the optimal power flow using simulated annealing," Electrical Power & Energy Systems (Elsevier), vol. 25,n°1, pp. 47-57, 2003.

[11] S. Frank, I. Steponavice, and S. Rebennak, "Optimal power flow: a bibliographic survey II, non-deterministic and hybrid methods," Int. J. Energy.System (Springer-Verlag), 2012.

[12] Victoire, T., & Jeyakumar, A. E. (2005c). "A modified hybrid EP–SQP approach for dynamic dispatch with valve-point effect," Electric Power Energy Systems, 27(8), 594–601.

[13] J.-C. Lee, W.-M. Lin, G.-C. Liao, and T.-P. Tsao, "Quantum genetic Algorithm for dynamic economic dispatch with valve-point effects and including wind power system," Electrical Power and Energy Systems, vol. 33, pp. 189–197, 2011.

[14] S. Hemamalini and S. P. Simon, "Dynamic economic dispatch using artificial immune system for units with valve-point effect," Electrical Power and Energy Systems, vol. 33, pp. 868–874, 2011.

[15] B. Panigrahi, P. V. Ravikumar, and D. Sanjoy, "Adaptive particle swarm optimization approach for static and dynamic economic load dispatch," Energy Convers Manage, vol. 49, pp. 1407–1415, 2008.

[16] X. Yuan, A. Su, Y. Yuan, H. Nie, and L. Wang, "An improved pso for dynamic load dispatch of generators with valve-point effects," Energy, vol. 34, pp. 67–74, 2009. August 18, 2011

[17] Y. Lu, J. Zhou, H. Qin, Y. Li, and Y. Zhang, "An adaptive hybrid differential evolution Algorithm for dynamic economic dispatch with valve-point effects," Expert Systems with Applications, vol. 37, pp. 4842– 4849, 2010.

[18] Z. J. Wang, Ying, H. Qin, and Y. Lu, "Improved chaotic particle swarm optimization Algorithm for dynamic economic dispatch problem with valve-point effects," Energy Conversion and Management, vol. 51, pp. 2893– 2900, 2010.

[19] R. Kumar, D. Sharma, and A. Sadu, "A hybrid multiagent based particle swarm optimization algorithm for economic power dispatch," Int Journal of Elect. Power Energy Syst., vol. 33, pp. 115–123, 2011.

[20] X. S. Yang, Nature-Inspired Meta-Heuristic Algorithms, Luniver Press, Beckington, UK, 2008.

[21] T. Apostolopoulos and A. Vlachos, "Application of the Firefly Algorithm for Solving the Economic Emissions Load Dispatch Problem", International Journal of Combinatorics, vol., pp.1-23, 2011.

[22] B. Mahdad, K. Srairi "Differential evolution based dynamic decomposed strategy for solution of large practical economic dispatch," 10th EEEIC International Conference on Environment and Electrical Engineering, Italy, 2011.

[23] B. Mahdad, Optimal Power Flow with Consideration of FACTS devices Using Genetic Algorithm: Application to the Algerian Network, Doctorat Thesis, Biskra University Algeria, 2010.

[24] Hingorani NG, Gyugyi L (1999), Understanding FACTS: Concepts and Technology of Flexible ac Transmission Systems. IEEE Computer Society Press.

[25] Acha E, Fuerte-Esquivel C, Ambiz-Perez (2004) FACTS Modelling and Simulation in Power Networks. John Wiley & Sons.

[26] M. A. Abido, "Multiobjective evolutionary algorithms for electric power dispatch problem," IEEE Trans. Power Systems, vol. 10, no. 3, pp. 315-329, 2006.

[27] K. Price, R. Storn, and J. Lampinen, Differential Evolution: A Practical Approach to Global Optimization. Berlin, Germany: Springer- Verlag, 2005.

[28] K. Zehar, S. Sayah, "Optimal power flow with environmental constraint using a fast successive linear programming algorithm: application to the algerian power system," Journal of Energy and management, Elsevier,vol. 49, pp. 3362-3366, 2008.

[29] B. Mahdad; Bouktir and T.; Srairi, K., "OPF with Envirenmental Constraints with SVC Controller using Decomposed Parallel GA: Application to the Algerian Network," Power and Energy Society General Meeting - IEEE, Page(s):1- 8. 2009.

[30] B. Mahdad,; Bouktir, T. and Srairi, K., "Optimal power flow of the algerian network using genetic algorithm/fuzzy rules", Power and Energy Society General Meeting - Conversion and Delivery of Electrical Energy in the 21st Century, IEEE, 20-24 July 2008 Page(s):1-8.

[31] B. Mahdad, T. Bouktir and K. Srairi, "Optimal Power Flow for Large-Scale Power System with Shunt FACTS using Efficient Parallel GA," Journal of Electrical Power & Energy Systems (IJEPES), Elsevier Vol.32, Issue 5, pp. 507-517, June 2010.

[32] E. Rashedi, H. Nezamabadi-pour, S. Saryazdi, 'GSA: A Gravitational Search Algorithm,' Information Sciences Journal, v. 179, pp. 2232-2248, 2009.

[33] S. Duman, U. Guvenc, Y. Sonmez, N. Yorukeren, "Optimal power flow using gravitational Search Algorithm," Energy Conversion and Management, vol. 59, pp. 86-95, 2012.

[34] E. Atashpaz-Gargari and C. Lucas, "Imperialist competitive Algorithm: An Algorithm for optimization inspired by imperialistic competition," in Evolutionary Computation, 2007. CEC 2007. IEEE Congress on, 2007, pp. 4661– 4667.

[35] P. K. Roy, S. P. Ghoshal, and S. S. Thakur, "Biogeography based optimization for multi-constraint optimal power flow with emission and non-smooth cost function," International Journal of Expert Systems with Apllications, vol. 37, pp. 8221-8228, 2010.

[36] T. Niknam, H. D. Mojarrad, H. Z. Meymand, B. B. Firouzi, "A new honey bee mating optimization algorithm for non-smooth economic dispatch," International Journal of Energy, vol. 36, pp. 896-908, 2011.

[37] D. Karaboga, B. Basturk, A powerful and efficient algorithm for numeric optimization: artificial bee colony (ABC) algorithm, J. Global Optim. 39 (3) (2007) 459–471.

Chapter 7

Harmonic Effects of Power System Loads: An Experimental Study

Celal Kocatepe, Recep Yumurtacı, Oktay Arıkan, Mustafa Baysal, Bedri Kekezoğlu, Altuğ Bozkurt and C. Fadıl Kumru

Additional information is available at the end of the chapter

http://dx.doi.org/10.5772/53108

1. Introduction

Today, electric power systems is spreading to a large area and wide variety of loads are connected to energy system. During the planning and management of power systems, accurate determination of load characteristics that connected to power system is very important. Thus, power system problems, that may occur, can be pre-determined and precautions may be taken against them.

Especially with the developing semi-conductor technology, harmonics have become one of the most popular issues in power system. In this study, the measurements for the harmonic effects of the loads in power system were carried out and also contribution of these loads to harmonic distortion was exhibited. Moreover, the effect of harmonics existing in power system on the performance of some equipment was analyzed experimentally. The obtained results were discussed and suggestions were given.

2. Power system harmonics

Harmonics can be defined as components with periodic waveforms having multiples of fundamental frequency. Harmonics, one of the most important issues of power quality, have recently come into prominence though they are known since the early time of the ac power systems. In 1893, only eight years after first ac power plant is built, engineers conducted harmonic analyses to identify and solve the motor heating problem [1]. A paper written by E.J.

INTECH
open science | open minds

© 2013 Kocatepe et al.; licensee InTech. This is a paper distributed under the terms of the Creative Commons Attribution License (http://creativecommons.org/licenses/by/3.0), which permits unrestricted use, distribution, and reproduction in any medium, provided the original work is properly cited.

Houston and A.E. Kennely in 1894 is one of the first documents in which the word harmonic is used [2]. The issues related to harmonics became widespread especially after power electronic devices are significantly penetrated into power systems. In order to enhance the power quality and to remove the negative effect of nonsinusoidal magnitudes on power system, harmonic magnitude levels are need to be specified and also harmonics have to be analyzed. In this part, the circuit quantities/components are identified and definitions related to harmonic distortion are introduced.

2.1. Electrical quantities for non-sinusoidal conditions

Electrical quantities, such as voltage and current, are usually defined for sinusoidal steady-state operating conditions. These electrical magnitudes are, however, needs to be redefined when there are harmonic components due to the nonlinear elements. In this case, instantaneous voltage and current can be represented as following;

$$v(t) = \sqrt{2} \cdot \left(V_1 + \sum_{n=2}^{\infty} V_n \cdot \sin\left(n\omega_1 t + \theta_n\right) \right) \tag{1}$$

$$i(t) = \sqrt{2} \cdot \left(i_1 + \sum_{n=2}^{\infty} i_n \cdot \sin\left(n\omega_1 t + \delta_n\right) \right) \tag{2}$$

where V_n and I_n are the effective values of voltage and current for nth harmonic level, respectively. θ_n and δ_n are respectively phase angels of voltage and current for nth harmonic component with respect to reference angle. $\omega_1 = 2\pi f_1$ is the angular frequency of the fundamental frequency f_1. The DC component is assumed to be zero for simplification.

A typical distorted voltage waveform and its harmonic components are shown in Figure 1. The voltage signal has the following function;

$v(t) = 220\sqrt{2} \cdot \sin\omega t + 20\sqrt{2} \cdot \sin(3\omega t + 90°) + 40\sqrt{2} \cdot \sin(5\omega t + 36°)$

Real power P can be represented by the following expression,

$$P = \frac{1}{T}\int_0^T p(t) \cdot dt = \sum_{n=1}^{\infty} V_n \cdot I_n \cos\left(\theta_n - \delta_n\right) \tag{3}$$

where T=1/f is known as period or cycle. Notice that voltages and currents having different frequencies have no effect on real power value calculated. Expression (e.g. multiplication of 3rd harmonic voltage by 5th harmonic current is not undefined in real power expression).

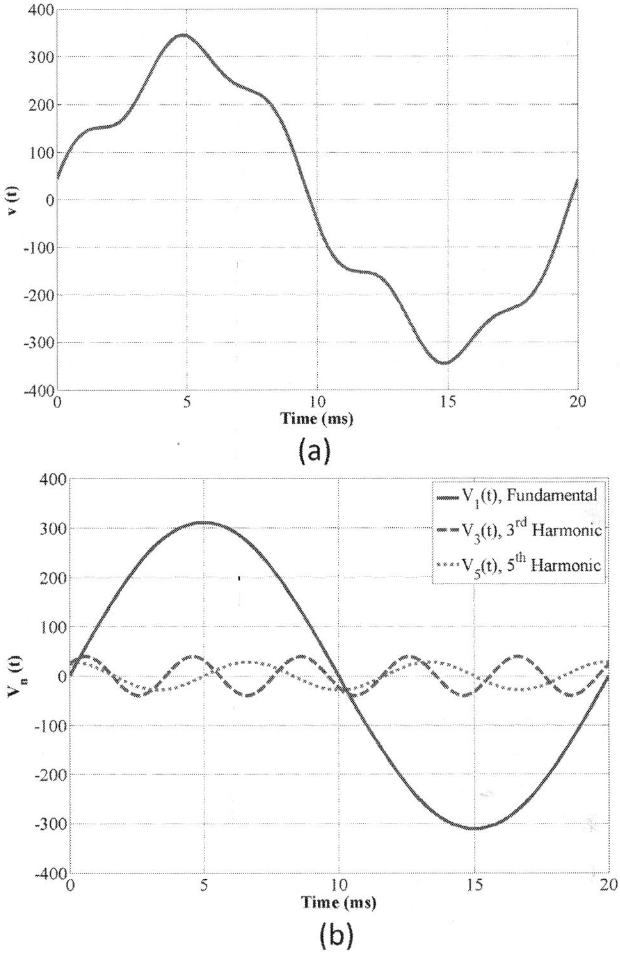

Figure 1. a) A typical distorted voltage waveform b) components of waveform

Apparent power S and distortion power D in power systems are defined as,

$$S = V \cdot I \tag{4}$$

$$D = \sqrt{S^2 - P^2 - Q^2} \tag{5}$$

Distortion power is not identical to real power and its value for sinusoidal conditions is zero.

Reactive power can be represented by,

$$Q = \sum_{n=1}^{\infty} V_n \cdot I_n \sin(\theta_n - \delta_n) \tag{6}$$

Power factor concept is used to determine how a current from AC power system is efficiently utilized by a load. In both sinusoidal and non-sinusoidal cases, power factor can be expressed as follows,

$$PF = \frac{P}{S} \tag{7}$$

2.2. Total Harmonic Distortion (THD)

Total harmonic distortion, THD, most-widely used index in related standards, is used to determine the deviation of the periodic waveform containing harmonics from the pure sinusoidal waveform. The total harmonic distortion of voltage and current waveform respectively can be expressed as following,

$$THD_V = \frac{\sqrt{\sum_{n=2}^{\infty} V_n^2}}{V_1} \tag{8}$$

$$THD_I = \frac{\sqrt{\sum_{n=2}^{\infty} I_n^2}}{I_1} \tag{9}$$

As seen from Eq. (7) and (8), total harmonic distortion is the ratio between rms values of harmonic components and rms value of fundamental component, and, is usually represented in percentage. THD value is equal to zero in a pure sinusoidal waveform.

2.3. Total Demand Distortion (TDD)

Total demand distortion, TDD, is related to particular load and defined as total harmonic current distortion,

$$TDD = \frac{\sqrt{\sum_{n=2}^{\infty} I_n^2}}{I_L} \tag{10}$$

where I_L is the maximum demand load current at the point of common coupling, PCC. This current value is the mean of the maximum currents which are demanded by the load through twelve months prior the measurement. TDD index is especially emphasized in IEEE Standard 519.

2.4. Crest factor (Cf)

The crest factor of a non-sinusoidal wave is represented,

$$Cf = \frac{\text{Peak Value}}{\text{RMS Value}} \tag{11}$$

This factor is the ratio between peak value and rms value of periodic wave which is the easiest way to indicate the harmonic components, and it is equal to $\sqrt{2}$ for a sinusoidal wave.

2.5. Transformer K-factor

The load current flowing through a transformer includes harmonic components when the transformer supplies nonlinear loads. Consequently the transformers under nonlinear loading cannot be run at their rating power. Transformer K-Factor index is used to determine the decrement quantity in nominal loading capacity of standard transformers running under harmonic conditions.

The transformer K-factor can be represented as follows,

$$K = \frac{\sum_{n=1} \left(n \cdot {I_n}/{I_1} \right)^2}{1 + THD_I^2} \tag{12}$$

where I is effective current and In is current of nth harmonic component.

To show how the concepts and definition related to harmonic distortion is changed in terms of signal waveform, a comparison is realized using several waveforms of current as shown in Figure 2. Peak values of pure sinusoidal waveform and square waveform are the same, 311 A, while pure sinusoidal waveform and distorted sinusoidal waveform have same rms value, 220 A. Distorted sinusoidal current signal has the identical harmonic components of waveform shown in Figure 2.

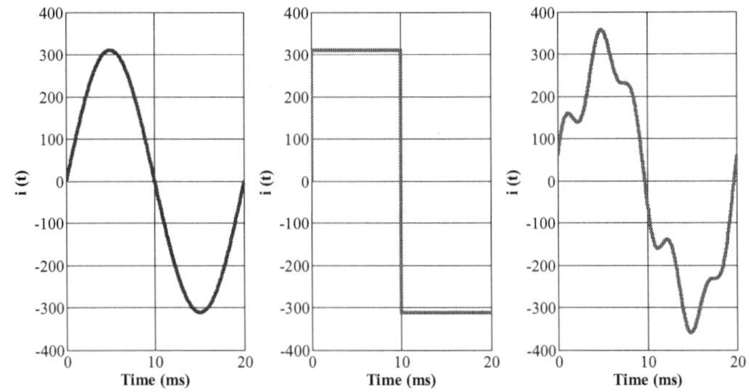

Figure 2. Several current waveforms

Quantities for the waveforms mentioned above are given in Table 1. The square shape current signal has the highest values for all harmonic indexes except crest factor. This is expected result since square waveform is the most distorted signal among the others. The highest crest factor value is obtained for distorted sinusoidal signal. Notice that, the peak value of the whole waveform would be different if the phase angles of the harmonic components are changed. The crest factor will have another value though the signal has the same harmonic components in that case. Maximum demand load current, I_L, is assumed to be 311 A for the calculations of TDD.

Electrical Quantity	Pure Sinusoidal	Square	Distorted Sinusoidal
I (A)	220	311	224,5
THD_I (%)	0	48,17	20,33
TTD (%)	0	43,37	14,38
Crest Factor	1,414	1	1,535
Transformer K-Factor	1	11,35	1,83

Table 1. Harmonic quantities calculated for several waveforms

2.6. Nonlinear elements

An electrical linear element has the constant ratio of voltage to current as illustrated in Figure 3. An electrical element which does not have linear relationship between voltage and current is defined as nonlinear element. When a nonlinear element is connected to a power system, it causes harmonic voltages and currents. These elements have either the electrical or magnetic circuit non-linearity characteristics.

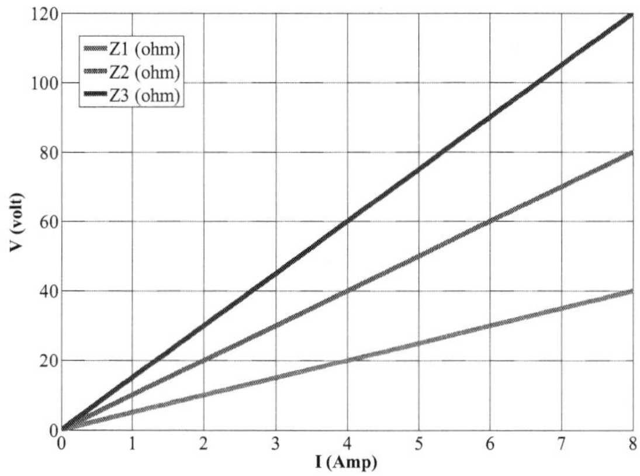

Figure 3. Linear electrical element characteristics

The relationship between voltage and current of a nonlinear element is mostly defined as,

$$I = K_1 + K_2 \cdot V + K_3 \cdot V^2 + K_4 \cdot V^3 + \ldots.. \tag{13}$$

where, I is the current through the nonlinear element and V is the voltage on the nonlinear element [3]. Constant values such as K_1, K_2, K_3 and K_4 can be different for every nonlinear element. These constant values given for every nonlinear element changes according to the current-voltage characteristic and acquired experientially. For example, the constant values of current function for a nonlinear element which flows 5,5 A at 100 V, 9 A at 150 V, 14,5 A at 200 V and 18 A at 230 V, are calculated as K_1=11,81, K_2=-0,19, K_3=1,5•10^3 and K_4=-2,44•10^{-6}, respectively. Some of them might be existent in a nonlinear element while some of them might not. Several nonlinear electrical element patterns are shown in Figure 4. Notice that the characteristic on the rightmost is nonlinear though it looks like linear element.

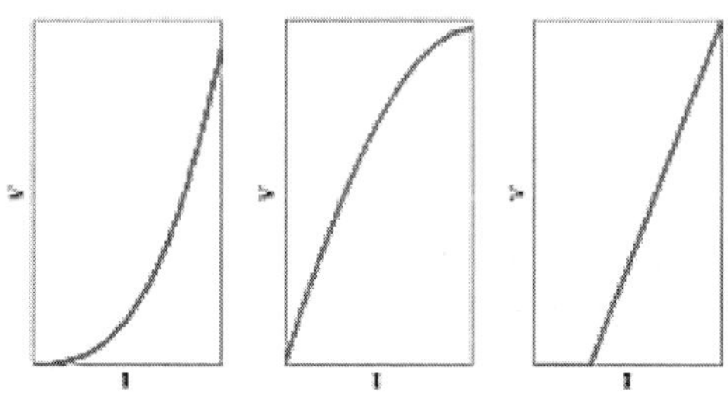

Figure 4. Nonlinear element patterns

3. Harmonic effects of power system loads

Due to increased diversity of the load that used in power systems, need to analyze load effects on power system occurred. The harmonic effects of the loads in power system were presented in this study. The loads in power system are considered in two different categories: distribution system loads and transmission system loads.

3.1. Distribution system loads

The harmonic effectiveness of the loads used frequently in power system was demonstrated by the performed measurements, in present study. The measurements were carried out for office equipment (computer, printer, scanner), air conditioner, lighting devices, motor drivers and kitchenware (refrigerator, microwave oven).

3.1.1. Office equipment

Many devices have been developed to facilitate professional life depending on technological developments. Growing number of these equipment occurred as power system loads. These office devices such as computer, printer, scanner etc. that involve power electronic components, have distortion effects on electrical power systems.

With the aim of seeing power quality distortions clearly, power quality measurements were performed for desktop computer, printer and scanner as office equipment. The real power, reactive power, total harmonic distortion in current (THD_I), power factor and cosϕ values obtained from the measurement on office equipment are given in Table 2 [4, 5];

	P (W)	Q(VAr)	THD_I(%)	PF	cosφ
Desktop Computer	140	100	70.3	0,82	1
Printer	174	154	36.9	0,76	0,92
Scanner	10	13	132.4	0,55	0,92

Table 2. Measurement Results from Office Equipment

As clearly seen in Table 2, THD_I values of all measured office equipment are exceed the standard values. Printer has worst power quality level with %132,4 THD_I value. Since power and current values of considered devices are very low, it is assumed that effects of this equipment may not harmful on power system. However, numbers of business centers have been increasing and very significant growth on collectively usage of these devices is observed. With including of individual user in this number, effects of office devices on power systems reach very high level that cannot be ignored.

All measured office equipment take attention with low power factor as seen in Table 2. In this case, it would be appreciate that implementation of compensation filter is realized especially on substations which feed business centers.

In Figure 5, voltage and current waveforms of measured equipment are shown.

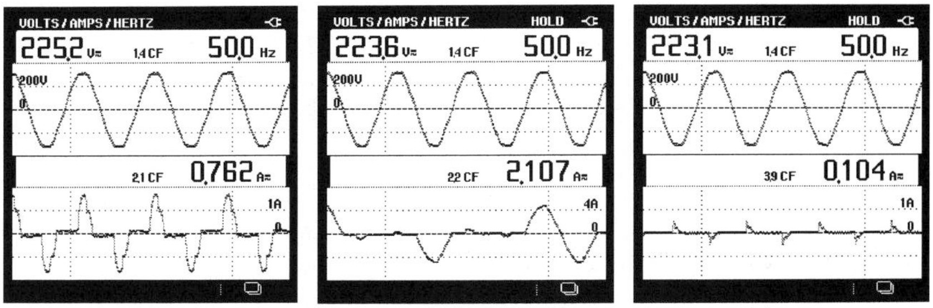

(a) Computer (b) Printer (c) Scanner

Figure 5. Voltage-Current Waveforms of Measured Equipment

3.1.2. *Air conditioning*

Air conditioning is used for both cooling and heating nowadays. Especially on summer, respectable increasement is shown on usage of air conditioning. Nevertheless, impact of air conditioning loads on electrical energy system increases. Particularly in warm regions, overload and disruptive effects can be shown on power system. This situation emphasizes the need to analyze the effects of air conditioning loads on power system. In line with this objective, power quality measurements of six commercial air conditioning units are realized and results are given in Table 3.

Air Conditioning	Btu/h	P_n (W)	I_n (A)	Power Factor	THD_v (%)	THD_i (%)
A	9000	1.000	4,0	0,98	2,600	17,042
B	9000	1.000	4,0	0,98	2,200	16,313
C	7000	680	3,0	0,98	3,702	17,879
D	12000	1230	5,7	0,99	3,502	14,861
E	21000	2100	9,5	0,57	3,201	19,881
F	11200	1250	6,0	0,99	1,600	13,419

Table 3. Measurement Results of Air Conditioners

By examination of Table 3, it can be clearly seen that, except air conditioner E, power factor values of all considered air conditioning unit are over 0,98. Similar with this situation cosφ values of measured air conditioning units are 0,99 or 1,00, except air conditioner E. Air conditioner E has very low power factor with value of 0,57. Therefore, power compensation should be done where these air conditioning units are used. However, considered air conditioning units are single-phase equipment and are used in places without an obligation to make compensation like houses, offices etc. In this case, reactive power demand is met by distribution transformers and voltage value will be change. This situation cause major problems for distribution transformers especially in summer, due to increased use of air conditioning.

Total harmonic distortion of air conditioner E is the worst one of considered air conditioning units with THD_I value of %19,881. Also, the power factor value of air conditioner E is very low. Voltage-current waveforms and harmonic spectrum of air conditioner E are given in Figure 6.

With examination of Table 3 values in general, total harmonic distortion values of all air conditioners are over 10%. Consequently, measured air conditioning units act as harmonic source and distort voltage waveform of power system.

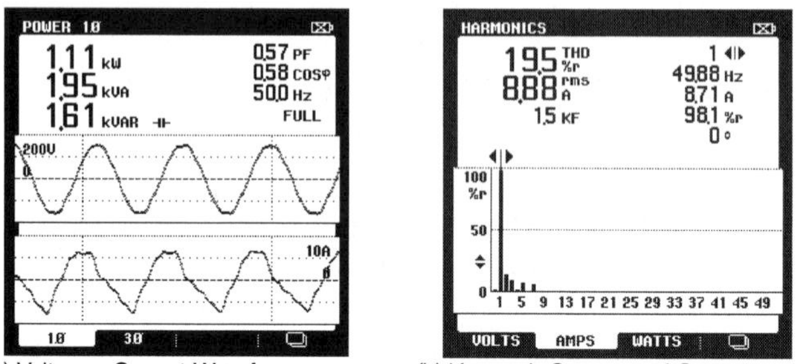

(a) Voltage – Current Waveform (b) Harmonic Spectrum of Current

Figure 6. Voltage-Current Waveforms and Harmonic Spectrum of Air Conditioner E

As a result of global warming, installed capacities of air conditioners are rising parallel with increased temperature. Correspondingly, harmonic distortion of low voltage power system is rising. In conclusion, filtered compensation is suggested on distribution substations that includes in a large number of air conditioning unit.

3.1.3. Lighting devices

Although the lighting loads are not taken into account generally, while the power system loads are ranked. They have an important place in power systems. The effects of lighting loads should be considered especially in the night hours.

Fluorescent lamps, which operate according to gas discharge principle and have high impact factor (effectiveness factor), are preferred instead of the incandescent lamps for lighting. Compact fluorescent lamps are widely used types of these devices. Block diagram of these lamps are shown in Figure 7. [6].

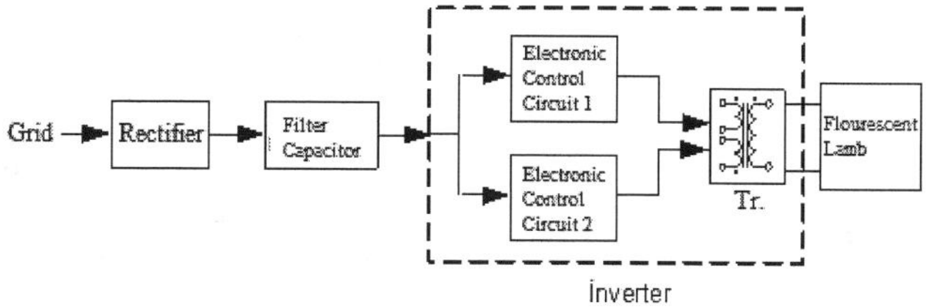

Figure 7. Block Diagram of Compact Fluorescent

The grid voltage with 50 Hz (or 60 Hz) main frequency is firstly directed by a rectifier and then filtered by a capacitor. The obtained DC voltage is converted to high frequency (20 kHz – 50 kHz) AC voltage by an inverter and applied to the fluorescent lamp. When the applied voltage frequency is going up, the luminous flux increases and impact factor of the device rises [7].

Fluorescent lamps are producing harmonic components and non-sinusoidal currents, because of their nonlinear current-voltage characteristics. Due to their characteristics, detailed analysis is required where they are used extensively. Power quality measurement results for commercially available devices are given in Table 4 [8].

	I_{rms} (A)	P (W)	Q (VAr)	Power Factor	cosφ	THD_I (%)
Compact Fluorescent A	0,077	10,3	14,0	0,59	0,89	108,8
Compact Fluorescent B	0,121	16,6	21,6	0,61	0,89	105,4
Compact Fluorescent C	0,170	22	34,0	0,54	0,94	123,3
Fluorescent Lamb	0,336	44	62	0,58	0,60	10,3

Table 4. Measurement Results of Fluorescent Lambs

According to measurement results THD_I values of compact fluorescent lamps reaches to % 123. Whereas, THD_I value of fluorescent lamp was % 10,3. Measurements show that all the lamps have very low power factor values.

The variation of THD_I according to voltage for compact fluorescent lamps is given in Figure 3.4. As shown in Figure 8, THD_I values of C lamp is particularly high over 220 V.

The harmonic currents injected into the network by a fluorescent lamp are negligible because of its low power. However, when the large number of fluorescent lamps used together, their effects are important for power quality distortion studies.

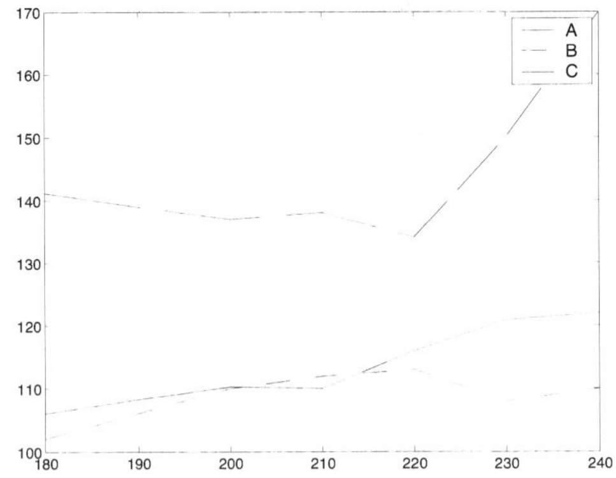

Figure 8. The variation of THD_I values with voltage for several compact fluorescent lamps.

3.1.4. *Motor drives*

Motor drives act an important role to ensuring control and efficient use of electrical machines. Due to technological developments, motor drives became more advances featured and reached a wider field of use. Many commercial facilities utilize motor drives for their motional systems. Motor drives become a significant power system load with increasement of usage.

Momentary or continuous variations on voltage and frequency can be occurred especially at the points that feed industrial plants. Motor drives are affected by these changes in addition to being a harmonic source. In Figure 9, total harmonic distortion of single-phase motor drive depending on variable voltage and frequency [9]. Voltage level is fixed to motor drives nominal voltage while frequency is variable.

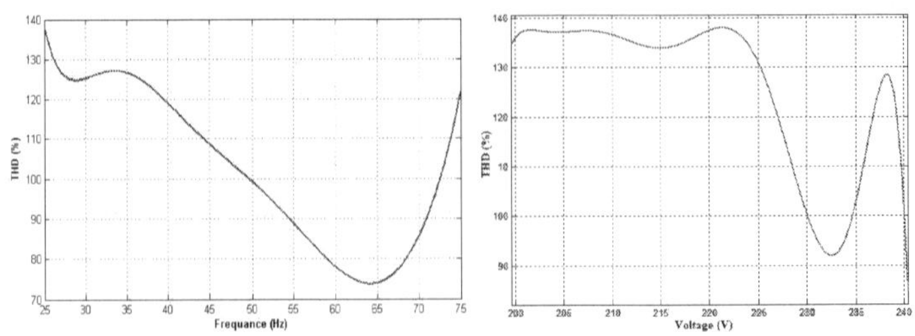

(a) THD_I variation by frequency (b) THD_I variation by voltage

Figure 9. THD_I characteristic of considered motor drive

Total harmonic distortion of motor drive current decreased to minimum value at nominal voltage and frequency levels, as seen in Figure 9. On the other hand, Total harmonic distortion reached very high values at low voltage and frequency levels. It can be suggested that usage of voltage and frequency regulator reduce total harmonic distortion in industrial plants, which includes motor drives.

3.1.5. Household equipment

Despite the conveniences that they create in daily life, household equipment take place in electrical energy system as a power system load. It is considering that household equipment is used several times per day; importance of their effects on power system is appeared clearly. Power quality measurements of an office type refrigerator and microwave oven are given in Table 5 [4, 5].

As seen in Table 5, THD_I values of all measured household equipment are exceed standard levels. Especially microwave oven has harmful effect on power system with %39,3 THD_I level. However, office type refrigerator have showed worst characteristic in terms of power factor. As a result of analyses, it can be said that other household equipment have similar effects on power system. In this case, characteristics of system loads must be considered and appropriate precautions must be operated during planning of power system.

	THD_I(%)	I_{rms}(A)	P (W)	Q(VAr)	Power Factor	cosφ
Refrigerator	13.3	0.608	97	98	0,70	0,72
Microwave Oven	39.3	5.39	1110	390	0,94	1

Table 5. Measurement Results of Household Equipment

Voltage and current waveforms of measured household equipment are shown in Figure 10.

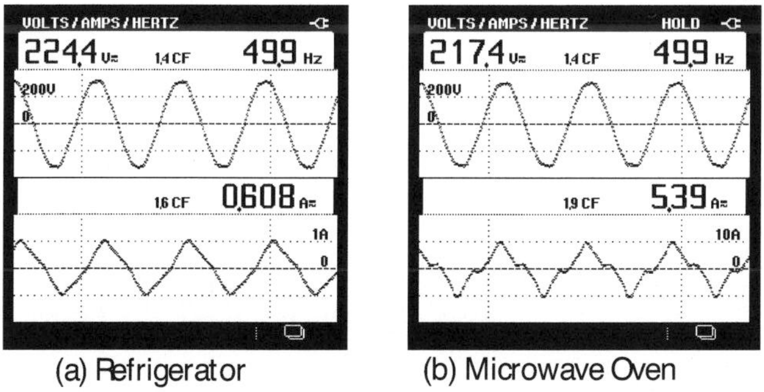

Figure 10. Voltage-Current Waveforms of Measured Household Equipment

3.2. Transmission system loads

The harmonic effectiveness of the loads which are directly connected to transmission system is introduced in this part. Within the scope of National Power Quality Project of Turkey, power quality measurements on more than 150 points were achieved. The results of measurements have been continuously carried out through 7 days and then processed.

In this study, harmonic effects of the loads which are directly connected to the measured transmission system are also demonstrated. In consequence of realized measurements, it has been seen that especially the iron-steel plants and large industrial facilities have considerable disturbance effect on the electrical network.

As an example THD_v variation with time measured in a distribution substation that feed industrial loads is shown in Figure 11.

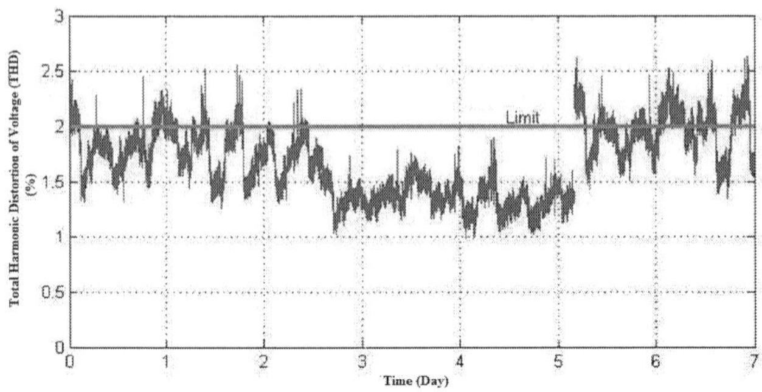

Figure 11. THD_v variation in time belonging to a distribution substation

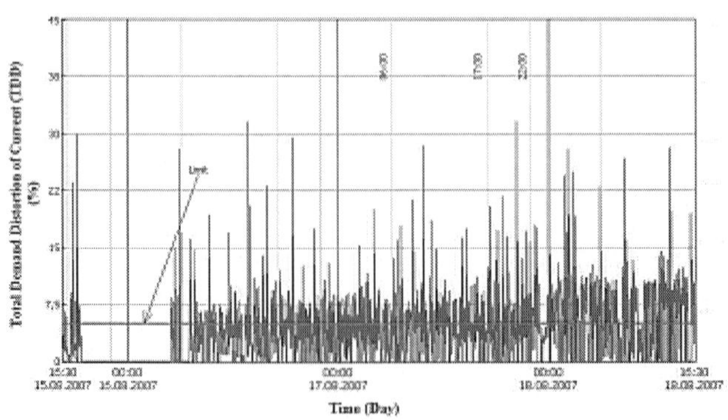

Figure 12. TDD_I variation of an iron-steel plant

Also, Figure 12 shows the TDD_I variation for an iron-steel plant. As it can be seen easily from the figure, notable distortions are observed at same substations. Because of that especially effects of large industrial plants and iron-steel plants have to be mentioned and sustainable solutions as harmonic filters, STATCOM etc. should be applied for substations which feed these loads. As a conclusion, it can be said that power quality problems could be seen in transmission system, besides distribution system.

4. Effects of harmonics on power system equipment

In power systems, protection relays are crucial elements for the system's reliability and protection. The overcurrent protection relays are used to protect the system's elements from over load and short circuit faults. Electromechanical, static and digital relays are the main types of over current relays. Generally, relay manufacturers design these relays for sinusoidal currents by giving them operating characteristics for sinusoidal conditions. The operation of these relays for non-sinusoidal currents including harmonics is not defined.

In this study, effects of harmonics on overcurrent relays are investigated for static inverse time overcurrent relay, electromechanical inverse time overcurrent relay, electromechanical definite time overcurrent relay.

4.1. Electromechanical Inverse Time Overcurrent Relay (EITOCR)

Electromechanical inverse time overcurrent relay (EITOCR) is used for overcurrent and short circuit protection of power systems elements (transmission lines, power transformers, generators etc.) The induction disc unit is the most important part of an EITOCR. The general structure of this relay is shown in Figure 13 [10]. The relay operates when the center coil is energized. As shown in Figure 13, the left pole is equipped with a lag coil while the right pole does not have any. The flux Φ is produced by the current of center pole coil. This flux passes through the air gap towards the disc and it reaches to the keeper.

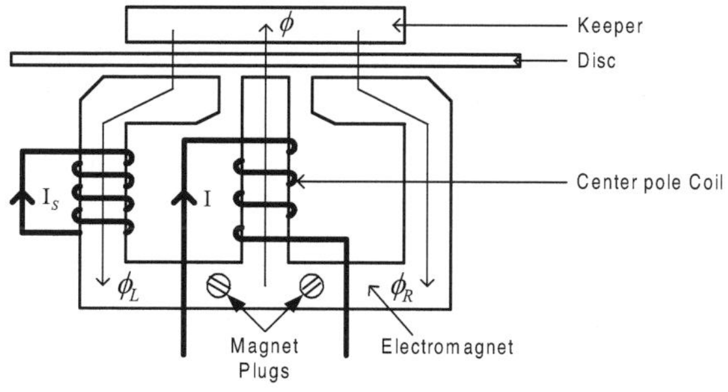

Figure 13. Induction disc unit [13]

The flux Φ consists of two parts: Φ_L through the left-hand leg and Φ_R through the right-hand leg. Φ is equal to summation of Φ_L and Φ_R. There is a short-circuited lagging coil on the left leg. This coil causes Φ_L to lag both Φ_R and Φ. When the fundamental pickup current applied to the center pole coil, a torque occurs on the disc. This causes the disc to begin to move. This torque results from the interaction between the disc currents produced by center pole flux and the other two pole fluxes. Directions of these torques are same [10, 11, 12].

In Figure 14, the standard current-time characteristic curves of EITOCR are given. [13].

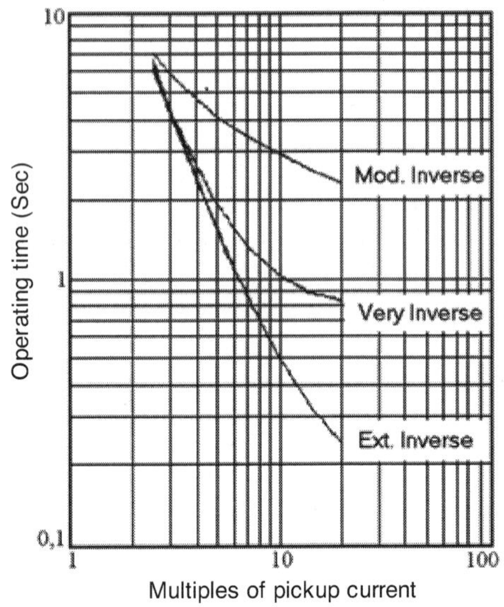

Figure 14. Time-current curves of an EITOCR [13]

In the lag coil circuit, increasing the frequency of the input current causes a little change in the current that is induced in the coil. The flux in this pole will decrease in inversely proportional to increasing of frequency. Because of this case, disc rotation slows down and the pick current of relay increases. Slowing down in disc rotation causes the operation time to increase. Frequencies of harmonic currents are different from the fundamental current's frequencies. Therefore, currents that include harmonics would have serious effects on pick up current and operation time of EITOCR.

In order to investigate the effects of harmonics on EITOCR, an experiment circuit was implemented. The experiment circuit for EITOCR is shown in Figure 15. Harmonics, pickup currents and operating times of the relay are measured and analyzed by means of data acquisition hardware and LabVIEW programme.

http://dx.doi.org/10.5772/53108

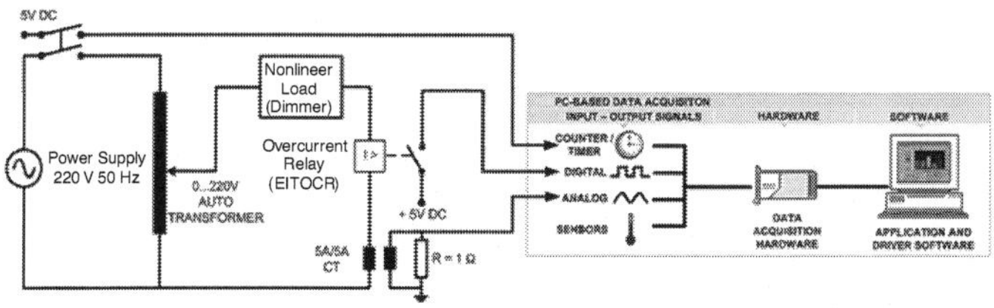

Figure 15. Experiment circuit for EITOCR [14]

The rms value of current (I_{rms}), total harmonic distortion of current (THD_I), rms value of fundamental current (I_1) and the relay's operating time (t) are measured for six modes. For all measurement modes, the relay's pickup current is set to 1 Amps for sinusoidal current and this pickup current value of relay is not changed during the experiments.

According to the experimental results, the pickup current of the relay and THD value of the non-sinusoidal current are given in Table 6. While THD value of current is increased, the pickup current of the relay increases. When THD_I is approximately 85%, although pickup current of the relay is set to 1 Ampere for sinusoidal current, it operates at 1.9 Ampere. Increasing in the pickup current of the relay shows that the relay cannot perform a suitable protection function and causes damage or heating up depending on the rms value of current in power system components such as transmission lines, motors and transformers.

Mode	α	I_1 (A)	I_{rms} (A)	THD_I(%)
0	0°	1.0975	1.10	6.0010
1	30°	1.1422	1.20	35.3101
2	60°	1.2177	1.30	46.0880
3	90°	1.3352	1.60	68.9976
4	120°	1.3888	1.68	70.6527
5	150°	1.4583	1.90	85.8677

Table 6. Pickup current values and THD_I values of EITOCR

The variation of relay's operating time versus the relay's rms current is shown in Figure 16.

The EITOCR has different operating time values for the same current value which is applied to the relay for six modes as shown in Table 7. While THD value of current increases, operating time of relay increases. According to this result, power system elements that are protected by this relay will be damaged because of the increase in the operating time of the relay.

Mode	I_{rms} (A)	THD_I (%)	t (s)
0	2.00	6.43	4.634
1	2.00	27.45	5.268
2	2.00	35.12	5.984
3	2.00	59.50	8.682
4	2.00	65.23	9.582
5	2.00	85.20	14.964

Table 7. Operating times of EITOCR for six modes

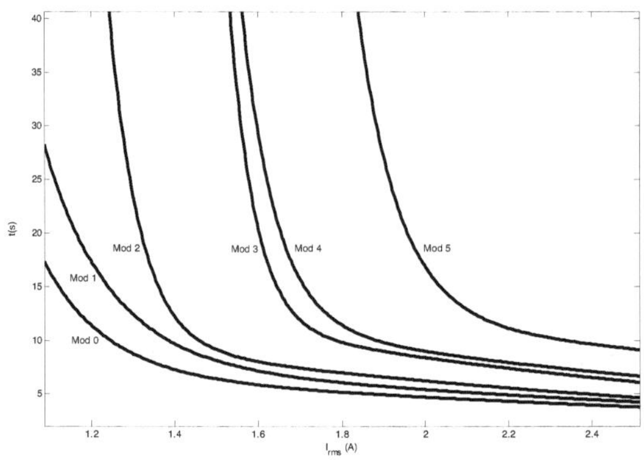

Figure 16. Operating time curves of relay versus rms values of current [14]

4.2. Elecromechanical Definite Time Overcurrent Relays (EDTOCR)

Non-sinusoidal load currents contain harmonic components. Each frequency component produces an independent and cumulative effect on relay operation. This effect appears as an increasing in pickup current of electromechanical definite time overcurrent relay. For very high frequencies, the pickup current of EDTOCR increases since the frequency increases [15]. Because of the unexpected increase of the pickup current value, this type of protection relays may not protect the power system elements reliably.

There are two types of time-current curves for overcurrent relays: inverse curve and definite curve. Operating time is inversely proportional to the current in inverse curve. As shown in Figure 17 [15, 16], operating time in definite curve is approximately constant if current is higher than several times of pickup current. Instantaneous relays have definite curve. These relays are used for short circuit protection. If an instantaneous relay have a time delayed unit, then this relay is suitable for overcurrent protection. Instantaneous overcurrent relays are also called as definite time overcurrent relay. Generally, time-current curves of these relays are given for sinusoidal current and they may be affected by current harmonics [15, 17].

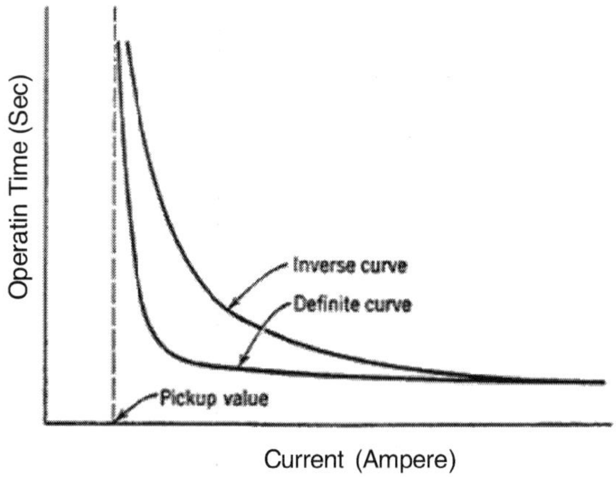

Figure 17. Operating time-current curves of overcurrent relays

In order to investigate the effects of harmonics on EDTOCR, an experiment circuit was implemented. The experiment circuit for EDTOCR is shown in Figure 18.

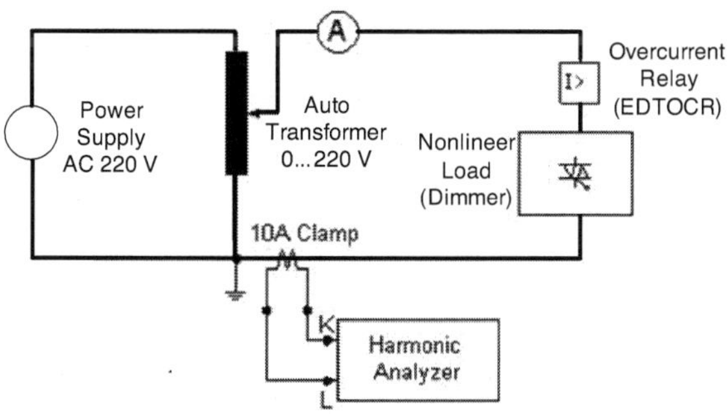

Figure 18. Experiment circuit for EDTOCR [17]

The pickup current value of EDTOCR (I_{pickup}), total harmonic distortion value (THD_I), rms value of fundamental current (I_1) and rms value of harmonic currents (I_n) for 3rd, 5th, 7th, 9th, 11th and 13th harmonic components are measured for each firing angle (α) value. For all measurements the pickup current of relay is adjusted to 1.095 Ampere for pure sinusoidal current and this setting is not changed during the experiments. The experimental results are given in Table 8.

THD_I (%)	I_1 (A)	I_3 (%)	I_5 (%)	I_7 (%)	I_9 (%)	I_{11} (%)	I_{13} (%)	I_{pickup} (A)
0.0	1.095	0.0	0.0	0.0	0.0	0.0	0.0	1.095
11.9	1.091	3.2	9.4	3.4	3.1	2.7	2.1	1.099
12.6	1.102	3.1	10.1	3.6	3.2	2.6	1.8	1.106
22.8	1.085	13.6	13.2	6.6	5.8	4.0	3.3	1.114
28.4	1.081	20.5	13.4	8.5	5.7	4.4	4.1	1.124
37.3	1.078	30.0	12.8	10.8	6.6	6.4	4.8	1.152
52.0	1.042	46.1	12.2	10.9	10.2	6.6	6.2	1.174
58.3	1.030	51.6	15.0	11.5	10.4	7.7	6.9	1.190
65.8	1.017	57.3	19.8	13.3	10.2	9.3	7.0	1.213
71.5	0.997	61.4	23.7	15.2	10.3	10.5	7.5	1.226
76.4	0.994	64.6	27.4	17.2	11.3	10.8	8.8	1.253
79.6	0.984	66.2	30.6	18.6	12.5	10.9	9.4	1.262
86.7	0.965	70.1	35.7	22.1	16.3	11.8	11.5	1.285
92.8	0.963	72.6	39.4	24.6	18.9	12.9	11.7	1.311
96.4	0.960	74.0	41.1	25.7	19.8	13.8	13.0	1.335
99.8	0.959	75.1	43.7	27.6	22.6	15.8	13.2	1.380

Table 8. Experimental results of EDTOCR [17]

According to the experimental results, the variation of relay's pickup current versus to THD_I is given in Figure 19. Pickup current value of the relay increases as long as THD_I values of relay current increases. Because of this problem, the relay is not suitable to protect the system and this circumstance causes damage or heating in power system elements.

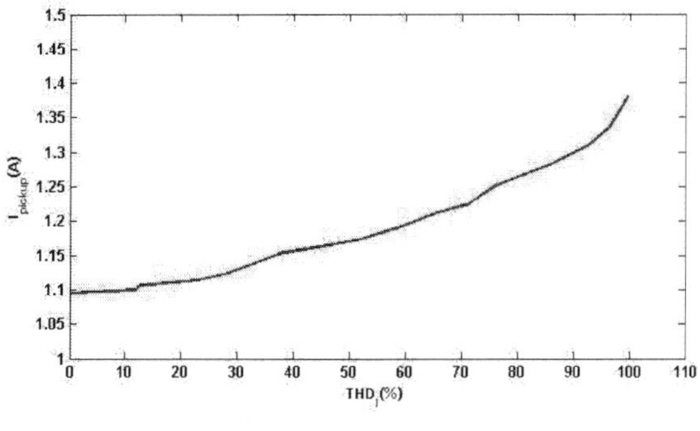

Figure 19. The pickup current values of EDTOCR versus % THD_I [17]

4.3. Static Inverse Time Overcurrent Relay (SITOCR)

Time-current curves of static over current relays are similar to the electromechanical relay's curves, but the structures of them are quite different. Static relays consist of analogue electronic circuits elements. In Figure 20, the general block diagram of a static inverse time over current relay is given [14]. When a short circuit fault occurs in the power system, the current increases to a very high value and the short circuit unit (I>>) of the relay operates instantly. In the event of a fault, if the current value is between nominal load current and short circuit current, the over current unit (I>) of the relay operates with a time delay. This time delay is called as operating time of relay [14].

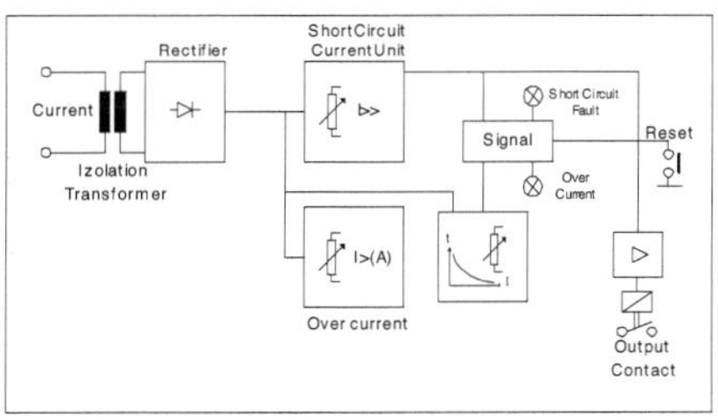

Figure 20. General block diagram of the SITOCR [14]

Time-current curve of a SITOCR is given in Figure 21. (pickup current of SITOCR is adjusted to 0.6 Amps). As shown in Figure 21, operating time of SITOCR is inversely proportional to the relay current.

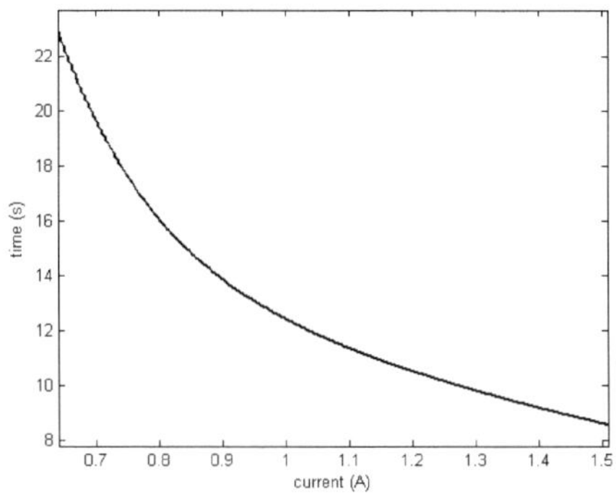

Figure 21. Current-time characteristic of the SITOCR

In order to investigate the effects of harmonics on SITOCR, an experiment circuit was implemented. The experiment circuit is shown in Figure 22.

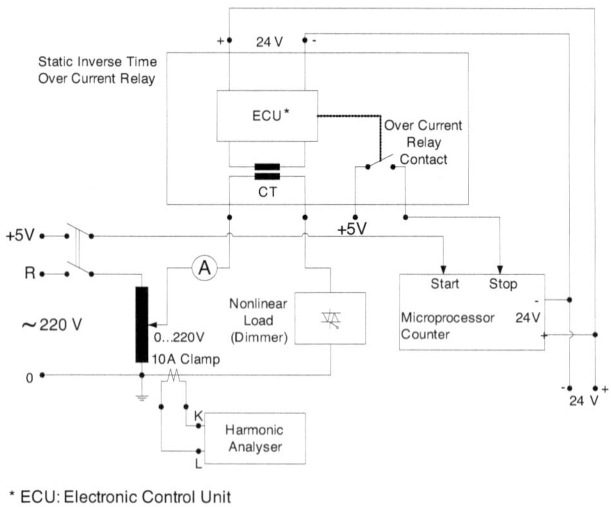

Figure 22. Experiment circuit for SITOCR [14]

The rms value of current (I_{rms}), total harmonic distortion of current (THD_I), rms value of fundamental current (I_1) and the relay's operating time (t) are measured for six modes. For all measurement modes, the relay's pickup current is set to 0.6 Amps for sinusoidal current and this pickup current value of relay is not changed during the experiments.

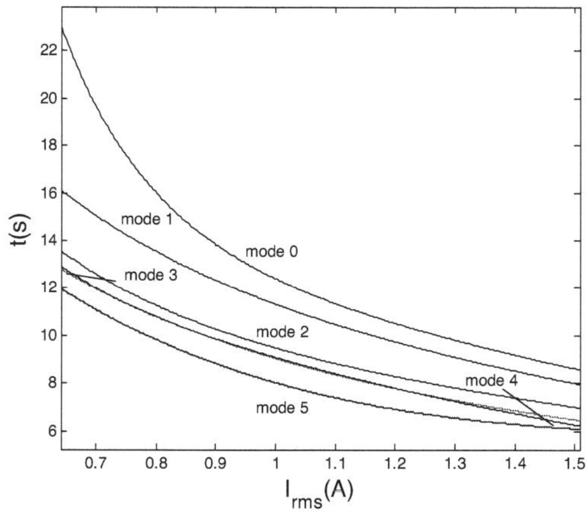

Figure 23. Operating time curves of relay versus rms current [14]

The experimental results for all modes are given Figure 23. Generally, the overcurrent relay characteristic for pure sinusoidal current (mode-0 curve) is given by the manufacturer of the relay. The other curves of modes (mode 1-5) are obtained for nonsinusoidal currents in the study. The most distorted case for current is mode-5. Although the RMS value of current is constant for all modes, the operating times of the relay is not. Operating time of relay decreases, whereas THD value of current increases as shown in Table 9.

Mode	I_{rms}(A)	THD_I (%)	t(s)
0	1,5	4,45	8,493
1	1,5	24,99	7,998
2	1,5	56,68	6,986
3	1,5	82,04	6,416
4	1,5	92,60	6,302
5	1,5	120,73	6,078

Table 9. Operating times of relay for all modes

As shown in Table 9 and Figure 23, while THD value of current increases, the operating time of relay decreases. Therefore, the selectivity cannot be provided in protection systems when harmonics exist in power systems.

5. Conclusion

Harmonic effects in power system are inspected in this study. This effect will further increase depending on the use of more devices which are produced by semiconductor technology. Furthermore, impact of iron-steel plants and large industrial areas on power quality disturbances cannot be ignored.

The representations of electrical quantities are modified due to harmonic components. These modified equations are given in subsection two. By using these equations, more realistic calculations could be realized. Especially when used in the systems with effective harmonic components, these equations lead to more reliable systems.

The distortion effects of distribution system loads and transmission system loads are given in subsection three. Influence of these loads can be easily seen from the given figures and tables. All of the office equipments and air conditioning units have significant harmonic distortion. In particular, this distortion effect increases in parallel to the size of the business center. In order to obtain a reliable and sustainable power system, these effects ought not be ignored. Furthermore, lighting devices, motor drives and household equipments are analyzed by using measurement results. As it can be easily seen from experimental results, all of them have harmful effects on distortion systems. According to the results of the studies, compact fluorescent lamps have more THD_I value when compared to fluorescent lamps. Use of a large number of these equipments could have an effective impact on the distortion of distribution systems. When focused on the motor drives, it can be said that frequency and voltage level variations have a considerable impact on the THD_I of these devices. By analyzing the measurement results, it has been seen that household equipments have harmful effects. Especially, the microwave oven, which has a greater power, has a remarkable THD_I value. As a result of analyses, it can be said that other household equipment have similar effects on power system. In this case, characteristics of system loads must be considered and appropriate precautions must be operated during planning of power system.

Furthermore, power quality disturbances on transmission lines are analyzed in this study. Turkey's power quality disturbance on transmission system is investigated by National Power Quality Project. According to the results of this project, it is seen that large industrial facilities and iron-steel plants have a disturbance effect on the network. Therefore, this impact has to be considered while planning and operating the power system. Electrical equipment as harmonic filter, STATCOM etc. should be used to eliminate the negative effects of nonlinear loads which have large application area on power systems. Thus, the quality of the power system would be at a better level.

In addition, influences of harmonics on power system equipment are very important. Results of an experimental study which performed on relays are given in subsection four in order to show these effects. According to the experimental results for over current protection relays, the operating time of static inverse time overcurrent relay decreases while THD value of current increases. Therefore, when harmonics exist in power systems, the selectivity cannot be achieved in protection systems. The operating time and pickup current of electromechanical inverse time overcurrent relay increase, while THD value of current increases. Similarly, electromechanical definite time overcurrent relay's pickup current increases, as THD value of current increases. According to the results it is seen that these relays cannot perform a suitable protection function and cause damage or heating up depending on the rms value of current in power system components such as transmission lines, motors and transformers. Finally, according to results of this study, the effects of harmonics on overcurrent relays are not same for all relays. These effects change depending on the type and the structure of overcurrent relays.

As a result of experimental studies, we can say that harmonic effects of power system loads and influence of harmonic components on power system equipment may create serious problems on system. Therefore, it can be clearly seen that power quality studies have great importance on the establishment and operation of the system.

Acknowledgement

Authors would like to thank the Public Research Support Group (KAMAG) of the Scientific and Technological Research Council of Turkey (TUBITAK) for full financial support of the project namely the National Power Quality Project of Turkey, Project No: 105G129.

Author details

Celal Kocatepe, Recep Yumurtacı, Oktay Arıkan, Mustafa Baysal, Bedri Kekezoğlu, Altuğ Bozkurt and C. Fadıl Kumru

Yildiz Technical University, Department of Electrical Engineering, Istanbul, Turkey

References

[1] Owen EL. A history of harmonics in power systems. IEEE Industry Applications Magazine 1998; 6–12

[2] Emanuel AE. Harmonics in early years of electrical engineering: a brief review of events, people and documents. Proceedings, IEEE ICHQP 2000; 1–7

[3] C. Kocatepe, M. Uzunoğlu, R. Yumurtacı, A. Karakaş, O. Arıkan, Elektrik Tesislerinde Harmonikler, Birsen Yayınevi, İstanbul, 2003.

[4] B. Kekezoğlu, O. Arıkan, C. Kocatepe, R. Yumurtacı, M. Baysal, A. Bozkurt, Elektrikli Ofis Donanımlarının Harmonik Etkilerinin İncelenmesi, 3e Electrotech, Sayı: 168, 150-156, Haziran 2008.

[5] B. Kekezoğlu, O. Arıkan, C. Kocatepe, R. Yumurtacı, M. Baysal, A. Bozkurt, Elektrikli Ofis Cihazlarında Güç Faktörü ve cosφ'nin İncelenmesi, 3e Electrotech, Sayı: 174, 110-114, Aralık 2008.

[6] Wenzel, E., AN1864 Application Note, 22W/120Vac Compact Fluoresant Lamp Driver with VK05CFL, ST Microelectronics Com, Italy, 2004

[7] S. Onaygil, Ö. Güler, Yüksek Frekanslı Elektronik Balastlar, 1. Ulusal Aydınlatma Kongresi, İstanbul, 1996

[8] R. Yumurtacı, O. Arıkan, A. Bozkurt, Kompakt Flourasant Lambaların Harmonik Kaynağı Olarak İncelenmesi, Enerji Verimliliği Kongresi – EVK'05, Kocaeli, 2005.

[9] B. Kekezoğlu, O. Arıkan, A. Bozkurt, C. Kocatepe, R. Yumurtacı, Analysis of AC Motor Drives as a Harmonic Source, 4th International Conference on Electrical and Power Engineering (EPE-2006), pp. 832- 83712-13 October, Iasi, Romania, 2006.

[10] Applied Protective Relaying, Westinghouse Electric Corporation Relay-Instrument Division, Newark, 1976.

[11] Horowitz, S. H., Phadke, A. G., Power System Relaying. John Wiley & Sons Inc., 1992.

[12] IEEE Standard Inverse-Time Characteristic Equations for Overcurrent Relays, Power System Relaying Committee of the IEEE PES, IEEE Std C37112., 1996.

[13] Yumurtacı R., Gulez K., Bozkurt A., Kocatepe C., Uzunoglu M., Analysis of Harmonic Effects on Electromechanical Instantaneous Overcurrent Relays with Different Neural Networks Models, International Journal of Information Technology, Vol. 11, No.5, 26-35, Aralık-2005.

[14] K.B. Dalci, R. Yumurtaci, A. Bozkurt, Harmonic Effects on Electromechanical Overcurrent Relays, Dogus University Journal, Vol. 6, pp. 202-209, Istanbul, 2005.

[15] Mason, C.R.: The Art and Science of Protective Relaying. John Wiley Eastern Limited New Delhi, 1991.

[16] Yumurtacı, R., Bozkurt, A., Gulez, K., Neural Networks Based Analysis of Harmonic Effects on Electromechanical Instantaneous Overcurrent Relays, SICE, The Society of Instrument and Control Engineers Annual Conference, Okayama/Japan, 1012-1015, 08-10 August 2005.

[17] R. Yumurtacı, A. Bozkurt, K. Gulez, Neural Networks Based Analysis of Harmonic Effects on Inverse Time Static Overcurrent Relays, INISTA, International Symposium on Innovations in Intelligent Systems and Applications, Istanbul/Turkey, 108-111, 15-18 June 2005.

Chapter 8

Bank Harmonic Filters Operation in Power Supply System – Cases Studies

Ryszard Klempka, Zbigniew Hanzelka and
Yuri Varetsky

Additional information is available at the end of the chapter

http://dx.doi.org/10.5772/53425

1. Introduction

Continuous technological development facilitates the increase in the number of nonlinear loads that significantly affect the power quality in a power system and, consequently, the quality of the electric power delivered to other customers. DC and AC variable speed drives and arc furnaces are ranked among the most commonly used large power nonlinear loads.

DC drives can be a significant plant load in many industries. They are commonly used in the oil, chemical, metal and mining industries. These drives are still the most common large power type of motor speed control for applications requiring very fine control over wide speed ranges with high torques. Power factor correction is particularly important for this drives because of relatively poor power factor, especially when the motor is at reduced speeds. Additional transformer capacity is required to handle the poor power factor conditions and more utilities are charging a power factor penalty that can significantly impact the total bill for the facility. The DC drives also generate significant harmonic currents. The harmonics make power factor correction more complicated. Power factor correction capacitors can cause resonant conditions which magnify the harmonic currents and cause excessive distortion levels. For the same reasons arc furnaces are very difficult loads for a supplier and for the customer they are very difficult objects of reactive power compensation and harmonics filtering.

One of the most common methods to prevent adverse effects of nonlinear loads on the power network is the use of passive filters. However, different configurations should be considered before making the final design decision. Among the performance criteria are current and voltage ratings of the filter components, and the effect of filter and system contingency conditions. Before any filter scheme is specified, a power factor study should be done to determine if any

INTECH
open science | open minds

© 2013 Klempka et al.; licensee InTech. This is a paper distributed under the terms of the Creative Commons Attribution License (http://creativecommons.org/licenses/by/3.0), which permits unrestricted use, distribution, and reproduction in any medium, provided the original work is properly cited.

reactive compensation requirements are needed. If power factor correction is not necessary, then a minimum power filter can be designed; one that can handle the fundamental and harmonic currents and voltages without consideration for reactive power output. Sometimes, more than one tuned filter is needed. The filter design practice requires that the capacitor and the reactor impedance be predetermined. For engineers not knowing the appropriate initial estimates, the process has to be repeated until all the proper values are found. This trialand-error approach can become complex as more filters are included in the systems.

While the effectiveness of a filter installation depends on the degree of harmonic suppression, it also involves consideration of alternate system configurations. As the supplying utility reconfigures its system, the impedance, looking back to the source from the plant's standpoint, will change. Similar effects will be seen with the plant running under light versus heavy loading conditions, with split-bus operation, etc. Therefore, the filtering scheme must be tested under all reasonable operating configurations.

The general procedure in analyzing any harmonic problem is to identify the worst harmonic condition, design a suppression scheme and recheck for other conditions. Analysis of impedance vs frequency dependencies for all reasonable operating contingencies is commonly used practice. A frequency scan should be made at each problem node in the system, with harmonic injection at each point where harmonic sources exist. This allows easy evaluation of the effects of system changes on the effective tuning. Of particular importance is the variability of parallel resonance points with regard to changing system parameters. This problem is illustrated by the practical example.

In a most classic cases all filter considerations are carried out under the following simplifying assumptions: (i) the harmonic source is an ideal current source; (ii) the filter inductance L_F and capacitance C_F are lumped elements and their values are constant in the considered frequency interval; (iii) the filter resistance can be sometime neglected and the filter is mainly loaded with the fundamental harmonic and the harmonic to which it is tuned e.g. [1]. The above assumptions allow designing simple filter-compensating structures. However, if a more complex filter structures or a larger number of filters connected in parallel are designed or their mutual interaction and co-operation with the power system (the network impedance), or non-zero filter resistances should be taken into account, these may impede or even prevent an effective analysis. An example of the new approach is the use of artificial intelligence methods, among them the genetic algorithm (AG) [2 - 4]. The usefulness of this new method is illustrated by examples of designing selected filters' structures: (a) a group of single-tuned filters; (b) double-tuned filter and (c) C-type filter.

2. Single-tuned single branch filter

Many passive *LC* filter systems, of various structures and different operating characteristics have been already developed [4 - 9]. Nevertheless, the single-tuned single branch filter (Fig. 1) still is the dominant solution for industrial applications, and it certainly is the basis for understanding more advanced filtering structures.

http://dx.doi.org/10.5772/53425

$$R_F \cong 0$$

$$\omega_r = n_r \omega_1 = \frac{1}{\sqrt{L_F C_F}}$$

$$L_F = \frac{1}{n_r^2 \omega_1^2 C_F} \tag{1}$$

$$C_F = \frac{n_r^2 - 1}{n_r^2} \cdot \frac{Q_F}{\omega_1 U^2}$$

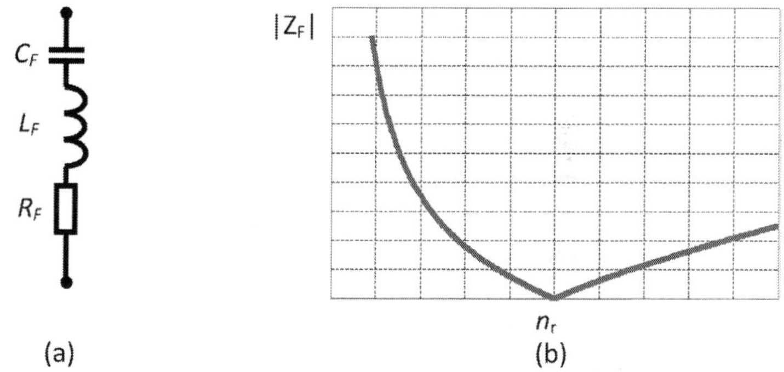

Figure 1. Single branch filter and the frequency characteristic

Where k single-tuned filters are operated in parallel in order to eliminate a larger number of harmonics then k voltage resonances (series resonances) and k current resonances (parallel resonances) occur in the system. These resonance frequencies are placed alternately and the series resonance is always the preceding one. In other words, each branch has its own resonance frequency.

The schematic diagram of an example group of filters in a large industrial installation and characteristics illustrating the line current variations and the 5th harmonic voltage variations in result of connecting ONLY the 5th harmonic filter are shown in Fig. 2. The figure also shows the 7th harmonic voltage variations prior to and after connecting the 5th harmonic filter. The 5th harmonic filter selectivity is evident — its connection has practically no influence on the 7th harmonic value.

Relations (2) allow determining parameters of a group of single-tuned filters taking into account their interaction, as well as choosing the frequencies for which the impedance frequency characteristic of the filter bank attains maxima, where (the filters' resistances R_i 0): C_i - the filters' capacitances; L_i - the filters' inductances; ω_{ri} - tuned angular frequency; n_{ri} - orders of filter tuning harmonics; m_i - orders of harmonics for which the impedance character-

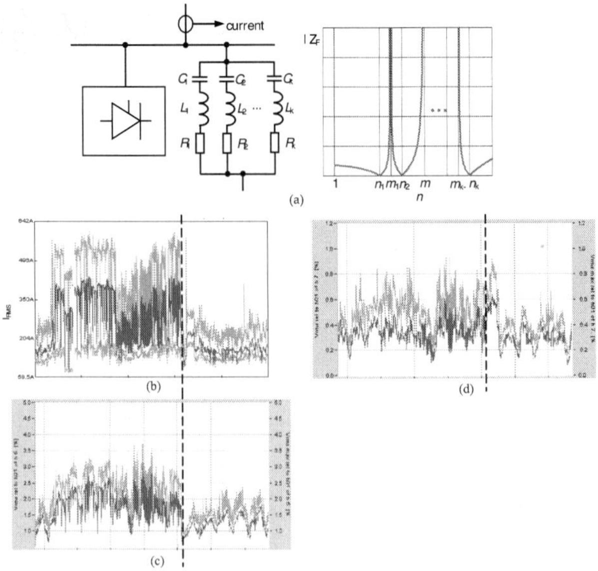

Figure 2. Groups of single branch filters: (a) schematic diagram; (b) current (*I*); characteristics of (c) 5th and (d) 7th voltage harmonic. The vertical line indicates the instance of the 5th harmonic filter connection

istic should attain maxima; Q_F - reactive power of the basic harmonic of the filter or group of filters and U – RMS operating voltage.

$$\begin{bmatrix} \frac{n_{r1}^2 m_1}{n_{r1}^2 - m_1^2} & \frac{n_{r2}^2 m_1}{n_{r2}^2 - m_1^2} & \cdots & \frac{n_{rk}^2 m_1}{n_{rk}^2 - m_1^2} \\ \frac{n_{r1}^2 m_2}{n_{r1}^2 - m_2^2} & & \cdots & \frac{n_{rk}^2 m_2}{n_{rk}^2 - m_2^2} \\ \vdots & & \ddots & \\ \frac{n_{r1}^2 m_{k-1}}{n_{r5}^2 - m_{k-1}^2} & & \cdots & \frac{n_{rk}^2 m_{k-1}}{n_{rk}^2 - m_{k-1}^2} \\ \frac{n_{r1}^2 \omega_1}{n_{r1}^2 - 1} & \frac{n_{r2}^2 \omega_1}{n_{r2}^2 - 1} & \cdots & \frac{n_{rk}^2 \omega_1}{n_{rk}^2 - 1} \end{bmatrix} \begin{bmatrix} C_1 \\ \vdots \\ C_k \end{bmatrix} = \begin{bmatrix} 0 \\ \vdots \\ \vdots \\ 0 \\ \frac{Q_F}{U^2} \end{bmatrix} \tag{2}$$

$$L_i = \frac{1}{n_{ri}^2 \omega_1^2 C_i} \qquad i = 1 \ldots k$$

2.1. Example 1

An example application of the method will be the design of single-tuned filters (two single-tuned filters) for DC motor (Fig. 3). The basis for design is modelling of the whole supplying system. The system may comprise nonlinear components and analysis of the filters can take into account their own resistance, which depends on the selected components values. Generally speaking, the model can be detailed without simplifications.

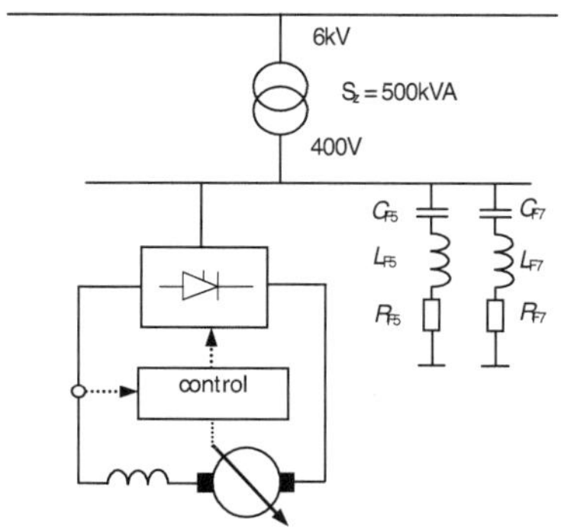

Figure 3. Diagram of the power system with the designed group of single-tuned filters for system with DC drive supplied by 6-pulse controlled rectifier:P_N = 22kW, U_N = 440V, I_N = 56,2A, J = 2,7kgm^2, R_t = 0,465Ω, L_t = 15,345mH, n_N = 1500 r/min, k = 2,62

Parameters of the single-tuned filters group were determined by means of the Genetic Algorithm minimising the voltage harmonic distortion factor with limitation of the phase shift angle between fundamental harmonics of the current and voltage$\phi_{(1)}>0$. Parameters of the applied Genetic Algorithm: (a) each parameter (C_{F5}, C_{F7}) is encoded into a 15-bit string; (b) range of variability from 1μF to 100μF; (c) population size 100 individuals; (d) crossover probability p_k = 0.7; (e) mutation probability p_m = 0.01; (f) Genetic Algorithm termination condition – 100 generations; (g) selection method Stochastic Universal Sampling (SUS); (h) shuffling crossover (APPENDIX A).

The genetic algorithm objective is to find the capacitance values of two single-tuned filters tuned to harmonics n_{r5} = 4.9 and n_{r7} = 6.9. It is worth pointing out that the genetic algorithm itself solves the problem of reactive power distribution between the filters. The voltage total harmonic distortion will be minimized and therefore power distribution between the filters will be achieved.

Basic characteristics of the power system, before and after connecting the filters, are tabulated in Fig. 4 (C_{F5} = 30.14μF; C_{F7} = 4.11μF; the phase shift angle between the voltage and current fundamental harmonics: prior to connection of filters – 11º, after connection of filters 0.2º).

In industry many of the supply systems consist of a combination of tuned filters and a capacitor bank. Depending on the system configuration the capacitor bank can lead to magnification or attenuation of the filters loading. Filter detuning significantly affects this phenomenon. Therefore, specifying harmonic filters requires considerable care under analysis of possible system configurations for avoidance of harmonic problems.

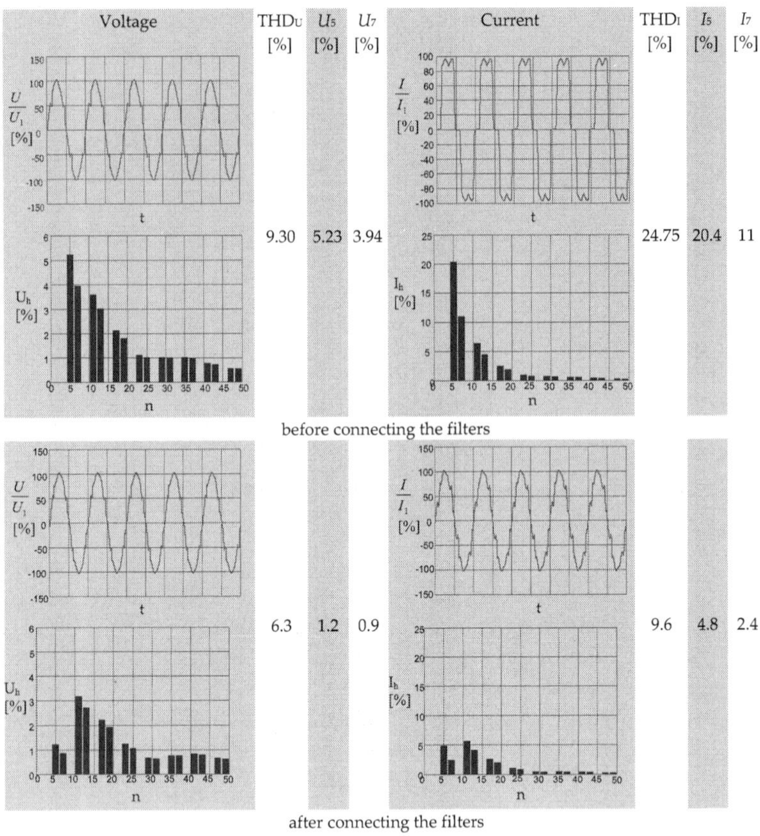

Figure 4. The voltage-current waveforms and spectrum before and after connecting the filters (U_1/I_1 – basic voltage/current harmonic; U_h/I_h – h. order voltage/current component

3. Parallel operation of filters

3.1. Example 2 – description of the system

Fig. 5 shows a one-line diagram of a mining power supply system which will be used to analyze operation characteristics of the single tuned harmonic filters in a power supply system including power factor correction capacitor banks. System contains two sets of powerful DC skip drives as harmonic loads connected to sections A and B. The drives are fed from six-pulse converters. As a result, there is significant harmonic current generation and the plant power factor without compensation is quite low.

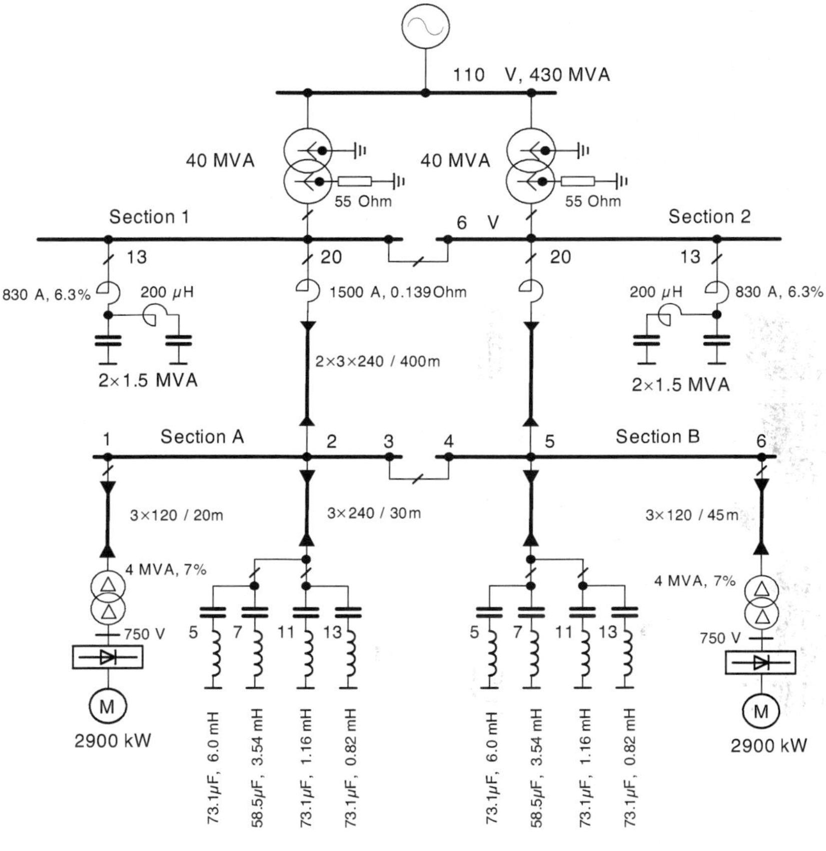

Figure 5. One-line diagram of a mining power supply system

Shunt capacitors 2×1.5 MVA connected to main sections 1 and 2 to partially correct the power factor but this can cause harmonic problems due to resonance conditions. The sections A and B can be supplying from the main section 1 or 2. Four single-tuned filters (5th, 7th, 11th, and 13th harmonic order) have been added to the sections A and B to limit harmonic problems and improve reactive compensation. Specifications of the harmonic filters are shown in the Table 1.

Allowable current limit for filter capacitors is 130% of nominal RMS value and voltage limit -110%. The iron-core reactors take up less space comparatively to air-core reactor and make use of a three-phase core. Reactors built on these cores weigh less, take up less space, have lower losses, and cost less than three single-phase reactors of equal capability. Reactors are manufactured with multi-gap cores of cold laminated steel to ensure low tuning tolerance. The primary draw back to iron-core reactors is that they saturate.

Filter, tuning	Capacitor bank			Reactor bank (three phase, iron-core)		
F5 $n_{r5}=4.81$	Bank rating	2×500	kvar	Nominal voltage	7.2	kV
	Nominal voltage	6.6	kV	Nominal current	120.0	A
	Nominal current	87.4	A	S.c. current	14.0	kA
	Capacitance	73.1	μF	Inductance	6.0	mH
	Cap. tolerance	-5...+10	%	Inductance tolerance	± 5	%
F7 $n_{r7}=6.98$	Bank rating	2×400	kvar	Nominal voltage	7.2	kV
	Nominal voltage	6.6	kV	Nominal current	100.0	A
	Nominal current	70.0	A	S.c. current	14.0	kA
	Capacitance	58.4	μF	Inductance	3.54	mH
	Cap. tolerance	-5...+10	%	Inductance tolerance	± 5	%
F11 $n_{r11}=10.94$	Bank rating	2×500	kvar	Nominal voltage	7.2	kV
	Nominal voltage	6.6	kV	Nominal current	130.0	A
	Nominal current	87.4	A	S.c. current	14.0	kA
	Capacitance	73.1	μF	Inductance	1.16	mH
	Cap. tolerance	-5...+10	%	Inductance tolerance	± 5	%
F13 $n_{r13}=13.02$	Bank rating	2×500	kvar	Nominal voltage	7.2	kV
	Nominal voltage	6.6	kV	Nominal current	130.0	A
	Nominal current	87.4	A	S.c. current	14.0	kA
	Capacitance	73.1	μF	Inductance	0.82	mH
	Cap. tolerance	-5...+10	%	Inductance tolerance	± 5	%

Table 1. Filter specifications

The saturation level is dependent upon the fundamental current and the harmonic currents that the reactor will carry. There is not standard for rating harmonic filter reactors and therefore, it is difficult to evaluate reactors from different manufacturers. For example, some reactor manufacturers base their core designs (cross sectional area of core) on RMS flux, while other will based it on peak flux (with the harmonic flux directly adding). There is a significant difference between these two design criteria. For evaluation purposes, reactor weight and temperature rise are a primary indication of the amount of iron that is used. The second feature of the reactors is considerable frequency dependency of eddy currents loss in the winding.

Equation (1) shows that the relative resonant frequency n_r depends on the power system frequency and filter inductance and capacitance. Any variation of these parameters causes deviation of the resonant frequency. So, possible deviation from the designed value can be obtained using (1) by the equation:

$$\frac{n_d}{(1+\Delta f_*)\sqrt{(1+\Delta L_*)(1+\Delta C_*)}} \le n_r \le \frac{n_d}{(1-\Delta f_*)\sqrt{(1-\Delta L_*)(1-\Delta C_*)}} \tag{3}$$

where: Δf_*- power system frequency variation, p.u.; ΔL_*, ΔC_*-filter inductance and capacitance variations, p.u.; n_d- designed relative resonant frequency (d = 5, 7, 11, 13).

Assuming $\Delta f_* \approx 0$, the possible deviation of relative resonant frequency n_r from the designed value for the investigated filter circuits can be defined using values of ΔL_*, ΔC_* from the Table 1:

$$0.93 n_d \le n_r \le 1.05 n_d \tag{4}$$

This means that the analysed filter circuits have the following possible ranges of relative resonant frequency n_r :

5-thorder filter - $4.3 \le n_r \le 5.1$;

7-thorder filter - $6.5 \le n_r \le 7.4$; (5)

11-th order filter - $10.2 \le n_r \le 11.5$;

13-th order filter - $12.1 \le n_r \le 13.7$.

It is obvious that the detuning of higher order filter is more sensitive for the same filter capacitance or inductance drift than detuning of lower order filter, as value of resonant frequencyω_r defines its deviation:

$$\Delta\omega_r \approx \frac{d\omega_r}{dC}\Delta C = -\frac{\omega_r}{2C}\Delta C \tag{6}$$

3.2. Filter characteristics analysis

In order to demonstrate filter circuits behavior under all reasonable operating configurations and get numerical results for comparison purposes, computer simulations have been performed using frequency and time domain software.

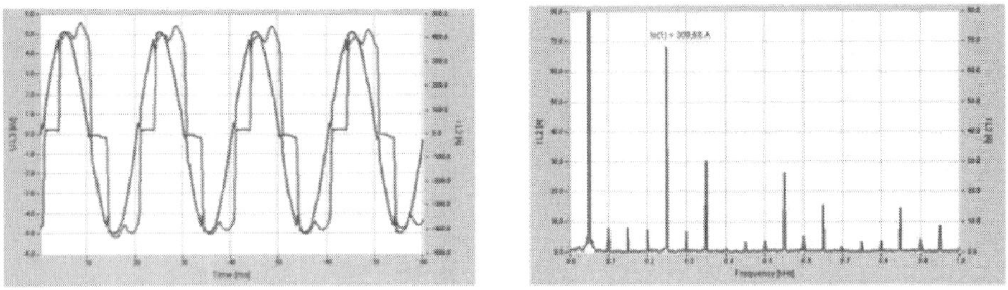

(a) (b)

Figure 6. Current and voltage waveforms of the fully loaded DC drive (a) and the current harmonic spectrum (b)

Measurements performed at the facility were used to characterize the DC drive load and obtain true source data for computer analysis of the filter characteristics. For example, Fig. 6 shows the DC drive current and its harmonic spectrum in the supply system consisting of 5th order filter under isolated operation of the section A.

Harmonic currents in the supply system components are listed in Table 2. There are obvious important findings from these measurements: 1) noncharacteristics current harmonics are present due to irregularities in the conduction of the converter devices, unbalanced phase voltages and other reasons; 2) there is resonance condition near 4th harmonic in the system configuration with 5th filter connected. Similar measurements also provided for the system with other filter sets.

Analysis of the system response is important because the system impedance vs frequency characteristics determine the voltage distortion that will result from the DC drive harmonic currents. For the purposes of harmonic analysis, the DC drive loads can be represented as sources of harmonic currents. The system looks stiff to these loads and the current waveform is relatively independent of the voltage distortion at the drive location. This assumption of a harmonic current source permits the system response characteristics to be evaluated separately from the DC drive characteristics.

In Fig. 7 are depicted the worst case of frequency scan for system impedance looking from the section A with several filters connected as concerns 5th harmonic filter loading. These conditions occur with upper limit (see (3)) of filter reactor and capacitor rating variations. Proximity of the frequency response resonance peaks to 4th and 5th harmonics produces significant magnification the harmonic currents in the 5th filter and feeder circuits.

Harmonic order	Feeder current, I_S		Drive current, I_D		5-th filter current, I_{F5}	
	A	%	A	%	A	%
1	241,49	100,0	309,68	100,0	157,87	100,0
2	8,49	3,5	8,04	2,6	0,61	0,4
3	9,75	4,0	8,03	2,6	1,77	1,1
4	41,62	17,2	7,51	2,4	39	24,7
5	27,02	11,2	68,09	22,0	42,54	26,9
6	4,31	1,8	6,75	2,2	2,42	1,5
7	25,43	10,5	30,21	9,8	5,28	3,3
8	0,74	0,3	0,85	0,3	0,28	0,2
9	2,81	1,2	3,36	1,1	0,57	0,4
10	2,92	1,2	3,49	1,1	0,57	0,4
11	22,09	9,1	26,36	8,5	4,33	2,7
12	4,16	1,7	5,04	1,6	0,89	0,6
13	12,7	5,3	15,43	5,0	2,77	1,8
14	1,21	0,5	1,49	0,5	0,27	0,2
15	2,79	1,2	3,37	1,1	0,59	0,4
16	2,84	1,2	3,41	1,1	0,59	0,4
17	12,14	5,0	14,57	4,7	2,48	1,6
18	3,52	1,5	4,28	1,4	0,73	0,5
19	7,35	3,0	8,74	2,8	1,38	0,9

Table 2. Harmonic currents for the system consisting of 5th harmonic filter

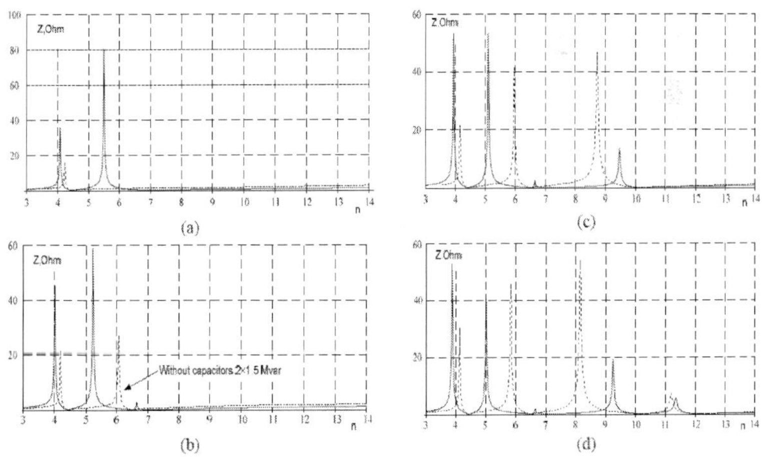

Figure 7. Frequency scans for the system impedance with 5th (a), (5+7)-th (b), (5+7+11)-th (c), (5+7+11+13)-th (d) harmonic filters

Harmonic current magnification in a filter circuit can be defined by the following factor:

$$\beta_{Fn} = \frac{|I_{Fn}|}{|I_{Dn}|} = \frac{|Z_n|}{|Z_{Fn}|} \tag{7}$$

and for the feeder circuit similarly:

$$\beta_{Sn} = \frac{|I_{Sn}|}{|I_{Dn}|} = \frac{|Z_n|}{|Z_{Sn}|} \tag{8}$$

where: I_{Dn}, I_{Sn}, I_{Fn} - the nth harmonic current of the harmonic source, feeder and filter, correspondingly; Z_n, Z_{Sn}, Z_{Fn}- the n-th harmonic impedances of the system, feeder and filter at the point of common connection, correspondingly.

The harmonic magnification factor allows estimating harmonic current in a filter or feeder circuit for several system configurations relative to source harmonic current. A value less than 1.0 means that only a part of the source harmonic current flows in the circuit branch.

Calculated values of harmonic magnification factors for analyse 5-th filter loading in the several system configurations are listed in Table 3. Column "Upper deviation limits" with 2×1.5 Mvar capacitors corresponds to the Fig. 7. The significant 4-th and 5-th harmonics magnification can be observed from the Table 3 in the 5-th filter and feeder circuits in the case of 2×1.5 Mvar capacitors connected. It can cause the filter overload and allowable system voltage distortion exceeding. On the other hand when lower deviation of the filter parameters the magnification factors are considerably less. Switching off the 2×1.5 Mvar capacitors reduces 5-th harmonic magnification in the circuits to acceptable levels, but 4-th harmonic is magnificated considerably more due to close to resonant peak.

System configuration	Upper deviation limits				Lower deviation limits			
	5th filter		Feeder		5th filter		Feeder	
With cap. 2×1.5 Mvar	β_{F4}	β_{F5}	β_{S4}	β_{S5}	β_{F4}	β_{F5}	β_{S4}	β_{S5}
F5	4.2	0.8	9.3	1.7	0.5	1.1	2.6	0.4
F5+F7	14.4	1.4	32.2	2.8	0.6	1.0	3.3	0.4
F5+F7+F11	4.8	2.8	10.2	5.5	0.8	0.9	4.5	0.4
F5+F7+F11+F13	2.4	18.6	5.2	36.7	1.4	0.8	7.4	0.3
Without cap. 2×1.5 Mvar								
F5	0.8	0.3	1.9	0.7	0.2	1.6	1.3	0.6
F5+F7	1.1	0.4	2.4	0.8	0.3	1.5	1.4	0.6
F5+F7+F11	1.5	0.5	3.2	1.0	0.3	1.4	1.6	0.5
F5+F7+F11+F13	2.1	0.6	4.5	1.1	0.3	1.3	1.8	0.5

Table 3. Harmonic current magnification factors β in the system

The calculated harmonic current magnification factors in filter circuits in the possible filter configurations are depicted in the Table 4. It is here noted that harmonic loading of the filters in the system without 2×1.5 Mvar capacitors depends on the filter configuration and filter detuning. It is well known that the series L-C circuit has the lowest impedance at its resonant frequency. Below the resonant frequency the circuit behaves as a capacitor and above the resonant frequency as a reactor. When a filter is slightly undertuned to desired harmonic frequency it has lower harmonic absorbing as a result of the harmonic current dividing between the filter and system inductances. If the filter is slightly overtuned than parallel resonant circuit created of the filter capacitance and system inductance will magnify the source harmonic current. Regularity of the phenomena for the analyzed system with multiply filter circuits one can see in the bottom part of the Table 4 for the system configuration without capacitors 2×1.5 Mvar.

Switching in capacitors 2×1.5 Mvar to the bus section changes the filters loading due to parallel resonant circuit created of the capacitors and system impedances. The resonant frequency of the system looking from the section A with several connected filters depends on the number of the filters and specifies the filter loading.

System configuration	Deviation limits							
	5th filter		7th filter		11th filter		13th filter	
With cap. 2×1.5 Mvar	Up	Lo	Up	Lo	Up	Lo	Up	Lo
F5	0.8	1.1	-	-	-	-	-	-
F5+F7	1.4	1.0	0.1	0.1	-	-	-	-
F5+F7+F11	2.8	0.9	0.1	0.1	0.6	1.9	-	-
F5+F7+F11+F13	18.6	0.8	0.1	0.1	1.3	1.3	0.6	3.2
Without cap. 2×1.5 Mvar								
F5	0.3	1.6	-	-	-	-	-	-
F5+F7	0.4	1.5	0.5	3.0	-	-	-	-
F5+F7+F11	0.5	1.4	0.7	1.9	0.7	1.3	-	-
F5+F7+F11+F13	0.6	1.3	0.8	1.5	2.5	1.0	0.7	2.1

Table 4. Harmonic current magnification factors β_{Fn} in the filter circuits

Figure 8 shows current waveforms and its harmonic spectrums for parallel 11th and 13th harmonic filters in the analyzed system obtain from time domain computer simulation of the system. The first observation of these two cases is significant harmonic overloading of the filters. In the case in question of filter iron-core reactor the phenomenon can cause the reactor temperature rise and its failure.

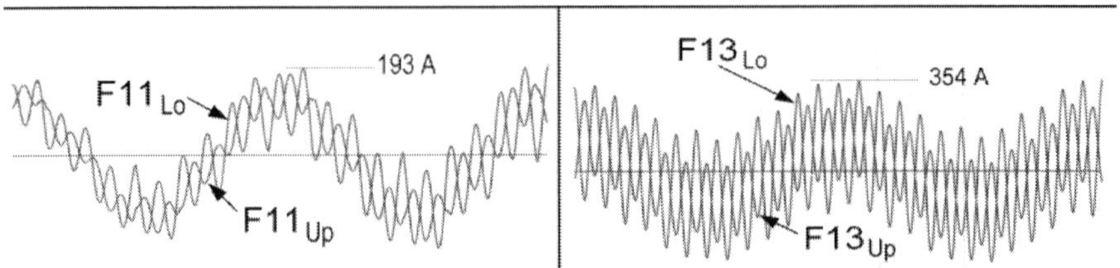

Circuit	1h	5h	7h	11h	13h	RMS
Drive, A	552	120	57,0	40,0	27,0	403
$F11_{Lo}$, A	107	15,0	3,0	110	3,0	110
$F11_{Up}$, A	127	15,4	2,2	63,0	3,4	101

(a)

Circuit	1h	5h	7h	11h	13h	RMS
Drive, A	552	120	57,0	40,0	27	403
$F13_{Lo}$, A	108	13,0	0,6	2,1	238	185
$F13_{Up}$, A	126	15,0	0,5	2,0	145	136

(b)

Figure 8. Current waveforms and its harmonic spectrums for parallel 11th (a) and 13th (b) filters

The most representative cases of the parallel filter configurations (e.g. when feeding sections A and B from section 1) are depicted in the Table 5. Two parallel the same order filters have opposite resonance detuning with upper and lower parameter deviation limits. From analysis of the Table 5 it is seen that opposite resonance detuning of the same order filters can cause considerable filter overload. As it has been noted earlier the higher order harmonic filters are more sensitive to filter component parameter variations from the detuning point of view. Furthermore, resonance detuning of the same order filters in the some system configurations can cause parallel system resonance peaks close to characteristic harmonic.

It should be quite clear from the above presented example that specifying harmonic filters and power factor correction requires considerable care and attention to detail. Main results of the investigation are follows:

- it is a bad practice to add filter circuits to existing power factor correction capacitors,
- improper design of the filter resonant point considering capacitor and reactor manufacturing tolerance and operation conditions can cause significant harmonic overloading of the filter,
- it is desirable to avoid the parallel operation of the same order filters in the system.

System configuration	Deviation limits							
	5th filters		7th filters		11th filters		13th filters	
With cap. 2×1.5 Mvar	Up	Lo	Up	Lo	Up	Lo	Up	Lo
2×(F5+F7)*	0.55	0.55	0.50	0.50	-	-	-	-
2×(F5+F7)	0.15	0.91	0.10	0.21	-	-	-	-
2×(F5+F7+F11+F13)	0.18	1.05	0.10	0.16	0.57	1.42	2.92	4.82
(F5+F7)+2×(F11+F13)	18.34	-	0.11	-	0.55	1.44	3.02	5.01
Without cap. 2×1.5 Mvar								
2×(F5+F7)*	0.41	0.41	0.50	0.50	-	-	-	-
2×(F5+F7)	0.13	0.77	5.40	9.44				
2×(F5+F7+F11+F13)	0.15	0.86	1.21	2.11	0.50	1.23	3.12	5.13
(F5+F7)+2×(F11+F13)	0.84	-	1.23	-	0.48	1.20	3.11	5.04

Note. *Both of 5th filters and 7th filters are fine-tuned.

Table 5. Harmonic current magnification factors β_{Fn} in the filter circuits (parallel operation)

4. Double-tuned filter

Double-tuned resonant filters are sometimes used for harmonic elimination of very high power converter systems (e.g. HVDC systems). Just like any other technical solution they also have their disadvantages (e.g. more difficult tuning process, higher sensitivity of frequency characteristic to changes in components values) and advantages (e.g. lower power losses at fundamental frequency, reduced number of reactors across which the line voltage is maintained, compact structure, single breaker) versus single-tuned filters. Such filters prove economically feasible exclusively for very large power installations and therefore they are not commonly used for industrial applications. There are, however, rare cases in which the use of such filter is justified. The double-tuned filter structure and its frequency characteristics are shown in Fig. 9. There are also the relations used to determine its parameters.

4.1. Example 3

As an example let us design a double-tuned filter (consider alternative configurations presented in Fig. 10) with parameters: $Q_F = 1$Mvar, $U = 6$kV, $n_1 = 5$, $n_2 = 7$, $n_R = 6$. Locations of the filter frequency characteristic extrema are determined using relations as in Fig. 9, whereas the genetic algorithm (APPENDIX A) determines the values of C_1 and C_2 for which the impedance-frequency characteristic attains the least value (at chosen harmonic frequencies) for the given filter power (R_{C1}, R_{C2}, R_{L1},R_{L2} –equivalent capacitor and reactor resistances; Fig. 9).

$$\begin{aligned}
&\omega_R = \frac{1}{\sqrt{L_2 C_2}} \quad \Rightarrow \quad L_2 = \frac{1}{\omega_R^2 C_2} \\
&\omega_S = \frac{1}{\sqrt{L_1 C_1}} \quad \Rightarrow \quad L_1 = \frac{1}{\omega_S^2 C_1} \\
&\omega_S = \frac{\omega_{n1}\omega_{n2}}{\omega_R} \\
&C_2 = \frac{\omega_S^2}{\omega_{n1}^2 + \omega_{n2}^2 - \omega_R^2 - \omega_S^2} C_1 \\
&C_1 = \left\{ \omega_1 \left(\frac{\omega_R}{\omega_{n1}\omega_{n2}} \right)^2 - \frac{1}{\omega_1} + \frac{\omega_1 \left[(\omega_{n1}^2 + \omega_{n2}^2 - \omega_R^2)\omega_R^2 - \omega_{n1}^2\omega_{n2}^2 \right]}{\omega_{n1}^2\omega_{n2}^2(\omega_R^2 - \omega_1^2)} \right\} \frac{U^2}{Q_F}
\end{aligned} \tag{9}$$

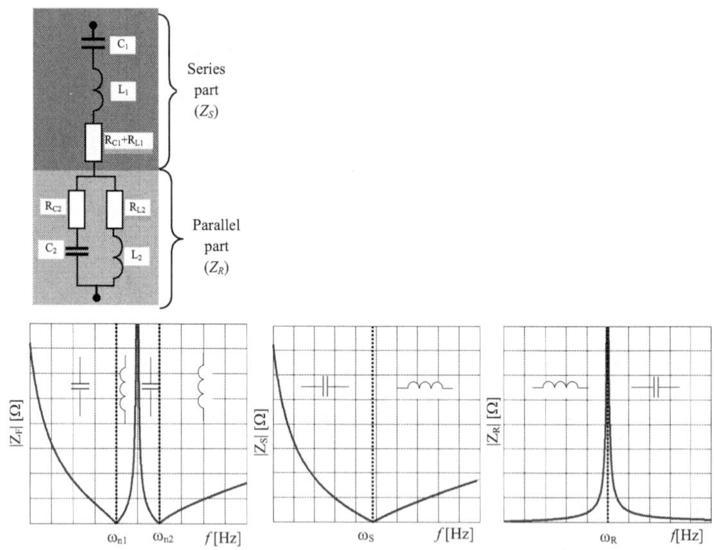

Figure 9. The double-tuned filter and its essential frequency characteristics;the basic configuration and frequency characteristics of: the series part, parallel part, and the whole filter (ω_R – angular resonance frequency of the parallel part; ω_S – angular resonance frequency of the series part; ω_{n1}, ω_{n2} – tuned angular frequencies of the double-tuned filter; equation 9 R 0).

Figure 11 shows graphic window of the programme developed by authors in the Matlab environment for optimisation of double-tuned filter. Ranges of filter parameters seeking are visible in the upper part of the widow, below the found characteristic is displayed, and basic parameters of the found solution are shown in the lowest part.

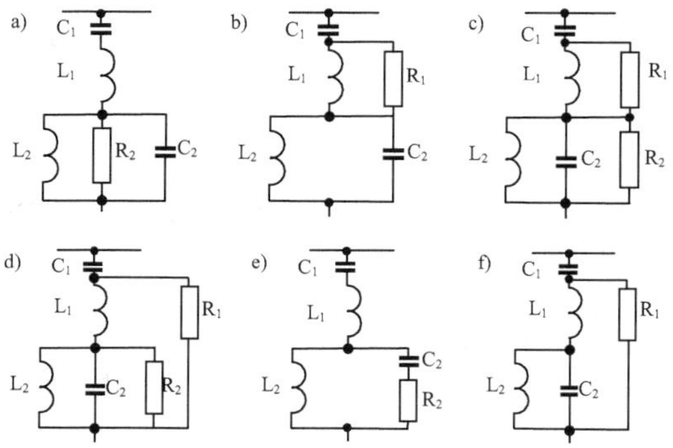

Figure 10. Alternative configurations of a double-tuned filter

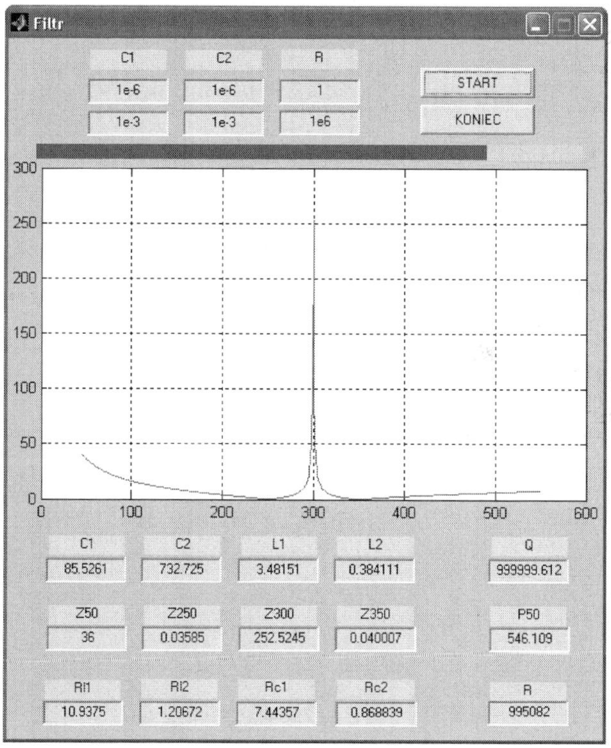

Figure 11. Graphic window of the program determining parameters one of the double-tuned filter in Fig. 10a (KONIEC = STOP)

The range of variability of decision variables: $C_1 = (10^{-6} - 10^{-3})$, $C_2 = (10^{-6} - 10^{-3})$. The Genetic Algorithm parameters: (a) each parameter is encoded into a 30-bit string, thus the chromosome length is 60 bits; (b) population size 1000 individuals; (c) crossover probability $p_k = 0.7$; (d) mutation probability $p_m = 0.01$; (e) Genetic Algorithm termination condition – 100 generations; (f) ranking coefficients $C_{min} = 0$, $C_{max} = 2$; (g) inverse ranking was applied in order to minimize the objective function; (h) selection method SUS and (i) shuffling crossover. The Genetic Algorithm goal was to minimize impedances for selected harmonics (n_1 and n_2) and maximize the impedance for the n_R harmonic.

Table 6 provides results of a double-tuned filter (Fig. 9 and 10) optimisation. The solutions are similar to each other (in terms of their values). It is noticeable that genetic algorithm is aiming to minimize the influence of additional resistances, that is to make the filter structures similar to the basic structure from Fig. 9. It means that additional resistances worsen the quality of filtering. The obtained result ensues from the applied optimisation method, i.e. optimisation of the frequency characteristic shape.

Parameter	Filter configuration						
	Fig. 10a	Fig. 10b	Fig. 10c	Fig. 10d	Fig. 10e	Fig. 10f	Fig. 9
C_1 [μF]	85.52	85.52	85.52	85.53	85.53	85.53	85.53
C_2[μF]	732.21	732.73	732.72	732.71	732.71	732.72	731.90
L_1[mH]	3.481	3.481	3.482	3.482	3.482	3.482	3.482
L_2 [mH]	0.384	0.384	0.384	0.384	0.384	0.384	0.385
R_{L1} [mΩ]	10.93	10.94	10.94	10.93	10.94	10.94	10.94
R_{L2} [mΩ]	1.207	1.207	1.207	1.207	1.207	1.207	1.208
R_{C1} [mΩ]	7.44	7.44	7.44	7.44	7.44	7.44	7.44
R_{C2} [mΩ]	0.869	0.868	0.868	0.868	0.868	0.868	0.870
Z_{50} [Ω]	36	36	36	36	36	36	36
Z_{250} [mΩ]	35.82	35.8	35.83	35.85	35.79	35,85	35.86
Z_{300} [Ω]	252.76	252.58	252.53	252.47	252.59	252.53	252.87
Z_{350} [mΩ]	40	40	40.04	40.01	40	40.01	40
Q_F[MVAr]	1	1	1	1	1	1	1
P_{50} [W]	546.09	546.06	546.10	546,11	546.07	546.11	546.11
R_1 [MΩ]	-	1	1	1	-	1	-
R_2 [MΩ]	1	-	1	1	0	-	-

Table 6. Basic parameters of filters from Fig. 10, designed using the genetic algorithm

5. C-type filter

The principal disadvantage of the majority of filter-compensating device structures is the poor filtering of high frequencies. To eliminate this disadvantage are usually used broadband (damped) filters of the first, second or third order; the C-type filter is included in the category of broadband filters [1, 2, 10]. Broadband filters have one more advantage, substantial for their co-operation with power electronic converters: they damp commutation notches more effectively than single branch filters - they have a much broader bandwidth. They also more effectively eliminate interharmonic components (in sidebands adjacent to characteristic harmonics) generated by static frequency converters. In the C-type filter in which the L_2C_2 branch (Fig. 12) is tuned to the fundamental harmonic frequency can be also achieved a significantly better reduction of active power losses compared to single branch filters. Thus the fundamental harmonic current is not passing through the resistor R_T, avoiding therefore large power losses.

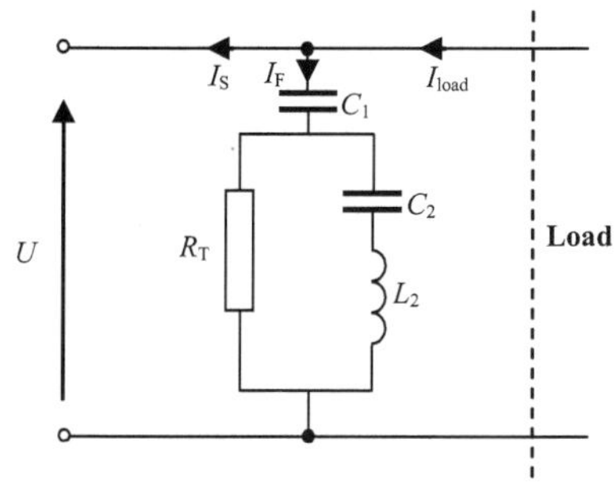

Figure 12. The C-type filter circuit

5.1. Example 4

In result of the arc furnace modernization (Fig. 13.) its power and consequently the level of load-generated harmonics have increased. It was, therefore, decided to expand the existing reactive power compensation and harmonic mitigation system. Prior to the modernization the system comprised two parallel, single-tuned 3rd harmonic filters that were the cause of a slight increase in the voltage 2nd harmonic.

Considering the system expansion the designed C-type filter should be tuned to the 2nd harmonic. Although currently the 2nd harmonic level in the existing system does not exceed the limit, connection of new loads may increase the 2nd harmonic to an unacceptable level.

5.1.1. Traditional approach

The filter impedance is given by (Fig. 12) [1]:

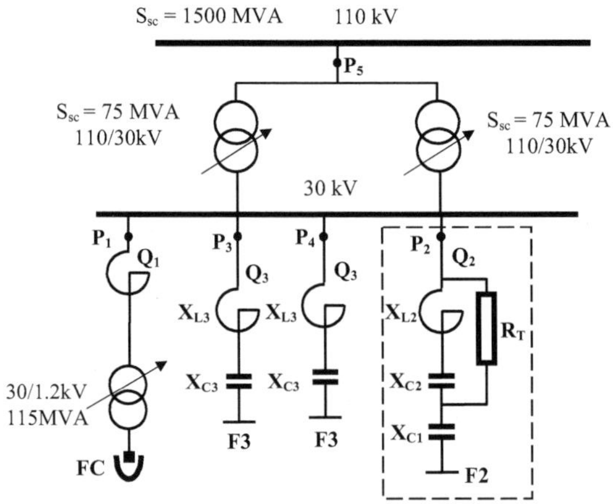

Figure 13. Single line diagram of the arc furnace power supply system

$$Z_F = \frac{\left(j\omega L_2 - j\dfrac{1}{\omega C_2}\right) R_T}{R_T + j\omega L_2 - j\dfrac{1}{\omega C_2}} - j\frac{1}{\omega C_1} \tag{10}$$

The L_2 and C_2 components are tuned to the fundamental frequency $_1$:

$$L_2 = \frac{1}{\omega_1^2 C_2} \tag{11}$$

hence

$$Z_F = \frac{jR_T\left(\omega^2 - \omega_1^2\right)}{R_T\omega\omega_1^2 C_2 + j\left(\omega^2 - \omega_1^2\right)} - j\frac{1}{\omega C_1} \tag{12}$$

The C-type filter is tuned to the resonance angular frequency $\omega_r = n_r\omega_1$

$$\omega_r \cong \frac{1}{\sqrt{L_2 \dfrac{C_1 C_2}{C_1 + C_2}}} \Rightarrow C_2 = C_1\left(n_r^2 - 1\right) \tag{13}$$

hence

$$Z_F = \frac{jR_T\left(\omega^2 - \omega_1^2\right)}{R_T\omega\omega_1^2 C_1\left(n_r^2 - 1\right) + j\left(\omega^2 - \omega_1^2\right)} - j\frac{1}{\omega C_1} \tag{14}$$

The filter reactive power (Q_F) for the fundamental harmonic is given by the relation:

$$Q_F = -\frac{U^2}{\mathrm{Im}\left(Z_F\left(\omega_1\right)\right)} \Rightarrow C_1 = \frac{Q_F}{\omega_1 U^2} \tag{15}$$

that is:

$$Z_F = \frac{jR_T U^2\left(\omega^2 - \omega_1^2\right)}{R_T\omega\omega_1 Q_F\left(n_r^2 - 1\right) + jU^2\left(\omega^2 - \omega_1^2\right)} - j\frac{\omega_1 U^2}{\omega Q_F} \tag{16}$$

Distribution of the load-generated harmonic current between the filter tuned to that harmonic and the system is:

$$\frac{\left|I_S\left(n_r\right)\right|}{\left|I_F\left(n_r\right)\right|} = k = \frac{\left|Z_F\left(n_r\right)\right|}{\left|Z_S\left(n_r\right)\right|} \Rightarrow R_T = \frac{U^2}{n_r^3 {Q_F}^2 k\omega_1 L_S}\sqrt{U^4 - n_r^4 {Q_F}^2 k^2 \omega_1^2 L_S^2} \tag{17}$$

Summarizing, the C-type filter parameters can be determined from above formulas. For the arc furnace power supply system (Fig. 13) and the design requirements:

Network	3rd harmonic filters	C-type filter
U = 30 kV	Q_F = 20 Mvar	Q_F = 20 Mvar
S_{sc}= 1500 MVA	L_3 = 18.48 mH	n_r = 1.9
L_S = 3.129 mH (supply network)	C_3 = 63 μF	q_{F2} = 10 (filter quality factor)
R_S = 30.0 mΩ (supply network)	R_3 = 30.0 mΩ	k = 1
	n_r = 2.95	

The C-type filter parameters are: C_1 =70.736μF, C_2 =198.24μF, L_2 =51.11mH, R_T =276.86 Ω.

Figure 14a shows frequency-impedance characteristics of: the power network, the resultant impedance of two single-tuned 3rd harmonic filters, and the C-type filter impedance. Fig. 14b shows frequency-impedance characteristics of: the network, the resultant impedance of the network and two 3rd harmonic filters, and the resultant impedance of the network, two 3rd harmonic filters and the C-type filter.

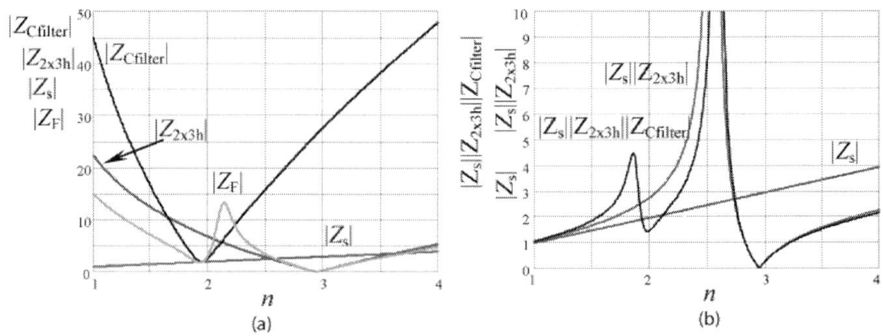

Figure 14. Frequency-impedance characteristics of: a) the power network equivalent impedance (Z_S), the resultant impedance of two 3rd harmonic filters (Z_{2x3h}), the C-type filter impedance ($Z_{Cfilter}$), the resultant impedance of two 3rd harmonic filters and the C-type filter (Z_F); b) the impedance seen from the load terminals: without filters (Z_S), the network equivalent impedance and two 3rd harmonic filters impedance connected in parallel ($Z_s||Z_{2x3h}$), and parallel connection of the network equivalent impedance, two 3rd harmonic filters and the C-type filter impedances ($Z_s||Z_{2x3h}||Z_{Cfilter}$)

Data listed in Table 7 demonstrate that connecting the C-type filter results in the expected reduction of the 2nd voltage harmonic in the supply system, whereas other harmonics are reduced to a small extent. Further reduction of the second harmonic can be achieved by improving the C-type filter quality factor q_{F2} and, consequently, reduction of the filter impedance for the filter resonant frequency and increasing the impedance for higher harmonics.

%	U_2	U_3	U_4	U_5	U_6	U_7	U_8	U_9	THD_U *
Busbars voltage harmonics without filters	1,76	3.01	1.66	2.88	1.12	1.75	1.00	1,12	5.87
Busbars voltage harmonics with two 3rd harmonic filters	2.47	0,27	0.95	1.87	0.78	1.24	0.72	0.81	4.07
Busbars voltage harmonics with two 3rd harmonic filters and the C-type filter	1.32	0.27	0.91	1.78	0.74	1.18	0.69	0.78	3.37

*Total harmonic voltage distortion factor THD_u determined from components up to 15th order.

Table 7. Voltage harmonics (at the 30kV side) without filters, with two 3rd harmonic filters, and with two 3rd harmonic filters and the C-type filter

Figure 15a shows the C-type filter frequency characteristics for different filter quality factors, figure 15b illustrates the relation between the resistance R_T and the coefficient k that indicates the distribution of the current harmonic to which the filter is tuned (Table 8).

Seemingly, the most advantageous solution is to increase the filter resistance R_T in order to ensure the largest possible part of the eliminated harmonic current flow through the filter

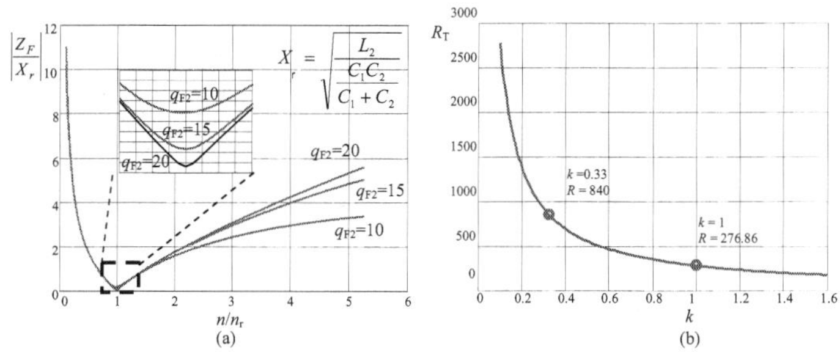

Figure 15. a) The C-type filter frequency characteristics for various quality factors $q_F = R_T/X_r$, b) the resistance R_T vs. the coefficient k

***I_F* [%]**	38.5	44.4	50.0	51.9	57.1	66.6	75.0	83.3	91.0
***I_S* [%]**	61.5	55.6	50.0	48.1	42.9	33.3	25.0	16.7	9.0
k	1.60	1.25	1.00	0.93	0.75	0.50	0.33	0.25	0.10
***R_T* [Ω]**	172	221	276.86	300	350	555	840	1111	2778

Table 8. The percentage distribution of the harmonic current between the filter tuned to that harmonic and the supply network; the corresponding R_T values and the designed filter quality factor.

instead of the supply network. But the increase in the resistance will reduce high harmonic currents through the filter. Thus a compromise between the filter ability to take over the harmonic the filter is tuned to, and its capability to mitigate other harmonics should found. Increasing the R_T resistance makes the C-type filter frequency characteristic similar to that of a single-branch filter.

5.1.2. Genetic approach

The goal of genetic algorithm is to seek the C-type filter capacitance (C_1) in order to compensate the system's reactive power, and determine the resistance value (R_T) to ensure a required distribution of the 2nd harmonic current. The filter parameters were computed by means of the Genetic Algorithm using the model from Fig. 13 in the Matlab environment. The arc furnace is regarded as an ideal harmonic current source and as a load for the fundamental harmonic with given active power (P) and reactive power (Q).

The range of variability of decision variables: $C_1 = (10^{-6} - 10^{-4}\text{F})$, $R_T = (1 - 10000)$. The Genetic Algorithm parameters: (a) parameter C_1 is encoded into 8-bit strings, and parameter R_T into a 12-bit string; (b) population size 200 individuals; (c) crossover probability $p_k = 0.8$; (d) mutation probability $p_m = 0{,}01$; (e) Genetic Algorithm termination condition – 30 generations; (f) ranking coefficients $C_{min} = 0$, $C_{max} = 2$; (g) inverse ranking was applied in order to minimize the objective function; (h) selection method SUS and (i) shuffling crossover. The optimization goal was to minimize total harmonic distortion of the supply network current THD_I and reduce the angle between fundamental voltage and current harmonics $\varphi(I_{(1)},U_{(1)})$ - (18).

$$F_{goal} = \begin{cases} THD_I + \sin\left(\varphi\left(I_{(1)},U_{(1)}\right)\right) & Q_F \leq 20\text{Mvar} \\ 100 \cdot \left(THD_I + \sin\left(\varphi\left(I_{(1)},U_{(1)}\right)\right)\right) & Q_F > 20\text{Mvar} \end{cases} \tag{18}$$

According with the achieved results the total capacitance $C_1 = 70.75\mu F$ and total capacitance $C_2 = 196.8\mu F$. The reactor L_2 inductance is 51.48mH. The resistor resistance is $R_T = 300\Omega \pm 10\%$.

Measurements in the power system, configured according to the above specification, were carried out in order to check the correctness of the system operation. The instruments locations were (fig. 13): P_1 – arc furnace, P_2 – C-type filter, P_3 – first filter of the 3rd harmonic, P_4 – second filter of the 3rd harmonic, and P_5 – at the 110kV side. Essential results of measurements are provided in Table 9.

Figures 16 – 18 illustrate voltage and current waveforms recorded at the 110kV side, the arc furnace supply voltage the arc furnace and the C-type filter currents and total harmonic voltage distortion factor THD_U at both: the 30kV and 110kV side. The measurements have demonstrated that the C-type filter performance has met the requirements, i.e. it attains the expected reduction of reactive power, ensures the second harmonic reduction in the power system and harmonic distortion THD_U reduction by means of high harmonics mitigation. The measurements verified the proposed method and the C-type filter designed using this method operates according to the requirements.

6. Conclusion

This chapter presents several selected cases of power electronic systems analysis with respect to high harmonics occurrence and reactive power compensation. For these cases are proposed classical solutions, i.e. power passive filters which still are a basic and the simplest method for high harmonics mitigation. Analytical formulas that enable to determine basic parameters of various filters' structures and a group of single-tuned filters are provided.

Also a method for passive filters' design employing artificial intelligence, which incorporates genetic algorithms, is presented. It has been proved that this method can be the attractive tool to solve some kinds of power quality problems. The results obtained using GA are very close to those obtained with the analytical method. Hence the conclusion that genetic algorithms can be an efficient tool for passive filters design. The advantage of the method employing genetic algorithms is the possibility of multi-criterial optimisation and taking into account at the design stage different (e.g. voltage or current) constraints. It also can be applied to filters of various structures and degrees of complexity and can account for filters' resistance that may influence the filter resonance frequency. In other words, genetic algorithm can be a useful design tool in cases where the system analysis is too complex or even not possible.

http://dx.doi.org/10.5772/53425

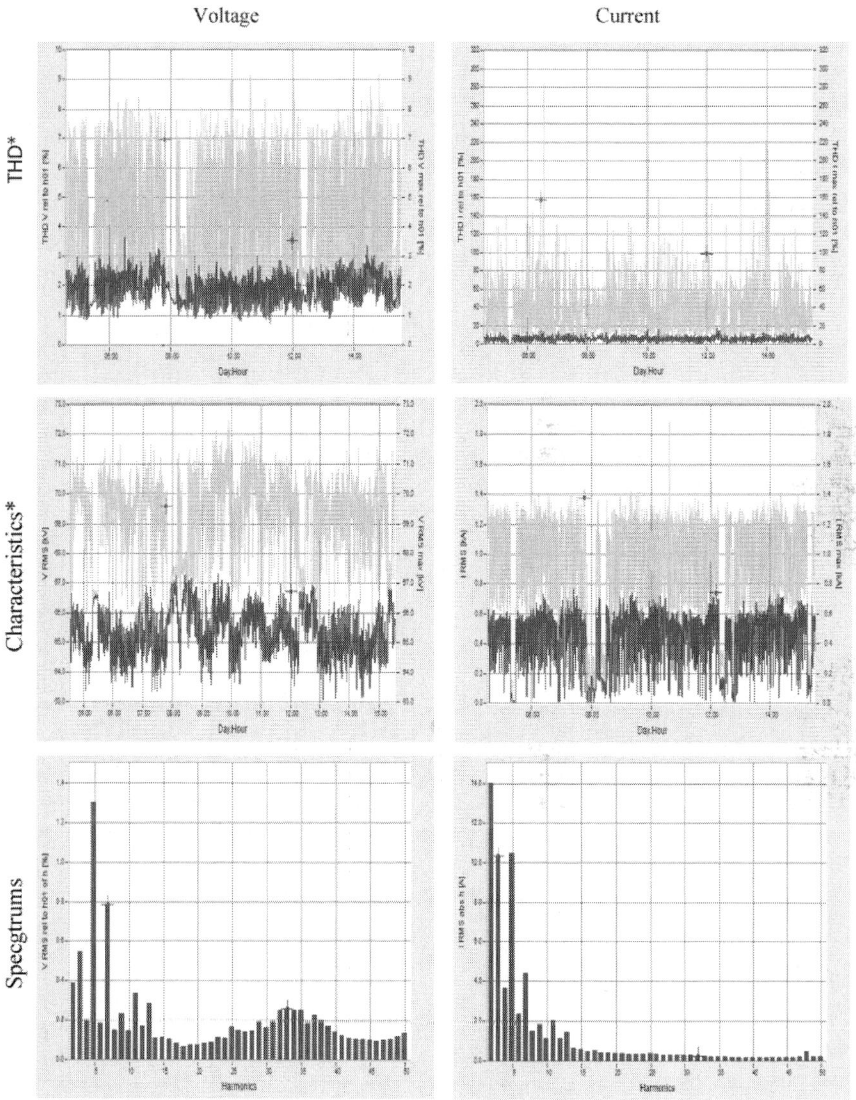

Figure 16. Voltage and current characteristics, spectrums and THD factors at 110 kV side. The graphs show two time characteristics: 10 min. averaged values (blue) and 10 ms maximum values (yellow)

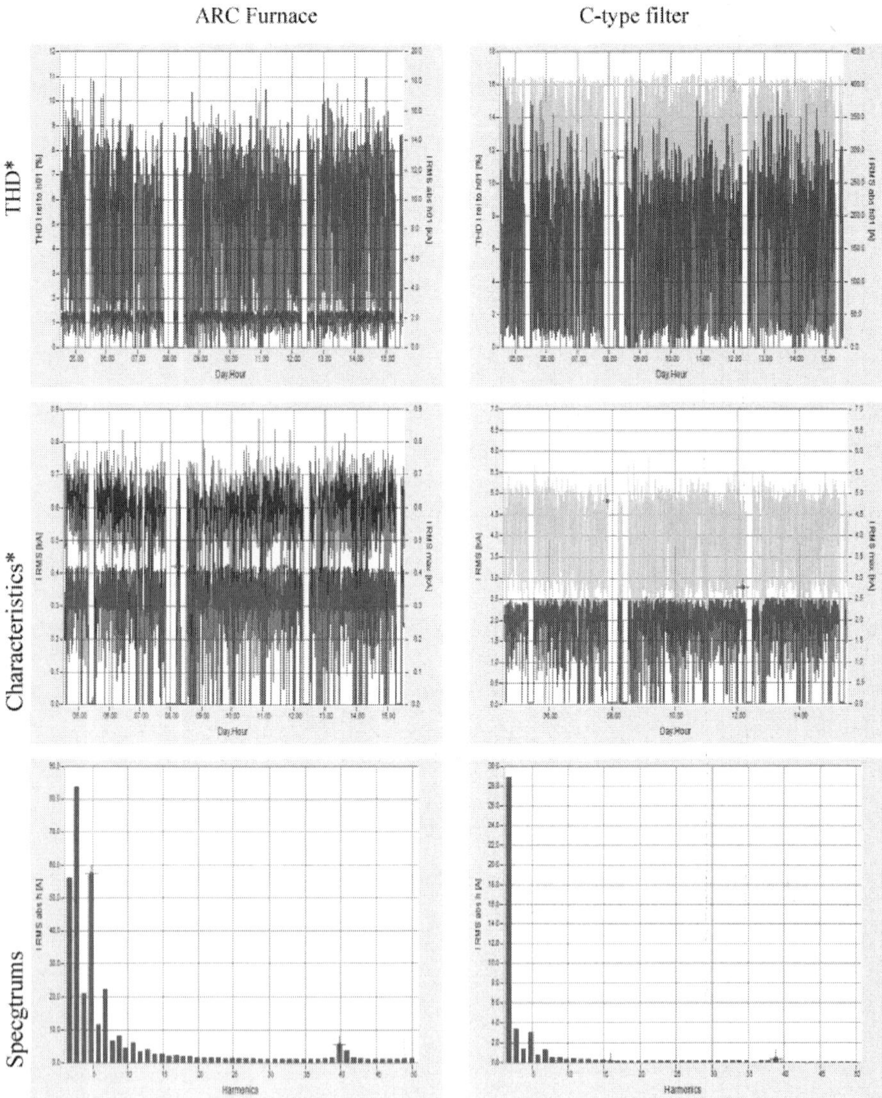

Figure 17. Current characteristics, spectrums and THD_I factors of Arc-furnace and C-type filter. The graphs show two time characteristics: 10 min. averaged values (blue) and 10 ms maximum values (yellow)

http://dx.doi.org/10.5772/53425

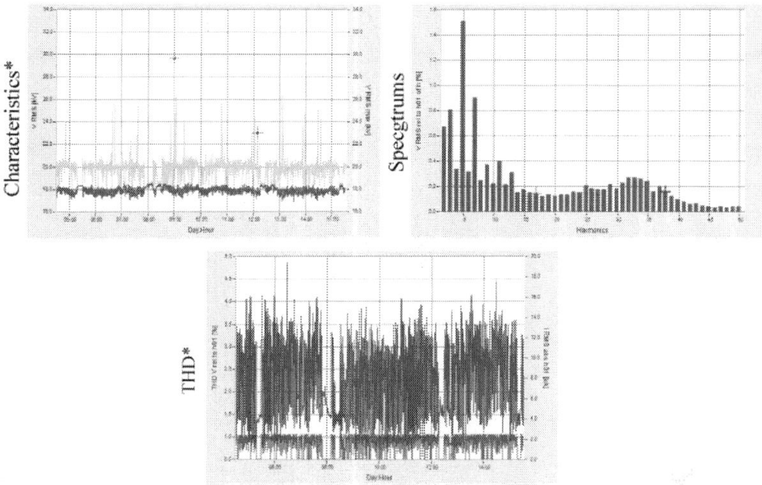

Figure 18. Current characteristics, spectrums and THD_U factors at 30 kV side. The graphs show two time characteristics: 10 min. averaged values (blue) and 10 ms maximum values (yellow)

Measurement point	P_1 (Furnace)	P_1 (Furnace)	P_2 (Filter C)	P_3 (3rd harm. Filter)	P_4 (3rd harm. Filter)	P_5(110kV)	P_5 (110kV)
Furnace and filters in operation	off	on	on	on	on	on	off
U_{RMS} [kV]	18.29		17.59			65.04	66.58
I_{RMS} [A]	-	2383	391	398	385	602	86.5
P [MW]	-	93.75	0.083	0.234	0.198	105	12.14
Q [MVAr]	-	71.19	19.55	19.84	19.68	28.2	6.52
S [MVA]	-	125.7	20.65	21.0	20.34	117.5	17.25
PF	-	0.744	0.0043	0.011	0.001	0.89	0.57
THD_U [%]	1.56		2.45			1.92	1.58
THD_I [%]	-	6.44	8.04	12.04	10.62	4.64	6.61
$I_{(1)RMS}$ [A]	-	2357	387	376	373	594	85.46
$U_{(1)RMS}$ [kV]	18.29		17.57			65.0	66.57
$U_{(2)RMS}$ [%]	0.06		0.73			0.42	0.07
$U_{(3)RMS}$ [%]	0.57		0.61			0.43	0.43
$U_{(4)RMS}$ [%]	0.04		0.34			0.19	0.04
$U_{(5)RMS}$ [%]	1.15		1.51			1.22	1.25
$I_{(2)RMS}$ [A]	-	58.7	32	10.8	9.8	15	2.21
$I_{(3)RMS}$ [A]	-	97.3	3.5	44.3	38.8	9	4.8
$I_{(4)RMS}$ [A]	-	23.1	1.5	5.1	4.5	3.7	0.6
$I_{(5)RMS}$ [A]	-	71.2	3.7	11.9	11.2	12.5	3.5
Pst [%]	1		16.66			9.17	1.02

Table 9. The Measurement results obtained over a period of 7 days

Appendix A - Genetic Algorithms

Genetic Algorithms (GA) are stochastic global search method, mimicking the natural biological evolution. It has been noted that natural evolution is done at the chromosome level, and not directly to individuals. In order to find the best individual, genetic operators apply to the population of potential solutions, the principle of survival of the fittest individual. In every generation, new solutions arise in the selection process in conjunction with the operators of crossover and mutation. This process leads to the evolution of individuals that are better suited to be the existing environment in which they live.

GA popularity is due to its features. They: (i) don't process the parameters of the problem directly but they use their coded form; (ii) start searching not in a single point but in a group of points; (iii) they use only the goal function and not the derivatives or other auxiliary information; (iv) use probabilistic and not deterministic rules of choice. These features consists in effect on the usability of Genetic Algorithms and hence their advantages over other commonly used techniques for searching for the optimal solution. There is a high probability that the AG does not get bogged in a local optimum.

An important term in genetic algorithms is the objective function. It is on the basis of all the individuals in the population are evaluated and on the basis of a new generation of solutions is created. Each iteration of the genetic algorithm creates a new generation. Figure 20. shows the basic block diagram of a Genetic Algorithm.

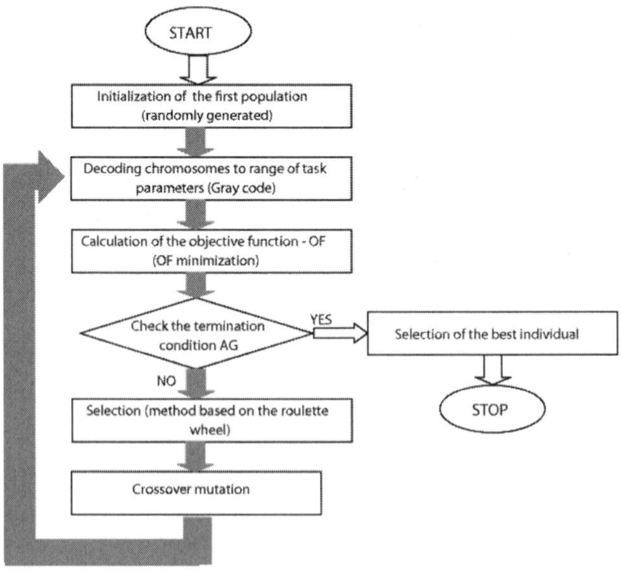

Figure 19. Block diagram of the basic Genetic Algorithm (GA)

Author details

Ryszard Klempka, Zbigniew Hanzelka* and Yuri Varetsky

*Address all correspondence to: hanzel@agh.edu.pl

AGH-University of Science & Technology, Faculty of Electrical Engineering, Automatics, Computer Science and Electronic, Krakow, Poland

References

[1] Dugan R., McGranaghan M., Electrical power systems quality, McGraw-Hill, 2002

[2] Yaow-Ming Ch., Passive filter design using genetic algorithms, IEEE Transactions on Industrial Electronics, vol. 50, no. 1, February 2003

[3] Younes M., Benhamida, Genetic algorithm-particle swarm optimization (GA-PSO) for economic load dispatch, Electrical Review 10/2011, 369-372

[4] Zajczyk R., Nadarzyński M., Elimination of the higher current harmonics by means of transverse filters, Electrical Review 10/2004, 963-966

[5] Hanzelka Z., Klempka R., Application of genetic algorithm in double tuned filters design, EPE01, Graz 27-29 VIII 2001

[6] Klempka R., Designing a group of single-branch filters, Electrical Power Quality and Utilisation, EPQU'03, September 17-19 2003, Krakow, Poland

[7] Klempka R., Filtering properties of the selected double tuned passive filter structures designed using genetic algorithm, EPE-PEMC 2002, Dubrownik

[8] Nassif A. B., Xu W., Freitas W., An investigation on the selection of filter topologies for passive filter applications, IEEE Transactions on Power Delivery, vol. 24, no. 3, July 2009

[9] Pasko M., Lange A., Influence of arc and induction furnaces on the electric energy quality and possibilities of its improvement, Electrical Review 06/2009, 67-74

[10] Badrzadeh B., Smith K. S., Wilson R. C., Designing passive harmonic filters for aluminium smelting plant, IEEE Transactions on Industry Applications, vol. 47, no. 2, March/April 2011

[11] Arillaga J., Watson N. R., Power system harmonics, John Wiley and Sons, 2003

[12] Chang S.-J., Hou H.-S., Su Y.-K., Automated passive filter synthesis using a novel tree representation and genetic programming, IEEE Transactions on Evolutionary Computation, vol. 10, No. 1, February 2006

[13] Eslami M., Shareef H., Mohamed A., Khajehzadeh M., Particle swarm optimization for simultaneous tuning of static var compensator and power system stabilizer, Electrical Review 09a/2011, 343-347

[14] Gary W. Chang, Shou-Young Chu, Hung-Lu Wang, A new method of passive harmonic filter planning for controlling voltage distortion in power system, IEEE Transactions on Power Delivery, vol. 21, no.1. January 2006

[15] Kolar V., Kocman St., Filtration of harmonics in traction transformer substations, positive side effects on the additional harmonics, Electrical Review 12a/2011, 44-46

[16] Piasecki Sz., Jasiński M., Rafał K., Korzeniewski M., Milicua A., Higher harmonics compensation in grid-connected PWM converters for renewable energy interface and active filtering, Electrical Review 06/2011, 85-90

[17] Pasko M., Lange A., Compensation of the reactive power and filtration of high harmonics by means of passive LC filters, Electrical Review 04/2010, 126-129

[18] Rivas D., Morán L., Dixon J. W., Spinoza J. R., Improving passive filter compensation performance with active techniques, IEEE Transactions on Industrial Electronics, vol. 50, no. 1, February 2003

[19] Świątek B. Hanzelka Z., Neural network-based controller for an active power filter. 10th International Power Electronics & Motion Control Conference, September 9-11, 2002, Cavtat & Dubrovnik, Croatia.

[20] Świątek B., Hanzelka Z., A neural network-based controller for an active power filter, 14th International Power Quality Conf. September 11-13, Rosemont, Illinois 2001

[21] Świątek B., Hanzelka Z., A single-phase active power filter with neural network-based controller, IASTED International Conf. Power and Energy Systems, July 3-6, 2001, Rhodes, Greece

[22] Świątek B., Klempka R., Kosiorowski S., Minimization of the source current distortion in systems with single-phase active power filters and additional passive filter designed by genetic algorithms, 11th European Conference on Power Electronics and Applications, EPE2005, September 11-14, 2005, Dresden

[23] Varetsky Y., Hanzelka Z., Klempka R., Transformer energization impact on the filter performance, 8th International Conference on Electrical Power Quality and Utilisation, September 21-23 2005, Krakow, Poland.

[24] Varetsky Y., Hanzelka Z., Filter characteristics in a DC drive power supply system, 13th International Conference on Harmonics & Quality of Power, Australia, 28th September–1st October 2008

[25] Vishal V., Bhim S., Genetic-algorithm-based design of passive filters for offshore applications, IEEE Transactions on Industry Applications, vol. 46, no. 4. July/August 2010

Chapter 9

Parameters Estimation of Time-Varying Harmonics

Cristiano A. G. Marques, Moisés V. Ribeiro,
Carlos A. Duque and Eduardo A. B. da Silva

Additional information is available at the end of the chapter

http://dx.doi.org/10.5772/53816

1. Introduction

The increasing use of power electronic devices in power systems has been producing significant harmonic distortions, what can cause problems to computers and microprocessor based devices, thermal stresses to electric equipments, harmonic resonances, as well as aging and derating to electrical machines and power transformers [1–3]. The most important problems that have been reported in the literature concerns to the difficulty of the frequency control within the micro-grids and the increase of the total harmonic distortion. These two factors may negatively impact on the protection system, power quality analysis and intelligent electronic devices (IEDs), in which digital algorithms assume that the fundamental frequency is constant. Based on this fact, there has been an increasing interest in signal processing techniques for detecting and estimating harmonic components of time-varying frequencies. Their correct estimation has become an important issue in measurement equipment and compensating devices. Although many methods have been proposed in the literature, it still remains difficult to detect and estimate harmonics of time-varying frequencies [4, 5]. The harmonic components (voltage or current) can change its frequencies due to continuous changes in the system configuration and load conditions, to the rapid proliferation of distributed resources, and to possibilities of new operational scenarios (e.g., islanded microgrids). Also the need for massive monitoring of networks is unquestionable within the concept of *smart grids*. An important line of research in the *smart grids* context is to identify and estimate time-varying harmonics that may appear in the current and voltage signals, and from this information, correct and adjust the digital algorithms that are part of protection equipments, power quality monitors and IEDs.

The concept of time-varying harmonics came recently to the vocabulary of power systems engineers, because more and more nonlinear loads, with dynamic behavior, are being

INTECH
open science | open minds

© 2013 Marques et al.; licensee InTech. This is a paper distributed under the terms of the Creative Commons Attribution License (http://creativecommons.org/licenses/by/3.0), which permits unrestricted use, distribution, and reproduction in any medium, provided the original work is properly cited.

connected to power systems and the fundamental frequency is experiencing a large range of variation. These factors have putting in check the traditional stationary spectral analysis methods, and many techniques for improving harmonics measurement have been proposed in recent years. Parametric and nonparametric methods that commonly have been used by the community of signal processing have been applied to power system harmonic estimation. These methods have in common that they need to estimate the fundamental frequency to adjust some internal parameters, like filter coefficients. The challenge is producing a harmonic estimator with high convergence ratio, high accuracy, low computational burden and immunity to the presence of interharmonic: conditions that are not ease to simultaneously deal with. The most used technique for harmonics estimation is based on the discrete Fourier transform (DFT) [8, 10]. The DFT algorithm is attractive because of its low computational complexity and its simple structure. However, DFT does not perform well if power system frequency varies around the nominal value. Several other techniques have been proposed in the literature for harmonic estimation. However, the DFT still appear to be the preferred algorithm mainly due to its simplicity.

2. Power quality and smart grid

The term "*smart grid*" has different definitions in the literature. Regardless of the precise definition, the term smart grid can be seen as a new paradigm, covering from conception to operation of the power systems, that makes intensive use of information and communication technologies, decentralized control approaches, powerful signal processing and computational intelligence techniques, renewable and distributed generation and storage energy facilities, self-generation, etc [11] to offer flexibility, robustness, and efficiency regarding generation, transmission, distribution, and consumption of electrical energy.

One of the promises of the *smart grid* is to improve the power quality. Therefore, the reliability or continuity of service is one of the consequences that will result from the implementation of the self-healing aspects of the Smart Grid. However power quality issues should not form an unnecessary barrier against the development of *smart grids* or the introduction of renewable sources of energy. The smart properties of future grids should rather be a challenge for new approaches in an efficient management of power quality [11].

An adequate power quality should guarantees electromagnetic compatibility between all equipment connected to the grid. Then, an important issue for the successful and efficient operation of smart grids is the introduction of advanced, flexible, robust, and cooperative set of signal processing and computational intelligent techniques for power quality analysis. With this set of techniques, an effective and extensive smart monitoring system can be devised and deployed. Such a monitoring system have to allow for the monitoring of such as voltage, current, bidirectional energy consumption at distribution transformers, substations transformers, smart meters, distribution feeders, distribution switching devices, and strategically installed power quality monitors in the power systems.

3. Harmonic estimation techniques: Before and after smart grid

Consider the monitored power line signal, after processed by the analog anti-aliasing filter, be expressed by

$$x(t) = \sum_{m=1}^{N_h} A_m(t)\cos[m\Omega_0 t + \phi_m(t)] + \eta(t), \tag{1}$$

where $A_m(t)$ and $\phi_m(t)$ are, respectively, the amplitude and phase of the m-th harmonic, N_h is the maximum harmonic order, $\eta(t)$ is the additive noise and Ω_0 denotes the angular fundamental synchronous frequency. Based on the definition of instantaneous frequency deviation [6, 7], the frequency of the m-th harmonic can be defined as

$$\psi_m(t) = \frac{d[m\Omega_0 t + \phi_m(t)]}{dt} = m\Omega_0 + \frac{d\phi_m(t)}{dt}. \tag{2}$$

Assuming Ω_0 constant, note that any variation in $\psi_m(t)$ can be expressed by the term $\frac{d\phi_m(t)}{dt}$. As $\psi_m(t)=m\psi_1(t)$ (the frequency of the m-th harmonic is equal to m times the fundamental frequency), from (2) we have

$$m\Omega_0 + \frac{d\phi_m(t)}{dt} = m\left[\Omega_0 + \frac{d\phi_1(t)}{dt}\right]$$

$$\therefore \frac{d\phi_m(t)}{dt} = m\frac{d\phi_1(t)}{dt}. \tag{3}$$

The goal of harmonic estimation techniques is to provide estimates of the parameters $A_m(t)$ and $\phi_m(t)$ using the discrete version $x[n]$ of the signal $x(t)$. Some of the existing techniques estimate these parameters considering that the fundamental angular frequency $\psi_1(t)$ is constant and nominal (steady) and some estimate the same parameters considering $\psi_m(t)$ time-varying.

Basically, we can say that before *smart grids*, the most used techniques for harmonic estimation assumed that $\psi_1(t)$ was constant and nominal. This is the reason that DFT is the standard algorithm adopted in international standards and implemented in the majority of equipments. It can be explained because, generally, in interconnected power systems the power frequency is high controlled and very near the nominal value. Thus, powerful harmonic estimation methods are not needed, except in especial applications. However, with the inclusion of new power generation technologies such as renewable energy source generation and distributed generation energy, the fundamental frequency of micro grids and isolated systems will suffer significant variations (this fact is already noted in actual power systems). In this new scenarios, very common in *smart grids*, new harmonic estimation techniques will be very needed and essential.

4. Methods for estimating steady-state harmonics

Methods for estimating steady-state harmonics are more simple than time-varying ones, and its algorithms do not use information of the fundamental frequency of the signal under estimation. In what follows, we describe four algorithms.

4.1. Discrete fourier transform

The most common and the most used technique for steady-state harmonic estimation is the discrete Fourier transform (DFT) [8-10]. The DFT method is simple and easy to be implemented in monitoring systems, but its application for time-varying harmonics is not recommended.

Given the discrete signal $x[n]$, the amplitude and phase of the k-th harmonic component can be straightforwardly estimated by the recursive equations:

$$\hat{A}_k[n] = 2\sqrt{Y_{c_k}{}^2[n](t) + Y_{s_k}{}^2[n]} \tag{4}$$

and

$$\hat{\phi}_k[n] = -\arctan\left(\frac{Y_{s_k}[n]}{Y_{c_k}[n]}\right), \tag{5}$$

respectively, where,

$$Y_{c_k}[n] = Y_{c_k}[n-1] + (x[n] - x[n-N])\cos(kw_0) \tag{6}$$

and

$$Y_{s_k}[n] = Y_{s_k}[n-1] + (x[n] - x[n-N])\sin(kw_0), \tag{7}$$

in which $w_0 = \Omega_0 / f_s$ is the discrete synchronous angular frequency, f_s is the sampling rate and N is the number of samples within a integer number of cycles of the fundamental power signal.

The DFT algorithm is very simple and its implementation is easy for real-time application. However, if the fundamental frequency is not nominal and constant, then the estimates can carry significant errors.

4.2. Demodulation

The demodulation technique presented in [15] can be used to estimate the parameters of harmonics as point out in [16]. In similar way to the DFT technique the demodulation technique can give erroneous results if its filter is fixed.

The k-th harmonic parameters can be estimated by

$$\hat{A}_k[n] = 2\sqrt{Y_{c_k}{}^2[n](t) + Y_{s_k}{}^2[n]} \tag{8}$$

and

$$\hat{\phi}_k[n] = -\arctan\left(\frac{Y_{s_k}[n]}{Y_{c_k}[n]}\right), \tag{9}$$

in which $Y_{c_k}[n]$ and $Y_{s_k}[n]$ are evaluated by

$$Y_{c_k}[n] = (x[n]\cos(kw_0)) * h[n] \tag{10}$$

and

$$Y_{s_k}[n] = (x[n]\sin(kw_0)) * h[n], \tag{11}$$

respectively, where $h[n]$ is the impulse response of a low-pass filter and $*$ denotes the linear convolution operator.

4.3. Goertzel

The Goertzel technique uses a second-order infinite impulse response filter to estimate the parameters of the *k*-th harmonic [17]. The Goertzel algorithm is more efficient than the Fast Fourier Transform (FFT) when the number of harmonics to be calculated is low.

The amplitude and phase of the k-th harmonic is estimated, respectively, by

$$\hat{A}_k[n] = 2\sqrt{Y_{c_k}{}^2[n](t) + Y_{s_k}{}^2[n]} \tag{12}$$

and

$$\hat{\phi}_k[n] = -\arctan\left(\frac{Y_{s_k}[n]}{Y_{c_k}[n]}\right), \tag{13}$$

in which $Y_{c_k}[n]$ and $Y_{s_k}[n]$ are evaluated by

$$Y_{c_k}[n] = \Re(X[k]) \tag{14}$$

and

$$Y_{s_k}[n] = \Im(X[k]), \tag{15}$$

where

$$X[k] = \exp(2\pi k)s[N-1] - s[N-2], \tag{16}$$

$$s[n] = x[n] + 2\cos(2\pi k / N)s[n-1] - s[n-2]. \tag{17}$$

4.4. Linear least squares

The linear least squares (LS) algorithm estimates several harmonics in one evaluation instead of a unique estimate [18]. Its advantage is the acquisition of several harmonics in only one evaluation. However the computational burden is high.

Basically, the LS algorithm has as a result the vector given by

$$\mathbf{v}[m] = [a_1[m]a_2[m] \ldots a_{N_h}[m]b_1[m]b_2[m] \ldots b_{N_h}[m]]^T \tag{18}$$

Thus, the amplitude and phase of the k-th harmonic (k $\in$ [1, 2,...,N_h]) are given by

$$\hat{A}_k[m] = 2\sqrt{b_k[m]^2 + a_k[m]^2} \tag{19}$$

and

$$\hat{\phi}_k[m] = -\arctan\left(\frac{b_k[m]}{a_k[m]}\right). \tag{20}$$

The vector $v[m]$ is evaluated as following

$$\mathbf{v}[m] = (\mathbf{H}^T[m]\mathbf{H}[m])^{-1}\mathbf{H}[m]^T\mathbf{y}[m], \tag{21}$$

where

$$\mathbf{y}[m] = [x[n]x[n-1]\ldots x[n-N]]^T \tag{22}$$

and

$$\mathbf{H}[m] = [\mathbf{H}_a[m]\mathbf{H}_b[m]] \tag{23}$$

$$\mathbf{H}_a[m] = \begin{pmatrix} \cos(w_0 n) & \cos(2w_0 n) & \cdots & \cos(N_h w_0 n) \\ \cos(w_0(n-1)) & \cos(2w_0(n-1)) & \cdots & \cos(N_h w_0(n-1)) \\ \cdots & \cdots & \cdots & \cdots \\ \cos(w_0(n-N)) & \cos(2w_0(n-N)) & \cdots & \cos(N_h w_0(n-N)) \end{pmatrix} \tag{24}$$

$$\mathbf{H}_b[m] = \begin{pmatrix} \sin(w_0 n) & \sin(2w_0 n) & \cdots & \sin(N_h w_0 n) \\ \sin(w_0(n-1)) & \sin(2w_0(n-1)) & \cdots & \sin(N_h w_0(n-1)) \\ \cdots & \cdots & \cdots & \cdots \\ \sin(w_0(n-N)) & \sin(2w_0(n-N)) & \cdots & \sin(N_h w_0(n-N)) \end{pmatrix} \tag{25}$$

In this kind of technique, the number of harmonic has to be known a priori. Otherwise, the performance can be considered reduced, also, the computational complexity is higher due to the matrix operations.

5. Methods for estimating time-varying harmonics

Methods for estimating time-varying harmonics consider that not only the amplitudes and phases of harmonics change, but also the fundamental frequency, and consequently, the harmonics frequencies. Thus, the frequency estimation is generally required to improve the algorithms.

5.1. Discrete fourier transform with sampling frequency control

The main weakness of the DFT is to estimate the harmonics when the sampling frequency is not synchronous with the fundamental frequency. In order to guarantee this synchronism we can control the sampling frequency as shown in Fig. 1. This method reduces and can also eliminate the errors caused by the mismatch between the fundamental frequency and the sampling frequency, however, it requires a robust and controllable ADC converter and a frequency estimation algorithm. As a result, its use is not recommended.

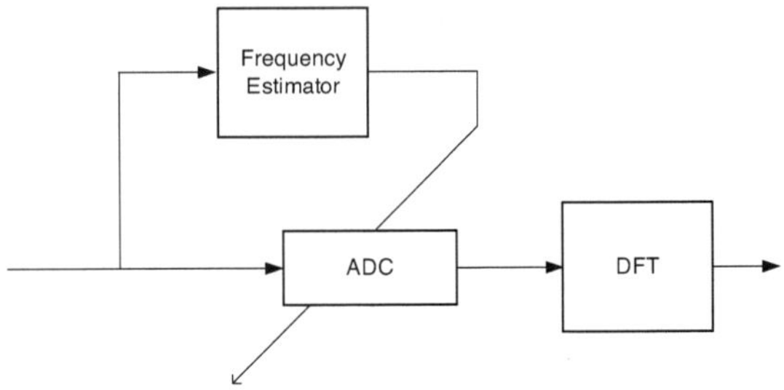

Figure 1. DFT with control of the ADC sampling frequency

5.2. Discrete fourier transform with signal resample

An alternative to the problem of synchronization of the sampling frequency when the sampling frequency is constant and not controllable is resampling the original signal before the harmonic estimation with the DFT. The drawback of this approach is the high computation complexity required by the resample process. Also, a frequency estimation technique is required to control the resampling process. Fig. 2 shows the block diagram of this strategy.

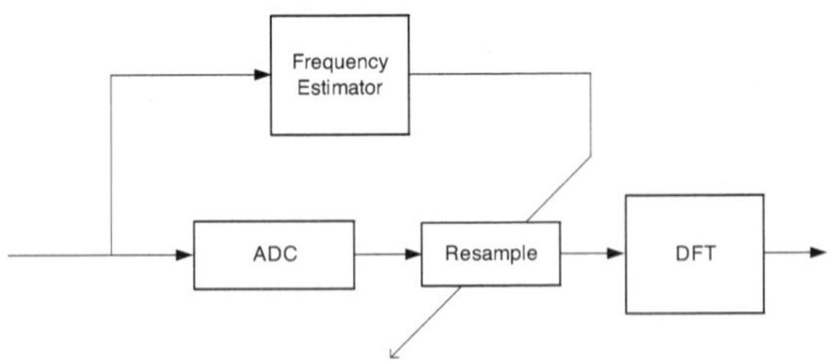

Figure 2. DFT with signal resample

5.3. Discrete fourier with window

An interesting way of improving the DFT algorithm is using a window in each block of data before the evaluation of the DFT. The windowing of the data can deal with the spectral leakage of the DFT caused by the frequency deviation by adding some computational burden to the algorithm. Some windows generally used are the triangular and Hanning. The coefficients of a Hanning window are computed from the following equation:

$$w[n] = 0.5\cos\left(1 - \cos\left(2\pi\frac{n}{N+1}\right)\right). \tag{26}$$

The triangular window has its coefficient given by

$$w[n] = \begin{cases} \frac{2n}{N}, & \text{if } 1 \le n \le \frac{N}{2}; \\ \frac{2(N-n)}{N}, & \text{if } \frac{N}{2} \le n \le N-1; \end{cases} \tag{27}$$

for N even, and

$$w[n] = \begin{cases} \frac{2n}{N-1}, & \text{if } 1 \le n \le \frac{N}{2}; \\ \frac{2(N-n)}{N-1}, & \text{if } \frac{N+1}{2} \le n \le N-1; \end{cases} \tag{28}$$

for N odd.

5.4. Demodulation

An interesting method based on demodulation technique for estimating time-varying harmonics is presented in [16]. This technique can provide very accurate estimates with a reasonable computational complexity.

The block diagram of the demodulation technique is depicted in Fig. 3. The LP blocks implement identical low-pass filters and the blocks COS and SIN implement the demodulation signals expressed by

$$d_{c_k}[n] = \cos(kw_0 n + \varphi_k[n]) \tag{29}$$

and

$$d_{s_k}[n] = \sin(kw_0 n + \varphi_k[n]), \tag{30}$$

respectively. The term $\varphi_k[n]$ control the instantaneous frequency of the demodulation signals (29)-(30). It is evaluated by

$$\varphi_k[n] = \varphi_k[n-1] + \frac{k}{f_s}(\psi_1[n] - \Omega_0), \tag{31}$$

where $\psi_1[n]$ is the estimated fundamental frequency in rad.

The blocks AMP and PHAS implement, respectively, the expressions

$$\hat{A}_k[n] = 2\sqrt{y_{cc_k}^2[n] + y_{ss_k}^2[n]} \tag{32}$$

and

$$\hat{\phi}_k[n] = -\arctan\left(\frac{y_{ss_k}[n]}{y_{cc_k}[n]}\right), \tag{33}$$

respectively, where $d_{c_k}[n]$ is the output of the low-pass filter at the top and $d_{s_k}[n]$ is the output of the low-pass filter at the bottom in Fig. 3.

In this technique, it is applied a approach to control the demodulation signals (blocks COS and SIN) and the frequency response of the low-pass filters (blocks LP) by the power frequency estimate, which is implemented by the block FREQ. The low pass filters are finite impulse response (FIR) filters which are controlled by a frequency estimator.

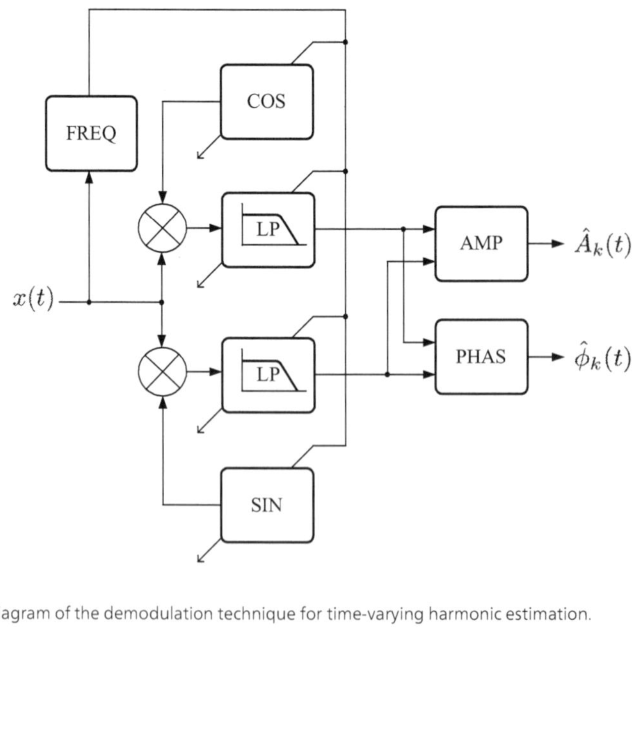

Figure 3. Block diagram of the demodulation technique for time-varying harmonic estimation.

5.5. Non linear least squares

The nonlinear least squares (NLS) uses the same expressions of the linear least squares for evaluate the harmonics [18]. The advantage of this approach is the improving of the estimates related to the linear version, however, the additional searching of the optimal frequency introduces additional delay in the technique and computational burden.

The NLS algorithm test several values of w_0 near its nominal value in order to minimize the euclidian norm of the following vector:

$$\mathbf{e}[m] = (\mathbf{I} - \mathbf{H}(\mathbf{H}^T[m]\mathbf{H}[m])^{-1}\mathbf{H}^T[m])\mathbf{y}[m] \tag{34}$$

Thus, with the optimal w_0, the harmonic parameters are evaluates as presented in section 4.4.

6. Performance analysis

In order to analyze the performance of the described techniques the following signal is considered:

$$\begin{aligned} x[n] = A_1[n]\cos(w_0 n + \phi_1[n]) + A_3[n]\cos(3w_0 n + \phi_3[n]) \\ + A_5[n]\cos(5w_0 n + \phi_5[n]) + A_7[n]\cos(7w_0 n + \phi_7[n]) \\ + A_9[n]\cos(9w_0 n + \phi_9[n]) + A_{11}[n]\cos(11w_0 n + \phi_{11}[n]) \\ + A_{13}[n]\cos(13w_0 n + \phi_{13}[n]) + A_{15}[n]\cos(15w_0 n + \phi_{15}[n]) \\ + A_{17}[n]\cos(17w_0 n + \phi_{17}[n]) + A_{19}[n]\cos(19w_0 n + \phi_{19}[n]) \\ + A_{21}[n]\cos(21w_0 n + \phi_{21}[n]) + A_{23}[n]\cos(23w_0 n + \phi_{23}[n]) \\ + A_{25}[n]\cos(25w_0 n + \phi_{25}[n]) + v[n], \end{aligned} \tag{35}$$

where $v[n] \sim \mathrm{N}(0, \sigma^2)$ is a white zero-mean Gaussian noise so that the signal-to-noise ratio (SNR) between the fundamental component and the additive noise is 60 dB (it should be noted that the SNR of the signal obtained from a power system usually ranges between 50 and 70 dB [19]).

Fig. 4 shows time estimations of the amplitude of the 3rd harmonic considering a 50% drop in the amplitude of signal when the fundamental frequency is equal to 60 Hz. Estimation delays of 2 cycles of the fundamental component are noted because the twocycles version of each technique was considered. However, when the fundamental frequency is set to 60.5 Hz such techniques exhibit significant errors in the estimates (time variations), as can be seen in Fig. 5. Otherwise, the time-varying techniques significantly improve the estimates. These results are depicted in Fig. 6. Table 1 shows the maximum of the absolute instantaneous error of all techniques after convergence of the algorithms (after the 50% drop in the amplitude of signal).

The best results are achieved with the demodulation and NLS techniques. Considering only these last two techniques, the errors were evaluated when the fundamental frequency varies between 59.5 Hz to 60.5 Hz for the 25th harmonic (See Fig. 7). Also, it is important to note that the improvement achieved by the DFT with hanning and triangular windows is significant compared with the standard DFT method as can be seen by Figs. 5 and 6. In order to better show this improvement, the errors, considering the DFT, DFT with triangular window and DFT with hanning window, were evaluated when the fundamental frequency varies between 59.5 Hz and 60.5 Hz for the 25th harmonic (Fig. 8).

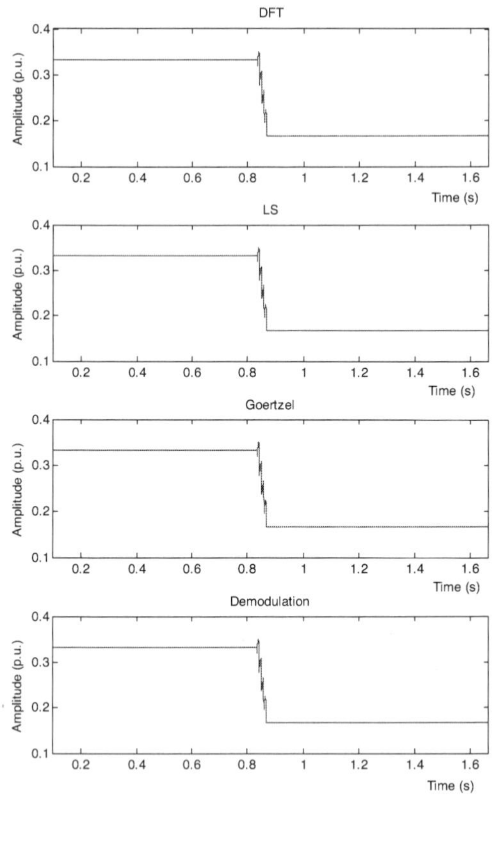

Figure 4. Estimation performance for the signal given by equation (35) in the case of a 50% drop in its amplitude for the *3rd* harmonic when the fundamental frequency is set to be 60 Hz considering the steady-state methods.

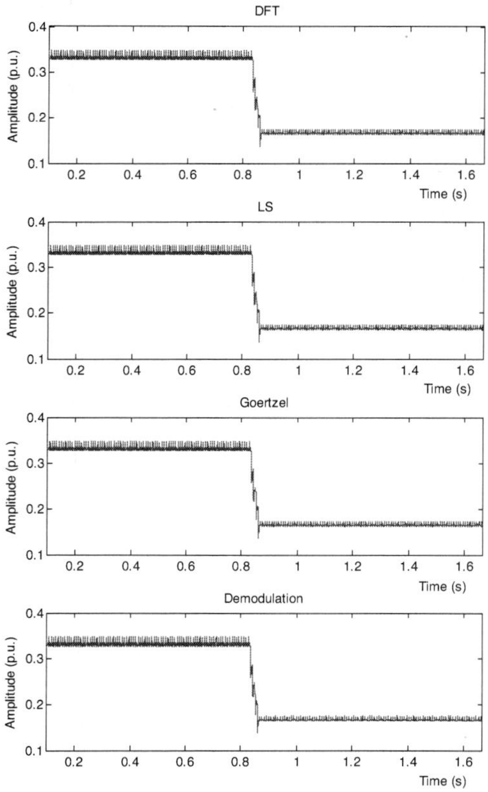

Figure 5. Estimation performance for the signal given by equation (35) in the case of a 50% drop in its amplitude for the 3*rd* harmonic when the fundamental frequency is set to be 60.5 Hz considering the steady-state methods.

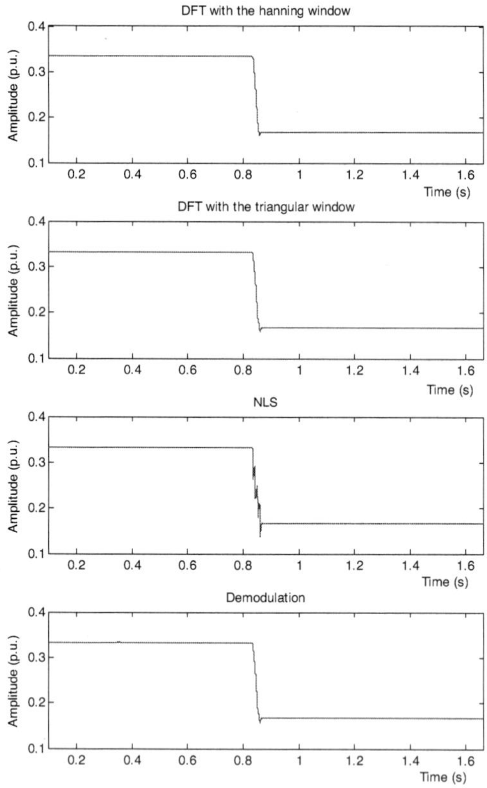

Figure 6. Estimation performance for the signal given by equation (35) in the case of a 50% drop in its amplitude for the 3*rd* harmonic when the fundamental frequency is set tobe 60.5 Hz considering the time-varying methods.

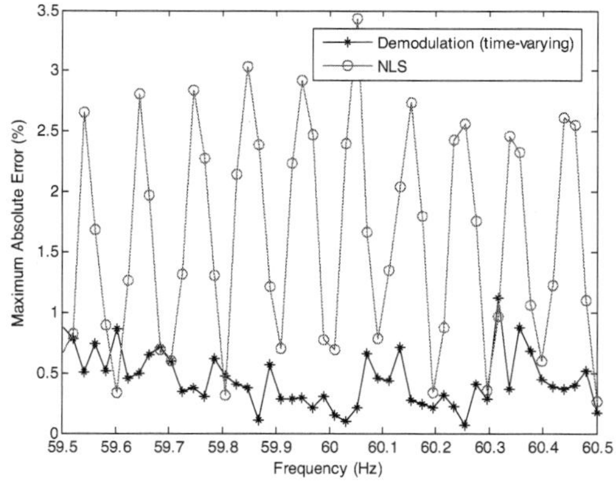

Figure 7. Maximum of the absolute instantaneous amplitude for the 25th harmonic when the fundamental frequency varies between 59.5 Hz to 60.5 Hz considering the NLS and Demodulaton method.

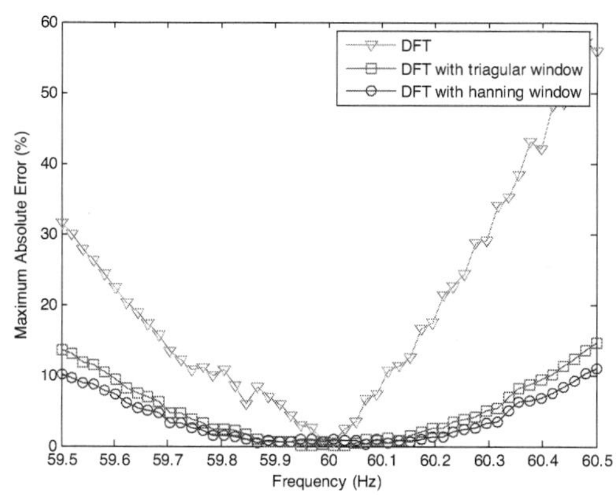

Figure 8. Maximum of the absolute instantaneous amplitude for the 25th harmonic when the fundamental frequency varies between 59.5 Hz to 60.5 Hz considering the DFT, DFT with triangular window and DFT with hanning window.

Technique	Maximum Instantaneous Error
DFT	4.7655
Goertzel	4.7655
LS	4.7655

Technique	Maximum Instantaneous Error
Demodulation (steady-state)	4.7655
DFT with hanning window	0.7479
DFT with triangular window	0.2296
NLS	0.0842
Demodulation (time-varying)	0.0660

Table 1. Maximum of the absolute instantaneous error of all techniques after convergence of the algorithms.

7. What is next and needed?

Most existing end-user equipment (computer, television, lamps, etc) emit almost exclusively at the lower odd integer harmonics, but there are indications that modern devices including certain types of distributed generators emit a broadband spectrum [11–14]. The measurement of these low levels of harmonics at higher frequencies will be more difficult than for the existing situation with higher levels and lower frequencies. This might require the development of new measurement techniques including a closer look at the frequency response of existing instrument transformers. Consequently, harmonic estimation of higher order harmonics will be very important and needed. In this case the sampling frequency should be increased to satisfy the Nyquist criterion and faster *analog to digital converter* (ADC) must be used to deal with this requirement.

Power electronic based photovoltaic solar and wind energy equipment may emit disturbances causing voltage fluctuations and unbalance. These types of electric sources will have large presence in the future grids very large. In order to deal with this new scenario, the harmonic estimation algorithms must be immune to higher voltage fluctuations and interharmonics.

An important issue associated with smart grid in regarding to harmonic estimation is the real time estimation of several harmonics instantaneously, including higher-order harmonics. Higher-order harmonics will be more and more important to estimate due its influence in sensitive electronics devices. Also, the dynamic and diversity of smart grid will demand different set of techniques to analyze the behavior of the time-varying harmonics. For deal with these issues, the use of reconfigurable hardware that allow the exchange of features between existing monitoring devices is of ultimate importance.

8. Concluding remarks

Several methods and techniques were developed so far for estimating steady-state and timevarying harmonics. Although several techniques can deal with time-varying harmonics, the implementation of them is incipient. However, the needs and demands related to *smart*

http://dx.doi.org/10.5772/53816

grids is pushing forward the development of new techniques as well as discussion of new measurement standards for time-varying harmonics.

Although *smart grids* offers the opportunity to improve the quality, efficiency and reliability for power systems, the increase of disturbances levels is inevitable. Thus, new challenges related to the power quality will be introduced.

Author details

Cristiano A. G. Marques[1], Moisés V. Ribeiro[1], Carlos A. Duque[1] and Eduardo A. B. da Silva[2]

1 Federal University of Juiz de Fora, Brazil

2 Federal University of Rio de Janeiro, Brazil

References

[1] Akagi, H. New trends in active filters for power conditioning, *IIEEE Trans. Ind. Appl.*, November (1996). , 32(6), 1312-1322.

[2] Watson, N. R, & Arrigala, J. Harmonics in large systems, *Electric Power System Research*, (2003). , 66, 15-29.

[3] Masoum, M. A. S, Moses, P. S, & Masoum, A. S. Derating of Asymmetric Three-Phase Transformers Serving Unbalanced Nonlinear Loads, *IEEE Trans. Power Delivery*, October (2008). , 23(4), 2033-2041.

[4] Lai, L. L, Chan, W. L, Tse, C. T, & So, A. T. P. Real-time frequency and harmonic evaluation using artificial neural networks, *IEEE Trans. on Power Delivery*, Jan. (1999). , 14(1), 52-59.

[5] Ribeiro, P. Time-Varying Waveform Distortions in Power Systems, Wiley-IEEE Press, (2009).

[6] Begovi, M. M, Djuri, c a. n. d P. M, Dunlap, c a. n. d S, & Phadke, A. G. Frequency tracking in power networks in the presence of harmonics, *IEEE Trans. on Power Delivery*, Apr. (1993). , 8(2), 480-486.

[7] Boashash, B. Estimating and interpreting the instantaneous frequency of a signal. I. Fundamentals, *Proceedings of the IEEE*, Apr (1992). , 80(4), 520-538.

[8] George, T. A. Harmonic power flow determination using the fast Fourier transform, *IEEE Trans. on Power Delivery*, Apr. (1991). , 2(2), 530-535.

[9] Thorp, J. S, Phadke, A. G, & Karimi, K. J. Real-time voltage phasor measurements for static-state estimation, *IEEE Trans. Power App. Syst.*, Nov. (1985). , PAS-104(11), 3099-3106.

[10] Computer Relaying for Power SystemsNew York: John Wiley and Sons, (1988).

[11] Bollen, M. H. J, Zhong, J, Zavoda, F, Meyer, J, Mceachern, A, & Opez, F. C. L. Power Quality aspects of Smart Grids, International Conference on Renewable Energies and Power Quality, Granada (Spain), 23th to 25th March, (2010).

[12] Bollen, M. H. J, Ribeiro, P. F, Larsson, E. O. A, & Lundmark, C. M. Limits for voltage distortion in the frequency range 2-9 kHz, *IEEE Transactions on Power Delivery*, July (2008). , 23(3), 1481-1487.

[13] Tentzerakis, S. T, & Papathanassiou, S. A. An Investigation of the Harmonic Emissions of Wind Turbines, *IEEE Trans. Energy Convers.*, March (2007). , 22(1), 150-158.

[14] Papathanassiou, S. A, & Papadopoulos, M. P. Harmonic analysis in a power system with wind generation, *IEEE Trans. Power Delivery*, pgs. 2006-2016, October 2006. Instantaneous phase tracking in power networks by demodulation, *IEEE Trans. On Instrumentation and Measurement*, vol. 41, no. 6, December (1992). , 21(4), 963-967.

[15] Djuric, P. M, Begovic, M. M, & Doroslova, M. ki, Instantaneous phase tracking in power networks by demodulation, *IEEE Trans. on Instrumentation and Measurement*, Dec. (1992). , 41(6), 963-967.

[16] Marques, C. A. G, Ribeiro, M. V, Duque, C. A, & Ribeiro, P. F. and E. A. B. da Silva, A Controlled Filtering Method for Estimating Harmonics of Off-Nominal Frequencies, *IEEE Trans. On Smart Grids*, March (2012). , 3(1), 38-49.

[17] Goertzel, G. An algorithm for the evaluation of finite trigonometric series, *The American Mathematical Monthly*, (1958). , 65(1), 34-35.

Chapter 10

Evaluation of the Distortion and Unbalance Emission Levels in Electric Networks

Patricio Salmerón Revuelta and
Alejandro Pérez Vallés

Additional information is available at the end of the chapter

http://dx.doi.org/10.5772/53373

1. Introduction

Electrical disturbances have important economic consequences for the consumer and the utilities, a fact which increases from the new electricity regulatory framework. Thus, Electric Power Quality, EPQ, has now become a priority within the field of electrical engineering. Power-quality deterioration is due to transient disturbances (voltage sags, voltage swells, impulses, etc.) and steady state disturbances (harmonic distortion, unbalance, and flicker). Among these are the so-called periodic disturbances such as distortion of the voltage and current waveforms, or unbalanced three-phase systems, [1-2]. This chapter is focused specifically on harmonic and unbalance phenomena.

These quality problems have entailed the need for measuring equipment to monitor the installation at the user side such as electric power quality analyzers, [3-7]. This equipment has a great number of quantities available relating to the harmonic distortion for each phase. Each measurement is usually composed of the voltage and current total harmonic distortion (THD) index, and RMS values of the total waveforms and the fundamental component. One problem with characterization of these data is representation of harmonic distortion when the installation has very different distortion levels in each phase. There are two choices: to consider the distortion level of each phase separately, managing three times more information, or to characterize the installation distortion using a global parameter which considers the whole three-phase system, [5, 7]. The first approach is more interesting for analyzing the cause of the problems in an installation. The second is more

INTECH
open science | open minds

© 2013 Revuelta and Vallés; licensee InTech. This is a paper distributed under the terms of the Creative Commons Attribution License (http://creativecommons.org/licenses/by/3.0), which permits unrestricted use, distribution, and reproduction in any medium, provided the original work is properly cited.

suited to characterizing the distortion level of the whole installation. This point of view is adopted in this chapter.

On the other hand, it is necessary to know the responsible for the production of periodic perturbations. That is, first, identification of distortion sources, and secondly, unbalances emission. Regarding on localization of sources producing distortion two different approaches can be distinguished: a) those based on measurements taken on the point of common coupling, PCC, [9-21], b) those based one measurement taken and processed simultaneously on different metering sections placed on the line connected to the same PCC, [22-24]. This chapter is developed within the first group. Thus, the objective to localization of sources producing distortion is to measure instantaneous values of current and voltage in each of the branches of the PCC, and from these measurements, establish which consumers are responsible for generating the distortion and quantify the distortion generated by each consumer. Regarding the issue of unbalance emission, it is measurement and evaluating the negative and zero sequence currents injected by loads of unbalanced structure at the PCC.

Originally, the problem was analyzed through the harmonic power sign with the objective of knowing the sense of harmonic power flow between source and one load in distorted systems. Nowadays, it has already been established in the technical literature that an analysis of this kind does not solve the problem, [10-11]. Recently, new indices have been introduced to evaluate a specific consumer distortion and unbalance level, [12, 15, 19]. In this chapter, a comparative analysis of these indices is carried out, having as reference different practical cases. The results obtained show that, in fact, these indices can help to valuate the periodic perturbation, although none of them solve the question definitively.

More recently, it has been found that the measure of quality indices of EPQ presents additional difficulties in the presence of capacitors. The capacitor do not produce harmonic but their presence contributes to the amplification of the harmonics existing in the electrical network. The capacitor behaviour makes that the indices proposed up now to identify distortion sources fail in the presence of this element, [25]. However, in [21] has proposed a new method that resolves the situation.

The chapter is structured as follows. Second section begins with the definitions of effective voltage and current contained in the IEEE Standard 1459 and continuous with the most suited distortion and unbalance indices for three-phase four-wire networks for characterizing the quality of the waveforms present in the PCC. In third section presents the theoretical basis of the main methods to identify sources of unbalance and distortion: harmonic powers, conforming and nonconforming currents, balanced linear currents and unbalanced non-linear currents. A comparative analysis of different PQ indices based on digital simulations has been performed. The fourth section analyzes the problems that present PQ indices in the presence of installed capacitor banks for power factor correction. The fifth section describes an experimental setup for measuring the PQ indices of a three-phase nonlinear load connected to the mains. Finally, a discussion of the results and conclusions are extracted.

2. Assessment of harmonic distortion and unbalance in power systems

In any system of measurement for power system, is usual to introduce a number of quantities characterizing the waveforms of voltage and current. This requiere to define the effective values of voltage, V_e, and current, I_e; here the definitions adopted in the IEEE Standard 1459-2000 (Std 1459), [26], are introduced. The definitions of these quantities are directly related to the definition of apparent power. Thus, in the 1459 Std apparent power is the maximum power that can be transmitted under ideal conditions (sinusoidal single phase or balanced three-phase sinusoidal systems) with the same impact of voltage (on the insulation and on the no-load losses) and the same impact of current (or line losses) from the PCC on the network. From this definition equivalent values of voltage and current which characterize the load impact on the power system are deduced.

2.1. Preliminary definitions

The explicit expression of the apparent power depends on how are characterized these voltage and current impacts. To determine the apparent power, we introduce an equivalent voltage and an equivalent current of a balanced system of positive sequence to produce the same impact on the network voltages and currents present in the system. In the following are determined equivalent quantities, effective voltage, V_e, and effective current, I_e.

A three-phase system consisting of an unbalanced load is supplied by a four-wire system wherein each of the lines has a resistance r and the neutral conductor has a resistance r_n. The effective current value, I_e, is

$$I_e = \sqrt{\frac{1}{3}\left(I_a^2 + I_b^2 + I_c^2 + \rho I_n^2\right)} \quad ; \quad \rho = \frac{r_n}{r} \tag{1}$$

The 1459 Std is ϱ = 1. On the premises of medium and low voltage typical ϱ = 0.2 - 4. Today, digital instrumentation can develop equipment that can adjust ϱ for any default.

The next step is to find an effective voltage V_e. This will take into account the non load power losses in magnetic cores of transformers and insulation 'upstream' of the load. Standard means that losses (non load losses voltage-dependent), P_Y, which are due to the line-neutral voltages and losses (non load losses voltage-dependent), P_Δ, which are due to the line-line voltages are equal. Thus, the expression for V_e:

$$V_e = \sqrt{\frac{3\left(V_{an}^2 + V_{bn}^2 + V_{cn}^2\right) + \xi\left(V_{ab}^2 + V_{bc}^2 + V_{ca}^2\right)}{9(1+\xi)}} \tag{2}$$

Where ξ is the power ratio,

$$\xi = \frac{P_{\Delta}}{P_{Y}} \tag{3}$$

Later, the concept of losses which depend on the voltage was abandoned in favor of whichever of loads that consume an equivalent active power, so that to determine the equivalent voltage V_e is assumed that the load is formed by a group resistors connected in Y, which consume a power P_Y and a group connected in Δ, which consume a power P_{Δ}. In any case the St 1459 considers $\xi = 1$. To a three-wire three-phase system where $I_n = 0$, the standard recommended simplified expressions,

$$V_e = \sqrt{\frac{V_{ab}^2 + V_{bc}^2 + V_{ca}^2}{9}} \quad ; \quad I_e = \sqrt{\frac{I_a^2 + I_b^2 + I_c^2}{3}} \tag{4}$$

referred to as the Buchholz-Goodhue, and original works proposed by the IEEE working group on nonsinusoidal situations.

As already stated above, the definition Std 1459 takes as apparent power or effective apparent power of the system, S_e, that originally suggested by F. Buchholz in 1922 and clarified by W. M. Goodhue in 1933,

$$S_e = 3 V_e I_e \tag{5}$$

Similarly for a total active power consumed by a load, P, the power factor is defined as,

$$PF = \frac{P}{S_e} \tag{6}$$

relating the minimum power loss and the current power loss.

The RMS values of the magnitudes of phase voltage and line current are determined as,

$$\begin{aligned} V_j^2 &= V_{j1}^2 + \sum_{\forall h \neq 1} V_{jh}^2 \\ I_j^2 &= I_{j1}^2 + \sum_{\forall h \neq 1} I_{jh}^2 \quad ; \quad j = a,b,c \end{aligned} \tag{7}$$

Likewise, it is desirable to split the effective voltage and current into two terms, one for the fundamental harmonic and the other for the remainder of the harmonics,

$$V_e^2 = V_{e1}^2 + V_{eH}^2 \quad ; \quad I_e^2 = I_{e1}^2 + I_{eH}^2 \tag{8}$$

where the subscript '1 'refers to the RMS values of the fundamental component,

$$V_{e1} = \sqrt{\frac{3\left(V_{an1}^2 + V_{bn1}^2 + V_{cn1}^2\right) + \left(V_{ab1}^2 + V_{bc1}^2 + V_{ca1}^2\right)}{18}} \quad ; \quad I_{e1} = \sqrt{\frac{I_{a1}^2 + I_{b1}^2 + I_{c1}^2 + I_{n1}}{3}} \tag{9}$$

and the subscript 'H' refers to the whole of the harmonic components other than the fundamental,

$$V_{eH} = \sqrt{\sum_{h \neq 1} \frac{3\left(V_{anh}^2 + V_{bnh}^2 + V_{cnh}^2\right) + \left(V_{abh}^2 + V_{bch}^2 + V_{cah}^2\right)}{18}} ; I_{eH} = \sqrt{\sum_{h \neq 1} \left(\frac{I_{ah}^2 + I_{bh}^2 + I_{ch}^2 + I_{nh}^2}{3}\right)} \tag{10}$$

2.2. Power quality indexes

The use of effective voltage and current values helps characterize the harmonic content of the three-phase system by using the so-called total harmonic distortion (THD) rates of voltage and current, [4]. Thus, for any phase ϕ (a, b, c) the voltage total harmonic distortion is,

$$VTHD_{\phi} = \sqrt{\frac{V_{\phi}^2 - V_{\phi_1}^2}{V_{\phi_1}^2}} = \sqrt{\frac{V_{\phi_2}^2 + V_{\phi_3}^2 + V_{\phi_4}^2 + \ldots + V_{\phi_N}^2}{V_{\phi_1}^2}} \tag{11}$$

and the current total harmonic distortion is,

$$ITHD_{\phi} = \sqrt{\frac{I_{\phi}^2 - I_{\phi_1}^2}{I_{\phi_1}^2}} = \sqrt{\frac{I_{\phi_2}^2 + I_{\phi_3}^2 + I_{\phi_4}^2 + \ldots + I_{\phi_N}^2}{I_{\phi_1}^2}} \tag{12}$$

where the second subscript represents the order of the harmonic. These indices THD, harmonic content of the waveform are compared with the fundamental harmonic. This is a definition in the standard IEC 61000 and is common in commercial network analyzers. However, a second definition is introduced where the harmonic content is compared to the RMS value of the waveform, [8]. Here we refer to this factor as the rate of total demand distortion, TDD. For voltage,

$$VTDD_{\phi}=\sqrt{\frac{V_{\phi}^2-V_{\phi_1}^2}{V_{\phi}^2}}=\sqrt{\frac{V_{\phi_2}^2+V_{\phi_3}^2+V_{\phi_4}^2+\ldots+V_{\phi_N}^2}{V_{\phi}^2}} \tag{13}$$

and to the current,

$$ITDD_{\phi}=\sqrt{\frac{I_{\phi}^2-I_{\phi_1}^2}{I_{\phi}^2}}=\sqrt{\frac{I_{\phi_2}^2+I_{\phi_3}^2+I_{\phi_4}^2+\ldots+I_{\phi_N}^2}{I_{\phi}^2}} \tag{14}$$

The values provided by THD and TDD factors are often very similar to waveforms with low distortion, yet their differences are more significant for high distortions. THD factors indicate a very high value, even infinite, when the waveform has a very small or zero value of the fundamental harmonic. Moreover, TDD values are always below 100% and this is somewhat larger errors in measurement instrumentation versus normal situations THD distortion.

The distortion characterization of the global systems requires extending the definition of harmonic distortion rates to three-phase systems. The rate of three-phase voltage total harmonic distortion, $VTHD_{3\phi}$, and the rate of three-phase current, $ITHD_{3\phi}$, is defined as,

$$ITHD_{3\phi}=\sqrt{\frac{I_e^2-I_{e1}^2}{I_{e1}^2}}\ ;VTHD_{3\phi}=\sqrt{\frac{V_e^2-V_{e1}^2}{V_{e1}^2}} \tag{15}$$

Similarly, is possible to define the three-phase total demand distortion of voltage, $VTDD_{3\phi}$, and the three-phase total demand distortion of current, $ITDD_{3\phi}$,

$$ITDD_{3\phi}=\sqrt{\frac{I_e^2-I_{e1}^2}{I_e^2}}\ ;VTDD_{3\phi}=\sqrt{\frac{V_e^2-V_{e1}^2}{V_e^2}} \tag{16}$$

According to their definition, the three-phase total demand distortion indices measure the lack of conformity of the line voltage and current waveforms with respect to sinusoidal waveforms, because of the harmonic content; without considering possible unbalances.

It is important to point out that these new indices may be calculated in a simple way with the harmonic measurements available from a commercial device. Usually, a power quality Analyzer supplies THD values, RMS values and fundamental component RMS values corresponding to each harmonic in each phase. So, the defined indices can be calculated for the measurements of each phase. For instance, the three-phase distortion indices can be calculated by means of the following expressions: to the voltage,

$$VTDD_{3\phi} = \sqrt{VTHD_a^2 \frac{V_{a_1}^2}{V_e^2} + VTHD_b^2 \frac{V_{b_1}^2}{V_e^2} + VTHD_c^2 \frac{V_{c_1}^2}{V_e^2} + VTHD_n^2 \frac{V_{n_1}^2}{V_e^2}} \tag{17}$$

and to the current

$$ITDD_{3\phi} = \sqrt{ITHD_a^2 \frac{I_{a_1}^2}{I_e^2} + ITHD_b^2 \frac{I_{b_1}^2}{I_e^2} + ITHD_c^2 \frac{I_{c_1}^2}{I_e^2} + ITHD_n^2 \frac{I_{n_1}^2}{I_e^2}} \tag{18}$$

To evaluate the unbalance conditions, a reference waveform may be established [2, 23]. In this paper a reference voltage waveform and a reference current waveform are defined in a similar way: they are sinusoidal and positive sequence waveforms whose RMS values, V_{e1+} and I_{e1+}, are the effective voltage and current defined in similar form as (16), respectively.

$$VTDD_{3\phi}^{+} = \sqrt{\frac{V_e^2 - V_{e1+}^2}{V_e^2}} \quad ; \quad ITDD_{3\phi}^{+} = \sqrt{\frac{I_e^2 - I_{e1+}^2}{I_e^2}} \tag{19}$$

These new factors act as indices of non-conformity of voltage and current waveforms with respect to sinusoidal waveforms of positive phase sequence of voltage and current, respectively. A comparison between the values obtained by applying the expressions (16) and (19) would determine whether the periodic disturbances are mainly due to the presence of distortion or the presence of unbalance. Finally, in [8] a weighted distortion index has been proposed. Their calculation is based on the RMS values of each phase. However, usually, the three-phase power system is not balanced. So, the power transferred by each phase differs considerably from the other phases. It affects the information supplied by the current indices in a relevant manner. For that reason, the definition of a new distortion index which takes this new effect into account may be interesting.

However, the indices defined in (17), (18), and (19) are not appropriated for establishing responsibility, between customer and supplier, for the lack of electric power quality.

3. Localization of sources producing distortion

The identification of the loads they produce distortion has been addressed by analyzing the direction of active harmonic power flow. Therefore, this section begins with the analysis of active harmonic power. Other indices are introduced with the aim of overcoming the limitations of the method of harmonic powers. Several practical cases using a simulation environment will be studied and a comparative analysis is performed.

3.1. Analisys of active harmonic powers

In a first approximation to the problem, balanced three-phase system are considered with several non-linear loads connected to PCC. Figure 1 shows the equivalent single-phase circuit, which includes n+ 1 branch, n corresponding to the loads and 1 to the supply. The grid is represented by the Thevenin circuit constituted by a voltage source Vs in series with inductive impedance Zs. Certainly, PCC in figure 1 represents a bus of a distribution system.

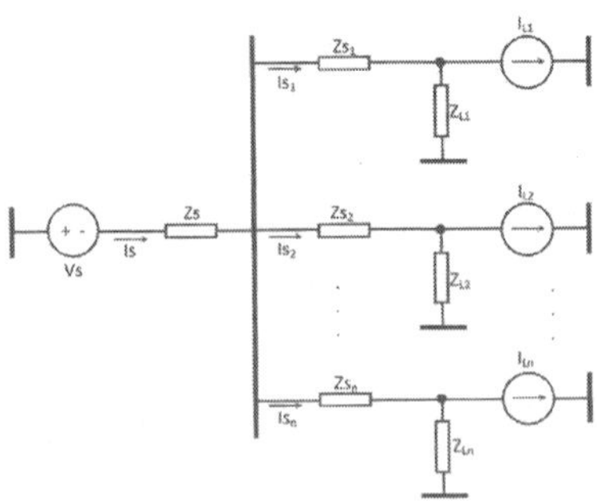

Figure 1. Non linear balanced three-phase system equivalent circuit.

Consumers are supplied by single impedances, where each load is modelled as impedance Z_{Lj} connected in parallel to a harmonic source I_{Lj}, for j branch. Due to the system topology, it is necessary the measurement of voltage and current in each branch connected to the PCC. From these measurements, the consumers responsible of distortion are established. Besides, distortion generated by each one is quantified. Thus, figure 1 can be reduced to figure 2.

Consumers are supplied by single impedances, where each load is modelled as impedance Z_{Lj} connected in parallel to a harmonic source I_{Lj}, for j branch. Due to the system topology, it is necessary the measurement of voltage and current in each branch connected to the PCC. From these measurements, the consumers responsible of distortion are established. Besides, distortion generated by each one is quantified. Thus, figure 1 can be reduced to figure 2.

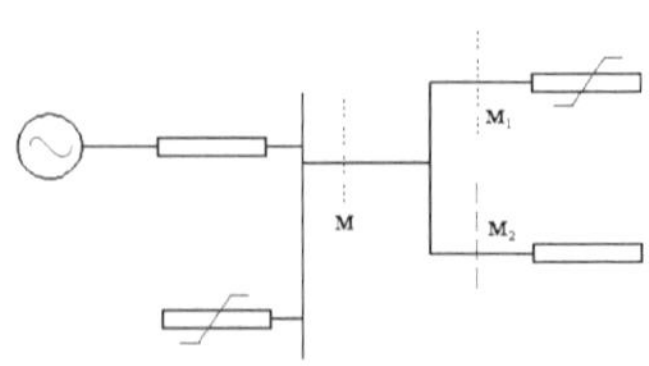

Figure 2. Sections of measurements in a power system simplified diagram with two loads connected to PCC.

From now on, diagram presented in figure 2 is considered as reference. There is a PCC where a sinusoidal voltage source is connected through source impedance. Besides, several linear and nonlinear loads are connected. Sections of measurements on M, M1 and M2 are considered.

A non-linear or time-varying load (measuring section M1) and linear load (measuring section M2) are considered to M. In M, bus voltage is non-sinusoidal due to the simultaneous effects of distorted caused by non-linear loads upstream and downstream M. The analysis of power systems which include non-linear loads can be carried out two subsystems. There is a power flow between them corresponding to each harmonic, [1, 9]. The system analysis in frequency domain needs the consideration of an equivalent circuit for each relevant harmonic. Thus, the system is simplified through a Thevenin equivalent circuit from the measuring section: a nonsinusoidal Thevenin voltage and frequency dependent Thevenin equivalent impedance.

In general, there are harmonics common to Thevenin voltage and current incoming non-linear loads. So, for each common harmonic there is a harmonic active power value corresponding to grid and consumer. Its sign depends on the subsystem responsible of the prevalent contribution. According to this, a harmonic analysis can not identify the distortion source from the active powers addition, but it only can identify the source which presents the prevalent contribution. It is because both harmonic currents as harmonic powers are due to the addition of two opposite contributions. Nevertheless, harmonic active and reactive powers measurements were the approach mainly adopted to identify the pollution generated by the consumer to the supply waveform quality.

3.1.1. An index to identify distorted loads based in harmonic powers

The analysis presented in above section suggests the necessity of finding an index which allows the evaluation of distortion generated by a specific load in the supply voltage system. According to this approach, the index should be based on the measurements of harmonic active powers corresponding to each load is needed. Several proposals within this approach have appeared in last years. Among them, named harmonic phase index, ξ_{HPI}, [15], defined as follows. A 3n current vector I is introduced where n is the maximum harmonic order considered. Vector I is built with RMS values of each phase of each harmonic load current. This is broken in two components, I_S and I_L whose elements corresponding to each harmonic are defined as follows;

$$I_{Sk} = \begin{cases} 0 & si \quad P_k \leq 0 \\ I_k & si \quad P_k \succ 0 \end{cases}$$

$$I_{Lk} = \begin{cases} 0 & si \quad P_k \geq 0 \\ I_k & si \quad P_k \prec 0 \end{cases} \tag{20}$$

The harmonic phase index here introduced is lightly different from the presented in [15],

$$u(t)=\sum_{k=1}^{n}\sqrt{2}U_k\, sin\left(k\omega_1 t+\theta_k\right) \quad ; \quad i(t)=\sum_{k=1}^{n}\sqrt{2}I_k\, sin\left(k\omega_1 t+\phi_k\right) \tag{21}$$

This index has the following significant characteristics for the purpose for which it is introduced. First, it is defined from the ratio of current RMS values that are the actual cause of disturbances generated by loads in the grid. Second, different harmonic values are not added, but in quadratic sum way. It avoids mutual cancellation between different harmonics.

3.2. Conforming and non-conforming currents

In [12], authors distinguish two kinds of loads, conforming loads and non-conforming loads. A conforming load does not cause a change in voltage waveform distortion or in symmetry of the phases. Any other load that changes the voltage waveform or symmetry is a non-conforming load. Thus, from the harmonic point of view, the author considers that a conforming load presents a current collinear to the voltage. A typical load will be constituted by a conforming part and a non-conforming part which may be modelled in a simple way by means of two parallel elements. Current measured in the input *i*(t) will be the sum of currents incoming to the conforming part i_n(t) and the non-conforming part i_d(t). The conforming current is the part of the current that presents the same distortion level as supply voltage. The rest of current is the non wished part of current; the incoming in non-conforming load. The split of power flow is obtained from the components established by the current.

Voltage and current measured in load terminals are expressed as the addition of fundamental harmonic and other components multiple of fundamental.

$$\mathbf{Y}\left(jkw_1\right)=\left|\mathbf{Y}\left(j\omega_1\right)\right|\angle k(\varphi_1-\theta_1) \quad ; \quad \mathbf{Y}\left(j\omega_1\right)=\left|\frac{\mathbf{I}_1}{\mathbf{V}_1}\right|\angle(\varphi_1-\theta_1) \tag{22}$$

The conforming current presents the same variation as voltage waveform and its phase may be lower or higher. On the other hand, due to the fact that the load can not generate power at fundamental frequency, conforming current evolves the complete fundamental active and reactive powers. Conforming current at fundamental frequency I_{n1} is equal to the total current at fundamental frequency, I_1. The rest of components to the conforming current for different frequencies are proportional to the corresponding harmonic voltage components. The proportional constant is a complex ratio:

$$\begin{aligned} i_n(t)&=\sum_{k=1}^{n}\frac{I_1}{V_1}\sqrt{2}V_k\, sin\left(k\omega_1 t+\theta_k+k(\phi_1-\theta_1)\right) \\ i_d(t)&=i(t)-i_n(t) \end{aligned} \tag{23}$$

Expressions in time domain for each line current are:

$$NC=\frac{I_d}{I}\times 100(\%) \tag{24}$$

Where V_1 and I_1 are fundamental voltage and current RMS values in PCC, θ_1 and ϕ_1 are voltage and current fundamental phases, θ_k is the k order harmonic phase of voltage and i(t) is the load current. The index of distortion proposed en [12] is the non collinear index, NC,

$$\begin{aligned} |\mathbf{Z}_{1a}| &= \frac{V_{1a}}{I_{1a}} \\ \angle \mathbf{Z}_{1a} &= |\theta_{1a}-\phi_{1a}| = \varphi_{1a} \end{aligned} \tag{25}$$

where I_d is i_d(t) RMS value.

3.3. Linear and nonlinear current

The problem of separating the contributions to the distortion of the supply and the consumer was approached was reasoned in [19] as follows. The deterioration of EPQ due to harmonics can be caused simultaneously in several points of network. Voltage and current waveforms measured in PCC are due to the combined effect of several polluting equipment connected in different places in the network. It is possible the determination of the specific load contribution to harmonic distortion, whereas all other is considered in the supply side which includes the rest of loads. The first step is fixing the ideal load conditions. Any load which shows a linear and balanced behavior represents and ideal load condition. In fact, if specific load is balanced and linear, the supply is the only responsible of harmonic distortion in PCC. Initially, it is necessary the identification of waveform of incoming current if an equivalent linear and balanced load is presented instead of the actual incoming load current. This load can be defined as the linear load that requires a fundamental active power equal to the fundamental active power actually flow by PCC. It requires a distorted current but it is not responsible of the distortion.

In order to model the three-phase balanced linear ideal load, three identical RL branches have been considered. Through the evaluation of R, L parameters, the part of load that represents the equivalent ideal load can be identified. So, from the consumer side, it is always possible the estimation of the linear load. This part of the actual load requires a balanced and linear which constitute an ideal current and represents one component of the load current in PCC. If ideal current is almost total current, the load is not responsible of distortion problems.

Equivalent balanced and linear load parameters R and L are estimated in phase 'a' according to the procedure indicated as follow, (the same method can be applied to phases 'b' and 'c'). If phase 'a' equivalent linear impedance is named by Z_{1a}, the R-L series circuit at fundamental frequency is:

$$\begin{aligned} R_a &= |\mathbf{Z}_{1a}| \cos\varphi_{1a} \\ X_{1a} &= |\mathbf{Z}_{1a}| sen\varphi_{1a} \\ L_a &= \frac{X_{1a}}{2\pi f_1} \end{aligned} \tag{26}$$

where V_{1a} and I_{1a} are voltage and current RMS values of fundamental frequency, respectively, in PCC, and θ_{1a} and ϕ_{1a} are phase angle of those magnitudes. So,

$$X_{ka} = 2\pi k f_1 L_a \quad , \quad k = 1, 2, \cdots n \tag{27}$$

where X_{1a} represents reactance of R-L series combination at fundamental frequency, f_1, and R_a and L_a are the corresponding parameters. Reactance values to voltage harmonic components are,

$$\begin{aligned} |\mathbf{Z}_{\mathbf{ka}}| &= \sqrt{R_a^2 + X_{ka}^2} \\ \varphi_{ka} &= tag^{-1}\frac{X_{ka}}{R_a} \quad , \quad k=1, 2, \cdots n \end{aligned} \tag{28}$$

The skin effect is neglected, that is, the resistance is assumed to be constant for all frequencies, and hence,

$$i_{La}(t) = \sum_{k=1}^{n} \frac{V_{ka}}{|\mathbf{Z}_{\mathbf{ka}}|} \sqrt{2}\, sen\left(2\pi k f_1 t + (\theta_{ka} - \varphi_{ka})\right) \tag{29}$$

where n is the most significant order of harmonic. As a consequence, current in the ideal linear load in phase 'a' is,

$$i_{nLa}(t) = i_a(t) - i_{La}(t) \tag{30}$$

This is the equivalent linear current and the supply system is the only responsible of its distortion. The difference between measured current and calculated ideal current is defined as non-linear current:

$$NL = \frac{I_{nLa}}{I_a} \times 100(\%) \tag{31}$$

The non-linear current indicates how much the actual load current in phase 'a' differs from the ideal linear load in terms of harmonic distortion. Thus, an index of non-linear current is defined with the intention to provide reliable information about distortion caused by the consumer:

$$V_+ = \frac{1}{3}\left(V_a + aV_b + a^2 V_c\right) \tag{32}$$

where I_{nLa} and I_a are the RMS values corresponding to non-linear and actual current, respectively.

3.4. Practical cases: Digital simulations

Harmonic powers and indices for assessing the harmonic distortion introduced in the previous sections have been applied to the system shown in Figure 3. This is a balanced system consisting of two AC / DC converters controlled firing angles 25° (Rectifier 1) and 45° (Rectifier 2).

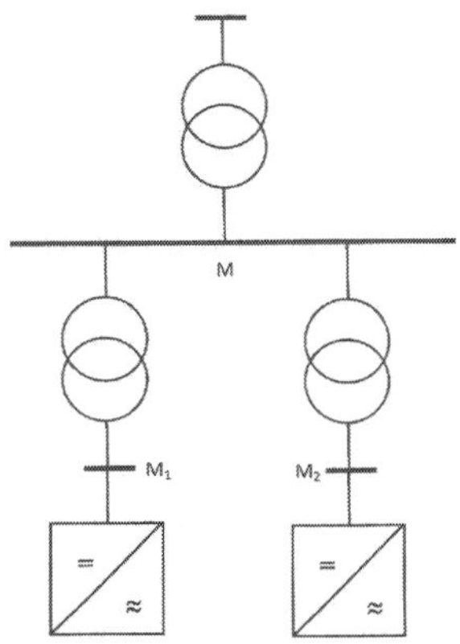

Figure 3. Circuit topology for the practical cases.

3.4.1. Case 1: Sinusoidal voltage

In the first situation has been considered a sinusoidal Thevenin voltage upstream of M and network parameters given by R = 0.005 Ω and L = 9,5x10^{-5} H. Under these conditions, the measuring point M, the VTDD is 15.06%.

Table 1 lists the powers of the main harmonics in the three measurement points M, M_1, M_2, Figure 3. The same table includes the values of total active power, fundamental and harmonic power in the three sections of measurement.

	Two loads	Load 1	Load 2
P1 (W)	$1.9129 \cdot 10^5$	$1.1377 \cdot 10^5$	$7.7512 \cdot 10^4$
P5 (W)	$-3.4427 \cdot 10^2$	$-1.2706 \cdot 10^3$	$9.2632 \cdot 10^2$
P7 (W)	$-1.2238 \cdot 10^1$	$-2.1575 \cdot 10^1$	$9.3370 \cdot 10^0$
P11 (W)	$3.7115 \cdot 10^1$	$1.5520 \cdot 10^2$	$-1.1809 \cdot 10^2$
P13 (W)	$1.1830 \cdot 10^2$	$2.7183 \cdot 10^2$	$-1.6761 \cdot 10^2$
P17 (W)	$4.7315 \cdot 10^1$	$2.7183 \cdot 10^2$	$-2.2452 \cdot 10^2$
PT (W)	$1.9127 \cdot 10^5$	$1.1365 \cdot 10^5$	$7.7620 \cdot 10^4$
PH (W)	$-1.4683 \cdot 10^1$	$-1.2237 \cdot 10^2$	$1.0769 \cdot 10^2$

Table 1. Active powers of the main harmonics and global results for the case 1.

The total harmonic power (point M) is -14.68 W in the direction of power flow from consumer to network. This is because the harmonic powers of certain orders are negative and larger absolute value than those of the harmonics with positive powers. Harmonic power measuring section M becomes a indicator of consumption nonlinear produced downstream of M. However, the presence of two non-linear loads of different consumption changes the situation and the harmonic power fail when their values are obtained in the measuring sections M_1 and M_2. Indeed, in the section of the M_2 measure harmonic power PH_2 is 107.69 W in the direction from network to load, while in M_1, the harmonic power PH_1 is -12.37 W in the direction from load to network. That is, the method of direction of power flow, when more than one source of distortion, determines the prevalent power for each harmonic, corresponding to the power flows in opposite directions, load to network versus network to load, into measuring point. Therefore, when more than one source of distortion, it is not possible to reliably identify the responsible of the disturbance through only the harmonic power. Table 2 presents four indices to assess the distortion. The first ITDD characterized the harmonic content of the current waveforms. The remaining three are defined to identify sources of distortion.

	Two loads	Load 1	Load 2
ITDD	23.59	31.14	37.68
HPI	22.08	26.37	16.70
NC	29.76	34.92	41.30
NL	25.92	33.17	38.97

Table 2. Current Total Demand Distortion, Harmonic Phase Index, Non Collinear Index and Non Linear Index values for the case 1.

HPI index account the current values prevailing at the measuring point. HPI indices in the three measuring points indicate currents prevalent from consumer to source direction, and are suitable for assessing the contribution of both loads. The same table 2 also includes the indices NC and NL. For measures in M, the three indices have consistent values. However, for each individual load, the HPI index assigns greater responsibility for the distortion to the load 1 before to the load 2, while NC and NL rates assigned greater responsibility to the load 2 before to the load 1. The values obtained by the latter two are consistent.

3.4.2. *Case 2: Non-sinusoidal voltage*

Tables 3 and 4 present the results for the same system of figure 3 but this time the TDD voltage at point M is 12.67%. In this case, the Thevenin voltage network is not sinusoidal but includes harmonics of order 5 and 7. Harmonic powers P_H are positive in all measurement points.

	Two load	Load 1	Load 2
PT(W)	$1.54371 \cdot 10^5$	$9.7743 \cdot 10^4$	$5.6623 \cdot 10^4$
P1 (W)	$1.4796 \cdot 10^5$	$9.5058 \cdot 10^4$	$5.2907 \cdot 10^4$
PH (W)	$6.4011 \cdot 10^3$	$2.6845 \cdot 10^3$	$3.7166 \cdot 10^3$

Table 3. Active powers of the main harmonics and global results for the case 2.

The indices NC and NL, Table 4, have values consistent with those obtained in the previous situation. But not so with HPI indices showing less stability against variations in network conditions. This time the HPI index shows an allocation of responsibility for the distortion between charges 1 and 2 contrary to case 1. By contrast, the results table further shows adequate stability of NL and NC indices versus distortion variations network.

	Two loads	Load 1	Load 2
ITDD	23.62	30.78	38.25
HPI	6.78	2.28	20.83
NC	30.67	35.57	42.97
NL	25.46	33.00	38.82

Table 4. Current Total Demand Distortion, Harmonic Phase Index, Non Collinear Index and Non Linear Index values in the case 2.

4. On the measurement of PQ indices in the presence of capacitor bank

In [21, 25] has shown experimentally that the presence of equipment power factor compensation capacitors based on amplified the distortion existing in the network. Moreover, as discussed later, the techniques proposed for assigning responsibility for harmonic distortion penalize capacitors like nonlinear loads. This is in clear contradiction with the vast majority of standards that require the use of capacitors for power factor correction to the fundamental frequency.

Indeed, an electronic load of the type power converter absorbs a current obtained through the sudden switching of electronic devices according to the on-off states. This current produces voltage drops in the inductive impedance of the grid, causing voltage peaks in the voltage waveform at the PCC. As a result, the current drawn by the linear load has a peak and RMS values higher because the load capacitors which tend to emphasize the high harmonics at the facility. This would be the case, for example, a set of discharge lamps (compensated) connected to the same PCC that a nonlinear load. The simplified equivalent circuit comprises an ohmic-inductive branch and a capacitive branch. High current harmonics are almost completely absorbed by the capacitive branch, since its impedance at these frequencies is very small compared with that of the other branch. This contributes to the increased distortion of the current absorbed by the linear load. Therefore, although the capacitors do not introduce new harmonics on the network, if they can dramatically amplify existing distortion produced by nearby non-linear loads.

To illustrate the situation has been considered a system consisting of a six-pulse rectifier controlled with a consumption of 91.77 kW in parallel with a line load consisting of 4 Ω resistor in parallel with a capacitor of 0.1 mF. The figure a) shows the waveform of the phase voltage at the PCC and figure b) the waveform of the current drawn by the RC load.

This has consequences for the effectiveness of the various indices introduced to identify sources of distortion. The situation is illustrated through two practical cases of simulation.

4.1. Case 1: Inductive and capacitive linear loads

Consider two linear loads for the same topology of Figure 3, the first consisting of a R = 2 Ω in series with an L = 0.01 H, and the second one R = 2 Ω in series with a C = 0.001 F; the supply voltage includes harmonics of orders 5 and 7. Tables 5 and 6 present the results.

	Two loads	Load 1	Load 2
P1 (W)	$4.4590 \cdot 10^4$	$2.2504 \cdot 10^4$	$2.2086 \cdot 10^4$
P5 (W)	$7.4478 \cdot 10^2$	$1.2859 \cdot 10^1$	$7.3192 \cdot 10^2$
P7 (W)	$9.0580 \cdot 10^1$	$7.7483 \cdot 10^1$	$8.9805 \cdot 10^1$
PT (W)	$4.5425 \cdot 10^4$	$2.2518 \cdot 10^4$	$2.2908 \cdot 10^4$
PH (W)	$8.3536 \cdot 10^2$	$1.3634 \cdot 10^1$	$8.2172 \cdot 10^2$

Table 5. Harmonic and total powers in the case 1.

	Two loads	Load 1	Load 2
ITDD	17.43	2.46	18.94
HPI	0.00	0.00	0.00
NC	7.34	12.74	20.58
NL	7.34	0.00	18.08

Table 6. PQ indices in the case 1.

Analysis of Table 6 shows how, in this case, only HPI rate indices identify the presence of linear loads in each measuring point. Although the HPI index identifies linear loads, however, then you will see that this index does not succeed when the system is unbalanced. The NL index can identify the RL load as linear but fails with RC load. The NC index fails to identify any such linear loads.

4.2. Case 2: Non linear load and capacitive load

Secondly, we have been considered a six-pulse controlled rectifier (load 1) in parallel with a linear load RC (load 2). This case is the same as above was used to obtain the waveforms presented in Figure 4. The table 7 includes the values obtained for the deferent indexes. In this case none of the identified indexes load 2 as linear load. The ITDD is the amplification of harmonics due to the presence of the capacitive load. Rates of identification of nonlinear loads HPI, NC, and NL, have high values for linear RC load.

	Two loads	Load 1	Load 2
ITDD	28.90	33.78	56.49
HPI	28.56	32.32	40.37
NC	31.95	34.84	56.04
NL	30.48	34.82	53.80

Table 7. Index values with the presence of capacitor banks.

5. Assessment of unbalance emission

From the point of view of electromagnetic compatibility we distinguish between immunity and emission of a particular disturbance. The emission of imbalance consists of measuring the current phase sequence different from the positive sequence injected by an unbalanced load into the PCC. This section introduces the most common index to assess the imbalance emission.

5.1. Conforming and nonconforming unbalance current

Regarding the issue unbalance, for three-phase phase voltages V_a, V_b, and V_c, and line currents I_a, I_b, and I_c, symmetrical components at the fundamental frequency are considered, V_+, V-, V_0, I_+, I- and I_0. These are defined for the voltages by the following expressions:

$$V_- = \frac{1}{3}\left(V_a + a^2 V_b + a V_c\right) \tag{33}$$

$$V_0 = \frac{1}{3}\left(V_a + V_b + V_c\right) \tag{34}$$

$$I_{n+} = I_+ \tag{35}$$

Where complex number a=exp(j2π/3). Similarly symmetrical components of currents are defined. The conforming current is the current that retains the same level of unbalance that the three-phase voltages, and coincides with the positive sequence current of the load current. Thus, the conforming current accomplishes all the active power and reactive power of positive sequence.

$$Z_+ = \frac{V_+}{I_+} \tag{36}$$

The ratio between V_+ and I_+ determines the impedance to the positive sequence current

$$I_{n-} = \frac{V_-}{Z_+} = I_+ \frac{V_-}{V_+} \tag{37}$$

The negative sequence component and zero sequence component of the conforming current will be in the same proportion as the sequence components of voltage, that is,

$$I_{n0} = \frac{V_0}{Z_+} = I_+ \frac{V_0}{V_+} \tag{38}$$

$$I_{d+} = 0 \tag{39}$$

The non conforming current is the current balance. In terms of symmetrical components,

$$I_{d-} = I_{-} - I_{n-} \tag{40}$$

$$I_{d0} = I_{0} - I_{n0} \tag{41}$$

$$i_{blj}(t) = \sum_{k=1}^{n} \frac{V_{kj}}{|\mathbf{Z}_{\mathbf{ka}}|} \sqrt{2} \sin\left(2\pi k f_1 t + (\theta_{kj} - \varphi_{kj})\right) \qquad j = b, c \tag{42}$$

Finally, the three phase values (I_{da}, I_{db}, I_{dc}) are obtained from the inverse transformation of Fortescue. Although the author does not indicate a specific index of unbalance, we have considered the ratio of norm of non-conforming currents with respect to the norm of the real current, NC_unb.

5.2. Balanced linear currents and unbalanced non-linear currents

To define an index that takes into account the charge imbalance, be chosen as the reference phase current phase extracted with the minimum RMS value. For further analysis this phase will be considered as phase 'a'. According to the procedure developed by the authors, [19], for estimating the parameters, it is assumed that the equivalent linear and balanced load is made up by three linear loads, balanced, equal to the estimated linear load for the phase 'a'. Thus, the current absorbed by the phases 'b' and 'c' for the same RL series load, estimated on phase 'a', are

$$i_{unlj}(t) = i_j(t) - i_{blj}(t) \quad ; \qquad j = b, c \tag{43}$$

These currents are called "balanced linear currents" and only responsible for the unbalance and distortion is the delivery system. The difference between the measured actual current in each phase and the calculated ideal current is called "unbalanced non-linear current",

$$i_{uj}(t) = i_{unlj}(t) - i_{nlj}(t) \quad ; \qquad j = b, c \tag{44}$$

These currents expresses how the phases 'b' and 'c' differ from ideal reference conditions in terms of phase unbalance and distortion. On the other hand, it is possible to calculate non-linear currents i_{nlj} (t) (j = b, c) for phases 'b' and 'c' in the manner provided in the previous section. Assuming that it is possible to separate the contribution of the distortion only by subtracting the nonlinear current i_{nlj} (t) of i_{unlj} (t) in the corresponding phase, then,

$$\frac{\|i_u\|}{\|i_{nl}\|}\times 100(\%) \tag{45}$$

These currents expresses how the phases 'b' and 'c' differ from ideal reference conditions only in terms of phase unbalance.

Once they have been introduced all these current components, it is possible to determine the degree of unbalance of a three phase load through the following definition of "unbalance current index",

$$\begin{matrix} \|i_u\|=\sqrt{\sum\limits_{j=b,c} I_{uj}^2} & \|i_{nl}\|=\sqrt{\sum\limits_{j=a,b,c} I_{nl_j}^2} \\ I_{uj}=\sqrt{\sum\limits_{\forall k} I_{kuj}^2} & I_{nlj}=\sqrt{\sum\limits_{\forall k} I_{kul_j}^2} \end{matrix} \tag{46}$$

Where RMS values I_{uj} and I_j and norms of the set phase currents are defined as follow,

$$V_e=\sqrt{\frac{V_{ab}^2+V_{bc}^2+V_{ca}^2}{9}} \tag{47}$$

I_{kuj} and I_{knlj} are the RMS values of corresponding harmonic components of i_{uj} (t) and i_{nlj}(t). Defined (46) has the disadvantage that for linear unbalanced loads takes very high values, even infinite. So, here has changed their definition by making the rate against the norm of the real current.

Finally it is also possible to evaluate the harmonic distortion on a three-phase basis by combining the values of the three different rates of each phase in a single index defined as the ratio between the norm of non-linear currents and the norm of the load currents.

5.3. Practical case: Digital simulation

As a case study a three-phase nonlinear unbalanced system is considered. A set of three loads are connected at the PCC. The load 1 is a balanced three-phase topology formed by three single-phase rectifier with a RL branch into DC side. The load 2 is an unbalanced non-linear load constituted by a single-phase rectifier with a RC branch into DC side in one phase and two resistors of different values in the remaining phases. Finally, the load 3 is a linear inductive and unbalanced load. Table 8 shows the total and harmonic powers, respectively, for each of the loads and for the whole of them. This time the harmonic power identifies the first two loads as non-linear and third load as linear.

	Three loads	Load 1	Load 2	Load 3
PT(W)	3.5433×10^4	1.2740×10^4	4.2502×10^3	1.8443×10^4
PH(W)	-3.4792	-2.4387	-3.7266	2.6861

Table 8. Harmonic and total active powers for unbalance loads.

Table 9 presents the usual distortion indices. They identify the sources of distortion except index HPI. This is based on the direction of power flows for each harmonic between network and load, and masks the power unbalance due to it. However, no information is provided on the issue of unbalance.

On the other hand, NC and NL indices identify the load 1 and the load 2 as non-linear, and the load 3 as linear.

	Three loads	Load 1	Load 2	Load 3
ITDD	25.86	59.53	86.35	1.27
HPI	25.97	39.67	86.56	35.59
NC	18.75	40.72	73.64	0.50
NL	16.80	34.02	83.41	0.20

Table 9. PQ indices in the practical case with unbalance loads.

Table 10 presents the values of the unbalance indices $ITDD^{+}$, NC_unb and UC. The first characterized the conformity of the current waveform of each load with the balanced sinusoidal waveform. The remaining indices identify unbalance responsability. Both identify the balanced load (load 1) and assign a value to unbalance loads.

	Three loads	Load 1	Load 2	Load 3
ITDD+	46.16	59.53	89.21	64.67
NC_unb	26.71	0.27	30.27	47.14
UC	41.19	0.29	25.38	68.99

Table 10. Indices of unbalance in the practical case with asymmetrical conditions.

One issue to note is that NC_unb index is obtained by applying the symmetrical components, while the UC not, this could lead to it not identify a supply voltage of negative sequence.

6. Practical cases: Experimental results

In the order to perform a validation of the results of simulations and verify the stability of the indices, has been built an experimental platform consists of three single-phase rectifiers connected in star and fed directly from the supply network. On the DC side of each rectifier has been connected a parallel RC branch, composed of a variable resistor and a capacitor 2200 μF, thus forming a nonlinear load, see Figure 5.

The measurement system consists of a data acquisition card (dspace-CP1104) and a signal conditioning system formed by three voltage sensors (LEM LV25-P) and three current sensors (LEM LA35-NP), thus the voltage and current signals are taken simultaneously, without introducing any phase change that could affect the accuracy of measurements. For data acquisition and processing, has developed a virtual instrument using Matlab and Control-Desk. This instrument stores the instantaneous values of each phase voltage and line current. The configuration of the virtual instrument was made following the recommendations EN 61 000-4-7 and EN 61 000-4-30, so that has been used window equal to five cycles of the fundamental component and a sampling frequency of 6400 Hz, thus avoiding problems of aliasing and leakage errors.

To evaluate the theory and the results obtained in simulations has been carried out daily measurements along a day at regular intervals of one hour. In this way has been possible to evaluate the changes into indices for two different cases, a non lineal balanced load(case A) and another case with non lineal unbalanced load (case B). In both cases the supply system network was used for feeding the loads.

6.1. Case A: Non linear and balanced load

In this case the load connected to the DC side of the three rectifiers consists of a resistance of 120 Ω in parallel with a capacitor 2200 μF. In Figure 6 (a) displays the voltage waveforms of the three phases whereas the Figure 6 (b) shows the waveforms of the currents taken from one of the measurements performed on the test system.

Table 11 shows the values of total active power (P_T) consumed by the load on the three different measurements made throughout the session, as well as fundamental harmonic (P_1) and the values of total active harmonic power (P_H).

	M 3	M 15	M 21
PT (W)	2134.00	2245.00	2210.00
P1 (W)	2169.00	2288.90	2254.40
PH (W)	-35.02	-43.85	-44.41

Table 11. Active powers of the main harmonics of the test system in Figure 4 with Non linear and balanced load (Case A).

Due to variations in the supply voltage throughout the day, small variations are produced in the active power consumed by the load. Analysis of Table 11, it follows that, as corresponding to the nonlinearity of the load, the total active harmonic power is negative. Furthermore. it can be seen as the values of total active harmonic power (P_H) are very small compared to the active power of fundamental harmonic. Moreover, as expected, there was a small value of VTDD throughout the day, about 3%, while the value of ITDD is considerably higher, around 78%. Figure 7 (a) shows the variations of VTDD throughout the day, whereas in Figure 7 (b) shows the variations of ITDD along test system and includes the results of the different rates to be evaluated.

6.2. Case B: Non lineal and unbalanced load

For unbalancing the load has changed the value of the resistances of each rectifier DC side, so that R1 = 80 Ω, R2=120 Ω and R3 = 484 Ω, while the capacitor is the same in the three loads, 2200 μF. In Figure 8 (a) shows the waveforms of voltages (supply network) whereas the figure 8 (b) shows the waveforms of currents resulting from the measurements taken.

Table 12 shows the values of total active power (P_T) consumed by the load on the three measurements made throughout the session, as well as the values of total active harmonic power (P_{TH}), and active power for the fundamental (P_{T1}).

Figure 9 shows how the measurement system includes all of the daily variation harmonic content and unbalance of the supply network voltage, whereas in Figure 9 (b) shows the daily trend of results unbalanced indices calculated for the same period.

	M3			M15			M21		
	Phase L1	Phase L2	Phase L3	Phase L1	Phase L2	Phase L3	Phase L1	Phase L2	Phase L3
P1 (W)	993.43	714.46	194.32	1041.80	752.64	202.11	1037.50	751.76	200.94
PTH (W)		-30.46			1958.80			1951.10	
PT (W)		1871.80			1996.60			1990.20	
PT1 (W)		1902.20			-37.82			-39.13	

Table 12. Active powers of the main harmonics of the test system in Figure 4 with Non lineal and unbalanced load (Case B).

The experimental results confirm the simulation results on the information provided by the PQ indices. The measurements show that for a typical load balanced/unbalanced nonlinear connected to the supply network, all the indices analyzed included some variations to changes of TDD voltage within a reasonable range. In any case, the index NC_unb respect the degree of unbalance, is the index experienced less variations. This index is defined from the fundamental harmonic symmetrical components, and therefore undergone fewer changes for a given load.

7. Discussion of the results and conclusions

The strong presence of waveforms of voltage and current distorted and/or unbalanced in electric power systems has driven the need to determine the contribution to the deterioration of the PQ of consumers connected to supply networks. The method of the direction of power flow has been widely used to identify the locations of harmonic sources. However, this method is unable to solve this task in all situations. This procedure does not locate a source of distortion in the case that there are multiple harmonics sources connected to the PCC. To overcome this situation, other procedures have been introduced under the constraint of seeking a solution, based on the realization of measures in only one section of the PCC. These methods have introduced new indices as they are, HPI, NC and NL, that advance but which do not solve the problem in all situations. Thus:

- The method of direction of harmonic power flow fails when there is more than a nonlinear load connected to the same PCC. The measurement of P_H can be positive at the terminals of a nonlinear load. This is because the harmonic powers up the sum of P_H are the contribution of power flow of two opposing, in the sense from load to network and another from network to load. The result for each harmonic corresponds to the prevailing power flow. Therefore, for the identification of sources of distortion, P_H is a not suitable indicator and would have to resort the study of the individual harmonics powers. However, this method is the only one that unambiguously identifies the linear loads, including those with capacitor banks for compensation.
- The index of harmonic phase, HPI, overcomes the drawbacks mentioned for the method of harmonic power direction, using the RMS values of current in one direction or another in the PCC. Moreover presents an adequate stability to changes in network distortion. However, result in errors for linear loads on systems with unbalanced and with the presence of the capacitor bank.
- The non-conforming current index, NC, while indicating the current component of the load presented to the voltage distortion front, is not useful from the practical point of view since it does not discriminate linear loads. In the case of the presence of capacitors is identified as distortion source. Moreover, this index shows more variation compared to other indices variations in the conditions of distortion of the mains.
- The non-linear current index, NL, it seems appropriate to characterize the current of a nonlinear load distorted. It also presents few variations to changes in the distortion of the mains voltage. However, the rate can result in errors in linear loads with different topologies including the RL parallel and/or capacitive branches.

Respect to the issues of unbalance loads produced by asymmetrical operation has been introduced ITDD+ index as a measure of non-conformity of the waveform of actual current

http://dx.doi.org/10.5772/53373

with respect to the balanced sinusoidal waveform positive sequence. This index, as the ITDD, characterized the set of waveforms of a three phase system but will not identify unbalanced loads. Two indexes to locate unbalance are introduced: Current Unbalance index, UC and Non-Conforming to the unbalance, NC_unb. Both identify the loads that are sources of unbalance, however, the UC would not detect the case of a supply voltage of negative sequence.

In conclusion, in the text above have reviewed the potential problems associated with the assessment of power quality in electrical installations. In particular, we have introduced harmonic distortion and unbalanced to characterize the voltage and current waveforms. A comparative analysis of the most common indices was made. Thus, to determine responsible for the generation of distortion through measurements made in a single measurement section, none of the indexes given is capable of resolving the issue reliably. However, since in practice all loads are nonlinear, the NL is an appropriate index for assessing the distortion source because it is little affected by the imbalance and distortion. For the determination of the issue of unbalance NC_unb index has shown a good performance.

Summary of indices

Effective Voltage, $V_e = \sqrt{\dfrac{V_{ab}^2 + V_{bc}^2 + V_{ca}^2}{9}}$

Effective Current, $I_e = \sqrt{\dfrac{I_a^2 + I_b^2 + I_c^2}{3}}$

Voltage Total Harmonic Distortion, $VTHD_{\phi} = \sqrt{\dfrac{V_{\phi}^2 - V_{\phi_1}^2}{V_{\phi_1}^2}} = \sqrt{\dfrac{V_{\phi_2}^2 + V_{\phi_3}^2 + V_{\phi_4}^2 + \ldots + V_{\phi_N}^2}{V_{\phi_1}^2}}$

Current Total Harmonic Distortion, $ITHD_{\phi} = \sqrt{\dfrac{I_{\phi}^2 - I_{\phi_1}^2}{I_{\phi_1}^2}} = \sqrt{\dfrac{I_{\phi_2}^2 + I_{\phi_3}^2 + I_{\phi_4}^2 + \ldots + I_{\phi_N}^2}{I_{\phi_1}^2}}$

Voltage Total Demand Distortion, $VTDD_{\phi} = \sqrt{\dfrac{V_{\phi}^2 - V_{\phi_1}^2}{V_{\phi}^2}} = \sqrt{\dfrac{V_{\phi_2}^2 + V_{\phi_3}^2 + V_{\phi_4}^2 + \ldots + V_{\phi_N}^2}{V_{\phi}^2}}$

Current Total Demand Distortion, $ITDD_{\phi} = \sqrt{\dfrac{I_{\phi}^2 - I_{\phi_1}^2}{I_{\phi}^2}} = \sqrt{\dfrac{I_{\phi_2}^2 + I_{\phi_3}^2 + I_{\phi_4}^2 + \ldots + I_{\phi_N}^2}{I_{\phi}^2}}$

Three-Phase Total Harmonic Distortion of Current and Voltage, $ITHD_{3\phi} = \sqrt{\dfrac{I_e^2 - I_{e1}^2}{I_{e1}^2}}$; $VTHD_{3\phi} = \sqrt{\dfrac{V_e^2 - V_{e1}^2}{V_{e1}^2}}$

Three-Phase Total Demand Distortion of Current and Voltage, $ITDD_{3\phi} = \sqrt{\frac{I_e^2 - I_{e1}^2}{I_e^2}}$; $VTDD_{3\phi} = \sqrt{\frac{V_e^2 - V_{e1}^2}{V_e^2}}$

Unbalance Three-Phase Total Demand Distortion of Voltage and Current, $VTDD_{3\phi}^{+} = \sqrt{\frac{V_e^2 - V_{e1+}^2}{V_e^2}}$; $ITDD_{3\phi}^{+} = \sqrt{\frac{I_e^2 - I_{e1+}^2}{I_e^2}}$

The Harmonic Phase Index, $HPI = \frac{\| I_L \|}{\| I_S \|} \times 100\,(\%)$

Non Collinear Index, $NC = \frac{I_d}{I} \times 100\,(\%)$

Unbalanced Non Collinear Index, $NC\_unb = \frac{\| i_d \|}{\| i \|} \times 100\,(\%)$

Non Linear Index, $NL = \frac{I_{nLa}}{I_a} \times 100\,(\%)$

Unbalanced Current Index, $UC = \frac{\| i_u \|}{\| i_{nl} \|} \times 100\,(\%)$

Acknowledgments

This work is part of the project "Measurements system to assessment the sources of imbalance and harmonic distortion in electric distribution networks", DPI2010-17709, sponsored by the Ministry of economy and competitiveness, Government of Spain.

Author details

Patricio Salmerón Revuelta and Alejandro Pérez Vallés

Departament of Electrical Engineering, E.T.S.I, Huelva University, Spain

References

[1] A. Menchetti, R. Sasdelli, *Measurement Problems in Power Quality Improvement*, ETEP, vol. 4, No. 6, Nov/Dec, 1994.

[2] A. Ferrero, A. Menchetti, R. Sasdelli, *Measurement of the Electric Power Quality and Related Problems*, ETEP, Vol. 6, No. 6, November/December 1996.

[3] G.T. Heydt, E. Gunther, 1996 "Post-measurement processing of electric power quality data", IEEE Trans. on Power Delivery, Vol. 11, No. 4.

[4] G.T. Heydt, W.T. Jewel, 1998 "Pitfall of electric power quality indices", IEEE Trans. on Power Delivery, Vol. 13, No. 2.

[5] D.D. Sabin, D.L. Brooks, A. Sundaram, 1999, "Indices for assessing harmonic distortion from power quality measurements: definitions and benchmark data", IEEE Trans. on Power Delivery, Vol. 14, No. 2.

[6] R. Sasdelli, G. Del Gobbo, G. Iuculano, 2000 "Quality management for electricity as a processed material". IEEE Trans. on Instrumentation and Measurement, Vol. 49, No. 2.

[7] F.J. Alcántara, P. Salmerón. *A New Technique for Unbalanced Current and Voltage Measurement with Neural Networks*. IEEE Trans. On Power Delivery, Vol. 20 No: 2, May 2005.

[8] P. Salmerón, R. S. Herrera, J. Prieto, A. Pérez. *New Distortion and Unbalance Indices based on Power Quality Analyzer Measurements*. IEEE Transaction on Power Delivery, Vol. 24, No. 2, April 2009, pp: 501-507.

[9] L. S. Czarnecki, T. Swietlicki, *Powers in Nonsinusoidal Networks: Their Interpretation, Analysis, and Measurement*, IEEE Trans. on Instr. and Measurement, Vol. 39, No. 2, April 1990.

[10] P. H. Swart, M. J. Case, J. D. Van Wyk, *On Techniques for Localization of Sources Producing Distorsion in Electric Power Networks,* ETEP, vol. 4, No. 6 Nov/Dec 1994

[11] P. H. Swart, J. D. Van Wyk, M. J. Case, *On Techniques for Localization of Sources Producing Distorsion in Three-Phase Networks*, ETEP, vol. 6, No. 6 Nov/Dec 1996.

[12] Srinivasan, K. *On Separating Customer and Supply Side Harmonic Contributions*, IEEE Trans. on Power Delivery, vol. 11, No. 2, April 1996.

[13] E. Thunberg y L. Söder. *A Norton Approach to Distribution Network Modeling for Harmonic Studies*. IEEE Transaction on Power Delivery, Vol. 14, No. 1, January 1999

[14] A. P. J. Rens y P.H. Swart. *On Techiques for the Localization of Multiple Distortion Sources in Three-Phase Networks: Time Domain Verification.* ETEP Vol. 11, No. 5, September/ October 2001.

[15] C. Muscas, *Assessment of Electrical Power Quality: Indices for Identifying Disturbing Loads*, ETEP, Vol. 8, No. 4, July/August 1998.

[16] J.P.V. Du Toit, J.H.C. Pretorius y W.A. Cronje. *Non Linear Load Identification under Non-Sinusoidal Conditions.* Sixth International Workshop on Power Definitions and Measurenments under Non-Sinusoidal Conditions, Milano, October 13-15, 2003.

[17] W. Xu, X. Liu y Y. Liu. *An Investigation on the Validity of Power-Direction Method for Harmonic Source Determination.* IEEE Transations on Power Delivery, Vol. 18, No. 1, January 2003.

[18] Ch. Chen, X. Liu, D. Koval, W. Xu y T. Tayjasanant. *Critical Impedance Method- A New Detecting HarmonicSources Method in Distribution Systems.* IEEE Transations on Power Delivery, Vol. 19, No. 1 January 2004.

[19] A. Dell´Aquila, M. Marinelli, V. G. Monopoli y P. Zanchetta. *New Power-Quality Assessment Criteria for Supply Under Unbalanced and Nonsinusoidal Conditions.* IEEE Transactions on Power Delivery, Vol. 19, No. 3, July 2004.

[20] Cataliotti and V. Consetino, *A New Measurement Method for the Detection of Harmonic Sources in Power Systems Base on the Approach of the IEEE Std. 1459-2000,* IEEE Transactions On Power Delivery, Vol.25, No. 1, pp. 332-340, January 2010.

[21] Reyes S. Herrera, Patricio Salmerón. *Harmonic disturbance identification in electrical systems with capacitor banks.* Electric Power Systems Research, Volume 82, Issue 1, January 2012, Pages 18-26.

[22] J.E. Farach, W.M. Grady y A. Arapostathis. *An Optimal Procedure for Placing Sensors and Estimating the Locations of Harmonic Sources in Power Systems.* IEEE Transations on Power Delivery, Vol. 8, No. 3, July 1993.

[23] L. Critaldi, A. Ferrero y S. Salicone. *A Distributed System for Electric Power Quality Measurement.* IEEE Transaction on instrumentation and Measurement, Vol. 51, No. 4, August 2002.

[24] C. Muscas, L. Peretto, S. Sulis, R. Tinarelli, *Investigation on Multipoint Measurement Techniques for PQ Monitoring,* IEEE Trans. Instrum. Meas. Vol. 51, pp. 1684, oct. 2006.

[25] N. Locci, C. Muscas, S Sulis, *On the Measurement of Power-Quality Indexes for Harmonic Distortion in the Presence of Capacitors,* IEEE Transactions on Instrumentation and Measurement, Vol 56, Issue 5, pp.1871 – 1876, Oct. 2007.

[26] *Definitions for the measurement of electric power quantities under sinusoidal, nonsinusoidal, balanced, or unbalanced conditions,* IEEE Std 1459-2000, January 2000.

Chapter 11

High Frequency Harmonics Emission in Smart Grids

Jaroslaw Luszcz

Additional information is available at the end of the chapter

http://dx.doi.org/10.5772/52874

1. Introduction

The term 'smart grid', is nowadays very often used in many publications and so far has not been explicitly defined, however it refers mainly to such an operation of electricity delivery process that allows to optimize energy efficiency by flexible interconnection of central and distributed generators through transmission and distribution system to industrial and consumer end-users [1], [3], [11], [13], [15], [17]. This functionality of power delivery system requires the use of power electronic converters at generation, consumer and grid operation levels. Harmonic pollution generated by power electronics converters is one of the key problems of integrating them compatibly with the power grid, especially when its rated power is high with relation to the grid's short-circuit power at connection point [18], [19], [20].

Contemporary power electronics converters has already reached rated power of several MW and are integrated even at the distribution level directly to medium voltage (MV) grid. Power electronics technologies used nowadays in high power and MV static converter increase the switching frequency significantly due to the availability of faster power electronic switches which allows to increase power conversion efficiency and decrease harmonic and inter-harmonic current distortion in frequency range up to $2\,kHz$. This trend significantly increases harmonic emission spectrum towards higher frequencies correlated with modulation frequency of switching conversion of power. Therefore typical harmonic analysis up to $2\,kHz$ in many power electronics application requires to be extended up to frequency of $9\,kHz$ which is the lowest frequency of typical electromagnetic interference analysis interest. Numerous problems related to current and voltage harmonic effects on contemporary power systems are commonly observed nowadays, also in frequency range $2-9\,kHz$. Levels and spectral content of current distortions injected into electric power grids are tending to increase despite the fact that the acceptable levels are determined by numerous regulations [2], [3], [7], [9], [12], [14], [16].

In recent years many of grid-side PWM boost converters of relatively high rated power have been introduced into power grid because of many advantages, like for example:

INTECH
open science | open minds

© 2013 Luszcz; licensee InTech. This is a paper distributed under the terms of the Creative Commons Attribution License (http://creativecommons.org/licenses/by/3.0), which permits unrestricted use, distribution, and reproduction in any medium, provided the original work is properly cited.

current harmonics limitation, reactive power compensation and bidirectional power flow. Implementation of smart grids idea will conceivably increase this tendency because of the need for bidirectional flow control of high power in many places of distribution and transmission power grid.

Typical carrier frequencies used in AC-DC PWM boost converters are within a range from single kHz for high power application up to several tens of kHz for small converters. Important part of conducted emission spectrum generated by those types of converters is located in frequency range below 2 kHz normalized by power quality regulations and above 9 kHz normalized by low frequency EMC regulation (especially CISPR A band 9kHz-150kHz). In between those two frequency ranges typically associated with power quality (PQ) and electromagnetic compatibility (EMC) respectively, where a characteristic gap of standard regulations still exists, the conducted emission of grid-connected PWM converters can be highly disturbing for other systems. Current and voltage ripples produced by grid-connected PWM converters can propagate through LV grids and even MV grids, where converters of power of few MW are usually connected. Filtering of this kind of conducted emission will require a new category of EMI filters with innovative spectral attenuation characteristic which is difficult to achieve by just adaptation of solutions that are already in use for current harmonics filtering for PQ improvement and radio frequency interference (RFI) filters used for EMC assurance.

2. Harmonic emissions of non-linear loads into power grid

Harmonics content defined for currents and voltages is an effect of its non sinusoidal wave-shape. Power electronics switching devices used in power conversion process like diodes, thyristors and transistors change its impedance rapidly according to line or PWM commutation pattern and produce non sinusoidal voltages and currents which are required to perform the power conversion process properly. Unfortunately, these non sinusoidal currents, as a results of internal commutation process in a converter, are also partly injected into the power grid as an uninvited current harmonic emission. Non sinusoidal load currents charged from power grid produce voltage harmonic distortions in power grid which can influence all other equipment connected to that grid because of the existence of grid impedance. This mechanism results that non-linear current of one equipment can be harmful for other equipment supplied from the same grid and also for the grid itself, like e.g. transformers, transmission lines.

A frequency spectrum range of harmonic distortions introduced into power grid can be exceedingly wide, nevertheless the maximum frequency range which is usually analysed is defined by CISPR standard as $30MHz$. Between $9kHz$ and $30MHz$ two frequency sub-bands are defined as CISPR A up to $150kHz$ and CISPR B above $150kHz$ (Figure 1). These two frequency rages are well known as conducted electromagnetic interference (EMI) ranges, where harmonic components of common mode voltages or currents are limited to levels defined by a number of standards.

In general, despite some specific cases, amplitudes of harmonic distortions observed in typical applications decrease with the increase of frequency, stating from several or tens percent in frequencies close to the power frequency and reach levels of only microvolts or microamps for the end frequency of conducted frequency band $30MHz$. Unfortunately, even so small voltage and current amplitudes can be really harmful, disturbing, and difficult to filter because of relatively high frequency which results with easiness of propagation by means of omnipresent parasitic capacitive couplings.

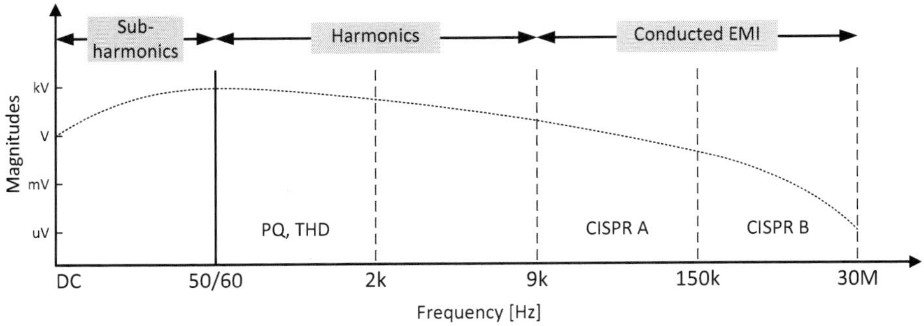

Figure 1. Harmonic distortions frequency sub-ranges.

On the other hand, typical harmonic distortion components which are usually recommended for analysing and solving PQ problems by standards are within frequency range up to $2kHz$. In this frequency range integer multiples of the fundamental power frequency ($50Hz$ or $60Hz$) harmonic components are defined usually up to 40th order.

Consideration of conducted emissions and harmonic distortions in these frequency bands: one up to $2kHz$ and second $9kHz - 30MHz$, were sufficient enough in last years in applications with classic line-commutated rectifiers and switch mode DC power supplies which are also fed by this type of rectifiers. During the last decades, with the increase of the rated power of single power supplies and increasing number of power supplies used the increased difficulties with acceptable current harmonic emission levels arise and other technologies like PWM boost rectifiers have been intensively introduced. The PWM modulation carrier frequency used in such applications is often within the range of $2 - 9kHz$ or adjacent ranges, which results with the significant increase of harmonic emission in this frequency range what will be discussed in the next sections of this chapter.

3. Harmonic distortions emission of grid-connected power electronics converters

Harmonic distortion emission is commonly understood as harmonics produced by non-linear loads, usually power electronics converters in the frequency range up to $2kHz$ which are strongly related to some of the power quality indices. From this point of view (PQ) harmonic distortion emission in the frequency range above $2kHz$ can be named as high frequency harmonics emission. On the other hand, from the EMC point of view, the conducted EMI emission below $9kHz$ is usually defined as low frequency EMC conducted emission.

The frequency map of different harmonic emissions, usually considered as conducted type emissions which are mainly propagating by conduction process along power lines, is presented in Figure 2. From this prospective we can distinguish three primary types of harmonic distortion emission of typical sources which can be associated to particular power electronics converters topologies and technologies. These are:

- classic PQ frequency range up to $2kHz$, where the main sources of harmonic distortions are usually line commutated rectifiers used in single- and multi-phase topologies using as power switches diodes or thyristors,
- high frequency harmonic distortion emission in the frequency range $2 - 9kHz$, where mainly PWM boost rectifiers, as a relatively new topology, are generating harmonic components correlated to the used PWM carrier frequency which depending on the topology and rated power of the converter is usually located between a few kHz and tens of kHz,
- conducted EMI emission in frequency range ($9kHz - 30MHz$), which is primarily an effect of DC voltage conversion by switching mode methods where power transistor switching processes are key sources of high frequency conducted emission which can easily propagate also towards AC power lines.

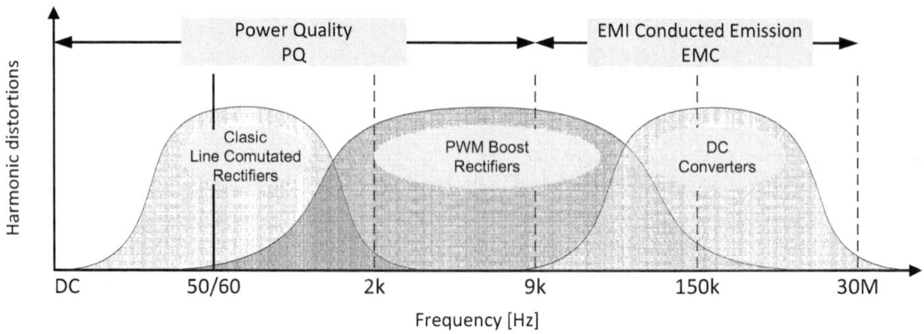

Figure 2. Characteristic distribution of harmonic emission spectra for different types of power electronics converters.

3.1. Low frequency harmonic emission of classic AC-DC converters

Classic, diode-based AC-DC converters (rectifiers) were successfully used for many years in multiple applications. Nowadays, because of extremely increasing number of such devices used in power system and significantly increasing its rated power, AC-DC converters for power of hundreds of kW are quite often used, its harmonic emission levels cannot be accepted by power grid operator demands. Significantly increasing problem of harmonic distortions in power grid led to legislation numerous of grid regulations which are set-up by grid operators and international standards. A typical configuration of six pulse three phase diode rectifier with DC link capacitor commonly used in medium power applications is presented in Figure 3 .

The exemplary input current waveform for this type of rectifier is presented in Figure 4 with correlation to input AC voltage. The maximum value of line current and its flow duration which is in six pulse rectifier always shorter than half cycle are accountable for the level of distortion. These parameters of current wave-shape are dependent of grid impedance and DC link capacitor parameters, especialy size, equivalent serial resistance (ESR) and equivalent serial inductance (ESL) In the evaluated case significant distortion of input current I_{AC} make a distortion effects slightly visible also at voltage waveform, where voltage deformations are correlated in time with the current pulses.

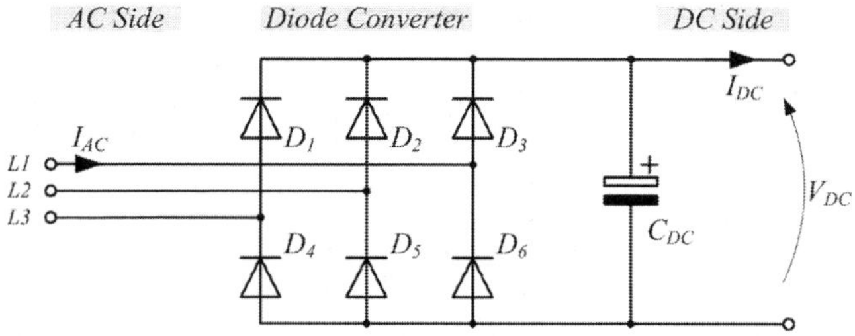

Figure 3. Six pulse diode rectifier with DC link capacitor.

The frequency domain representation of input current, calculated for 10 cycle period with rectangular widowing as a discrete Fourier transform (DFT) product [8] is presented as harmonic amplitudes I_k with $5Hz$ resolution in frequency range up to $2kHz$ in Figure 5 and up to $25kHz$ in Figure 6 . The characteristic harmonics for six-pulse rectifier are non nontriplen odd harmonics (5th, 7th, 11th, 13th etc.) and its amplitudes decrease with frequency

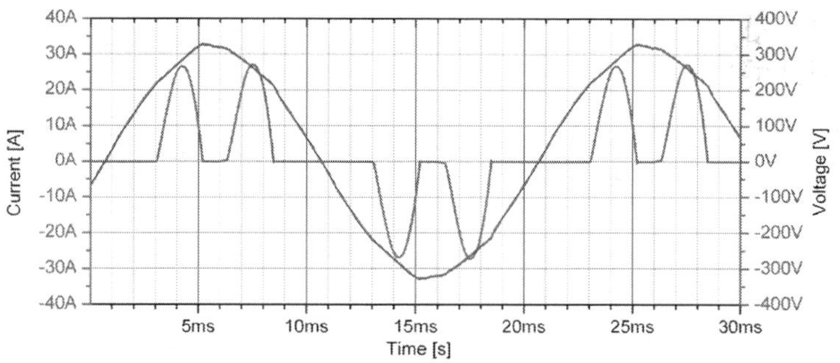

Figure 4. Six pulse diode rectifier - typical input current and voltage waveforms.

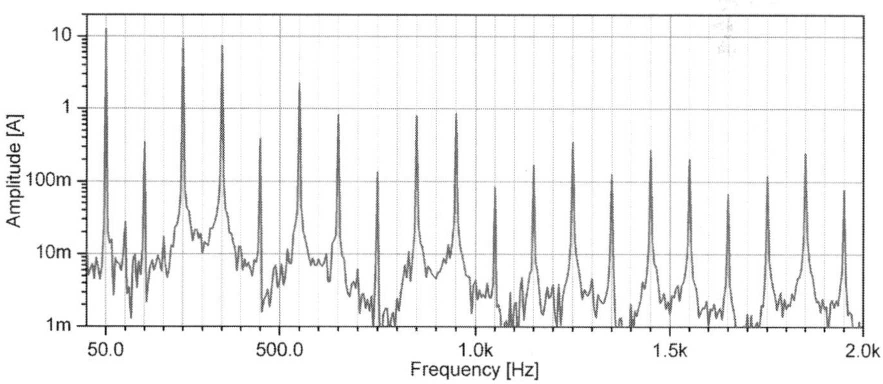

Figure 5. Six pulse diode rectifier - typical input current harmonics spectrum up to 2kHz.

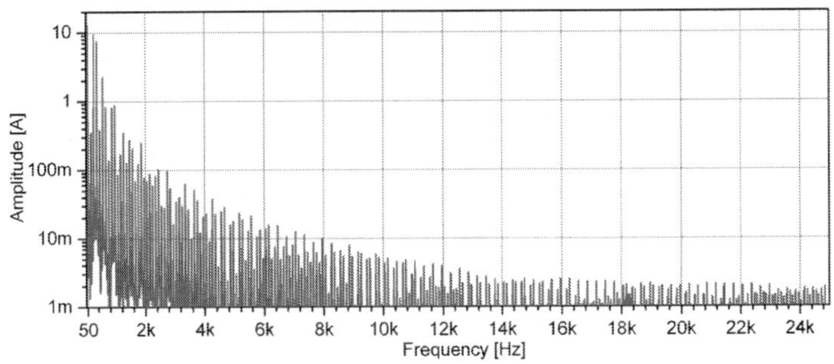

Figure 6. Six pulse diode rectifier - typical input current harmonics spectrum up to 25kHz.

The total harmonic distortion (THD) content of input current can be calculated using formula (1) where each harmonic group I_n is determined according to formula (2) . In the analysed example presented in Figure 4 the obtained THD was over 95%. To reduce so high harmonic emission number of passive filtering techniques can been introduced. AC reactors (L_{AC}) and DC chokes (L_{DC}) (Figure 7) are typically used and allow to decrease input current THD below 30%. Adequate input current waveform and its frequency domain representation for diode rectifier with passive filtering are presented in Figure 8, 9 and 10.

$$THD(I) = \frac{\sqrt{\sum_{n=2}^{40} I_n^2}}{I_1} \tag{1}$$

$$I_n = \sqrt{\frac{I_{k-5}{}^2}{2} + \sum_{i=k-4}^{k+4} I_i^2 + \frac{I_{k+5}{}^2}{2}} \tag{2}$$

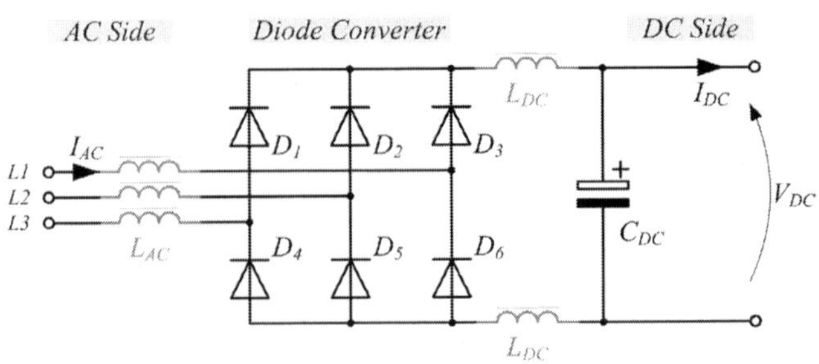

Figure 7. Six pulse diode rectifier with passive filtering of line current harmonic distortions.

http://dx.doi.org/10.5772/52874

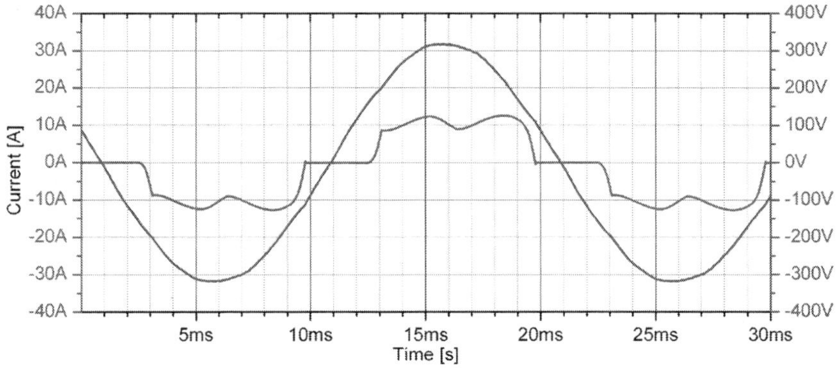

Figure 8. Six pulse diode rectifier with passive filtering - input current and voltage waveforms.

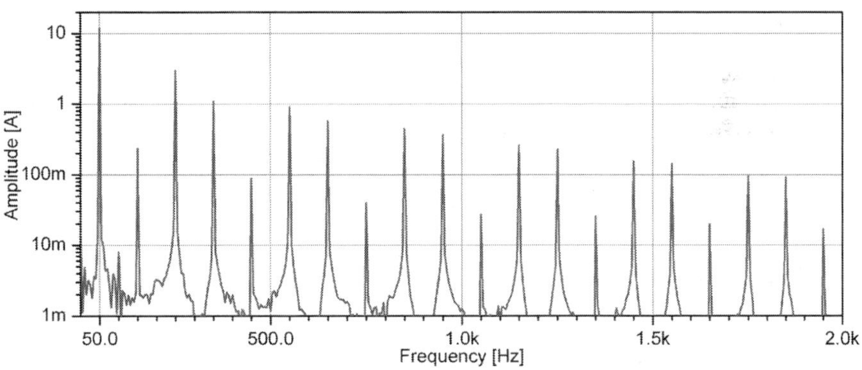

Figure 9. Six pulse diode rectifier with passive filtering - input current harmonics spectrum up to 2kHz.

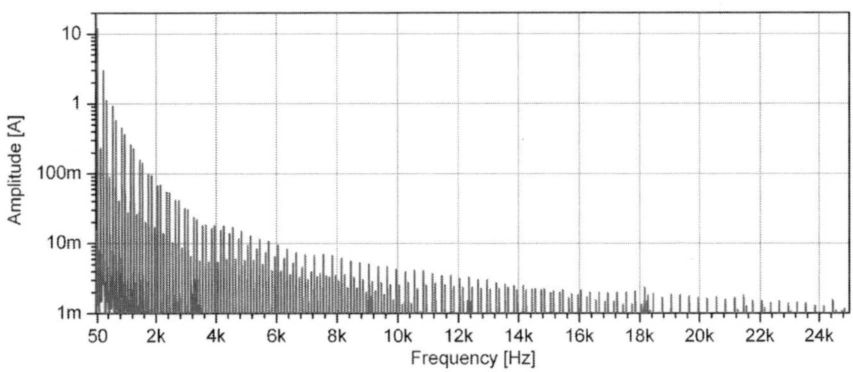

Figure 10. Six pulse diode rectifier with passive filtering - input current harmonics spectrum up to 25kHz.

3.2. High frequency harmonic emission of modern AC-DC converters

Severe limitations of the line current harmonic performance improvement possibilities of classic diode rectifiers stimulate introducing fully controlled switches in AC-DC converters. Accompanying significant increase of IGBT transistor performance during the last decade allows to obtain successful implementation of PWM boost AC-DC three phase converter topology in many applications where harmonic distortion emission has to be limited. PWM boost type AC-DC converters besides line current harmonic distortion significant reduction in frequency range up to $2kHz$ have a number of other advantages [4], [5], [10], like for example:

- ability to transform energy bidirectionally, which significantly increases the range of applications especially in energy saving purpose and renewable and distributed energy systems,
- possibility to control line current phase, which allows to maintain reactive power consumption within required limits and also stand-alone operations as a power factor correction system,
- autonomous operation as a harmonic distortion compensator for other non-linear loads working in the power grid.

PWM boost rectifier basic topology is based on the six pulse power transistors bridge which is connected to power grid through AC line reactor (Figure 11). AC line reactor L_{AC} allows to control line current freely using suitable PWM strategies, which results in a possibility to considerably decrease the line current harmonic emission level in frequency range below $2kHz$. Essential problem, tightly related to the current harmonic distortion emission in the frequency range close and above PWM modulation carrier frequency are input current ripples which are an effect of line and DC bus voltage commutations over the AC line reactor inductance L_{AC} (Figure 12).

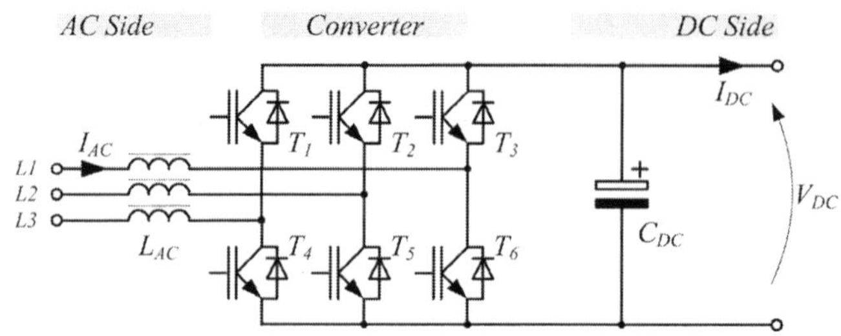

Figure 11. Three-phase grid connected PWM boost converter topology.

The exemplary line current waveform obtained using this method is presented in Figure 13, where nearly sinusoidal current can be seen with low harmonic content in frequency band below $2kHz$ (Figure 14), however with some noticeable distortions in higher frequency range which are an effect of existing limitations of the used PWM control method. To minimize the PWM carrier frequency related harmonic emission low pass filtering methods are used,usually based on the LCL filter topology. Nevertheless, the harmonic emission effect correlated with PWM carried frequency is observable in most of applications (Figure 15). The

http://dx.doi.org/10.5772/52874

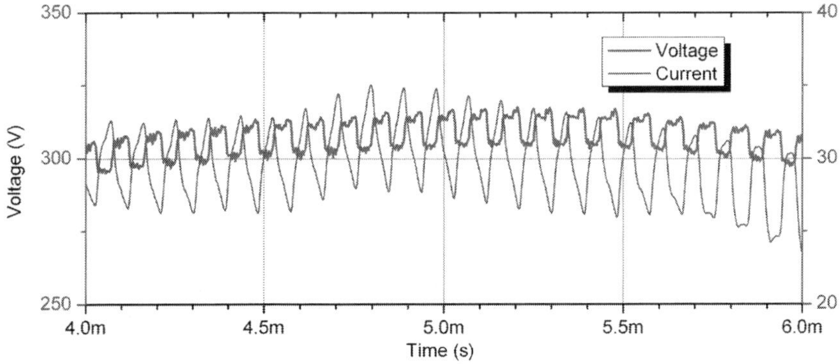

Figure 12. Line current and voltage ripples generated by PWM boost converter

maximum emission is observed around modulation frequency which in the tested converter was set to $15kHz$. In higher frequency range, close to integer multiples of PWM carrier frequency harmonic products of modulation are usually also observed. Perfect elimination of this PWM-related emission is not possible and became more difficult to realize by using passive filters with the increase of frequency. An example of input current and voltage waveforms and its harmonic content in frequency domain representation recorded in PWM boost converter are presented in Figure 13, 14 and 15.

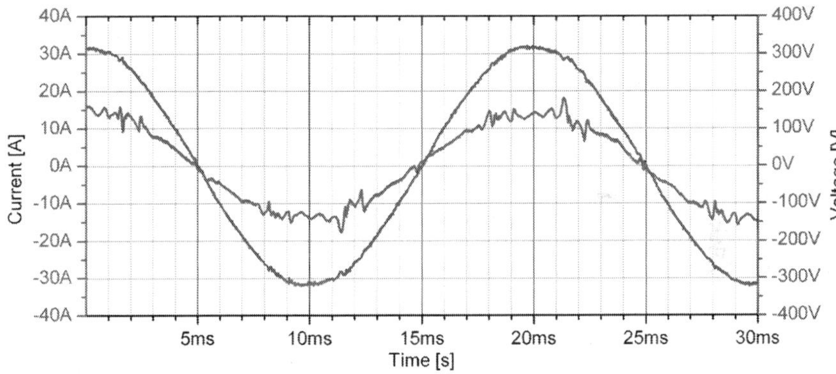

Figure 13. PWM boost converter – input current and voltage waveforms.

3.3. Comparison of line current harmonic distortion of diode and PWM boost rectifiers

Detailed comparison of current harmonic distortion emission has been done for three phase six pulse diode rectifier and PWM boost converter with the three phase IGBT transistor bridge. Both converters has been tested in similar supply condition and using similar load, which allows to minimize the influence of line impedance and DC load level on the obtained results. Comparison of input current harmonic distortion emission should be carried out

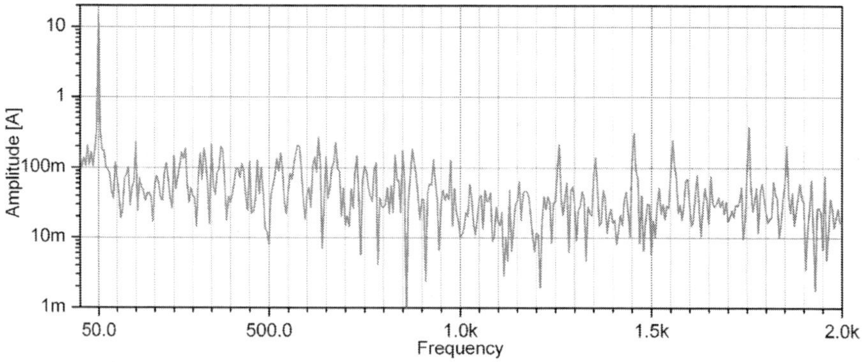

Figure 14. PWM boost converter – input current harmonics spectrum up to 2kHz.

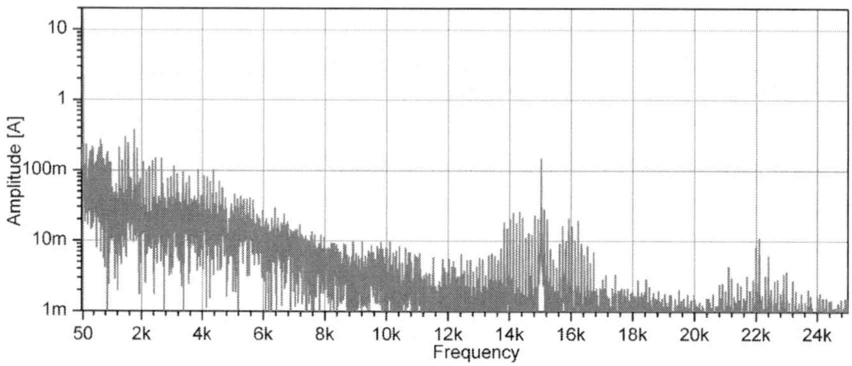

Figure 15. PWM boost converter – input current harmonics spectrum up to 25kHz.

separately for different frequency sub-ranges presented in Figure 1 and 2, because of different evaluation methods which have to be used in each particular sub-range.

3.3.1. Frequency range up to 40th harmonic order

Typical analysis, important from the total harmonic distortion (THD) limitations point of view, consider frequency range up 40th harmonic order of power frequency (in $50Hz$ system it is up to $2kHz$). In this frequency range classic line commutated rectifiers generate dominating characteristic harmonics orders $n * (p \pm 1)$ correlated with number of pulses p depending on rectifier topology. According to this rule, for six pulse rectifiers harmonics of order H5, H7 and H11, H12 and H17, and 19 etc. are dominating (Figure 16 blue line).

The use of PWM boost conversion technology allows to decrease harmonic emission for this specific orders significantly (about tens of times for H5 and H7, about ten times for H11, H13 and H17, H19). Unfortunately, use of PWM boost conversion technology introduces extra harmonic components emission for frequencies values in between integer multiplies

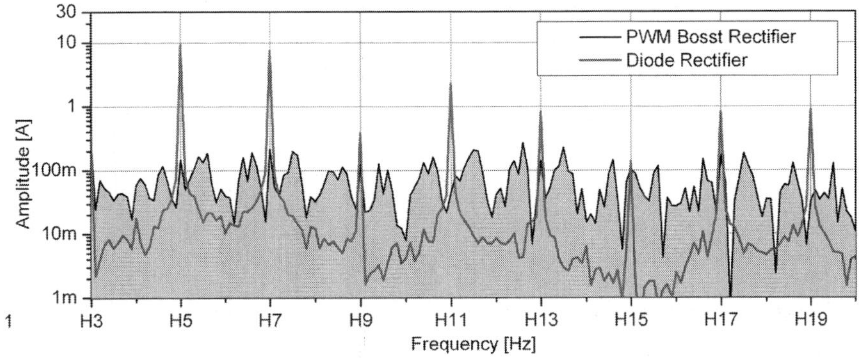

Figure 16. Harmonic emission of three phase six pulse diode rectifier with comparison to PWM boost converter emission

of power frequency: inter-harmonics (Figure 16 grey area). Detailed calculation results are presented in Table 1, where the decrease of harmonics content from 97% down to 3.5% and increase of inter-harmonics from 2% up to 10% are listed. The final effect of obtained harmonic reduction using PWM boost technology is the decrease of THD from 97% to 11% whereas inter-harmonic content is considerable: around 10%.

AC-DC Converter topology	Harmonics	Inter-harmonics	THD	PWHD
Diode rectifier	97%	2%	97%	48%
Diode rectifier with passive filtering	29%	0.5%	29%	28%
PWM boost converter	3.5%	10%	11%	34%

Table 1. Comparison of current harmonic distortion emission spectra of different AC-DC converters topologies

Higher order harmonic distortion of line current is particularly important in several applications because of its disturbing potency in power grid. To asses the certain limitation levels in standard [6] partial weighted harmonic distortion (PWHD) is extra defined using formula (3). According to this rule, harmonics above 14th order up to 40th order are considered with the weighting factor increasing with harmonic order. The best performance in terms of PWHD index, have been observed for diode rectifier with passive filters: only 28% (Table1). For the PWM boost converter and diode rectifier without passive filter PWHD index is substantially higher: 43% and 48% respectively.

$$PWHD(I) = \frac{\sqrt{\sum_{n=14}^{40} n I_n^2}}{I_1} \tag{3}$$

3.3.2. Frequency range $2 - 9kHz$

Harmonic distortions in frequency range up to $9kHz$ are characterized in standard [8] as a result of grouping of harmonics DFT product obtained for 5 cycles of observation within $200Hz$ sub-bands using rectangular window. Proposed grouping method results with 35 sub-bands (groups) H_g, 200Hz wide each with the center frequencies of the band starting from 2.1kHz up to 8.9kHz. Grouping algorithm is represented by formula 4.

$$H_{g(200Hz)} = \sqrt{\sum_{k=g-90Hz}^{g+100Hz} I_k{}^2} \tag{4}$$

For the purpose of comparison of current harmonic emission of the diode rectifier and PWM boost converter the FFT analysis has been done according to [8]. To demonstrate more clearly the effect of harmonic emission character the raw DFT product in frequency range $2-9kHz$ is presented in Figures 17 and 18.

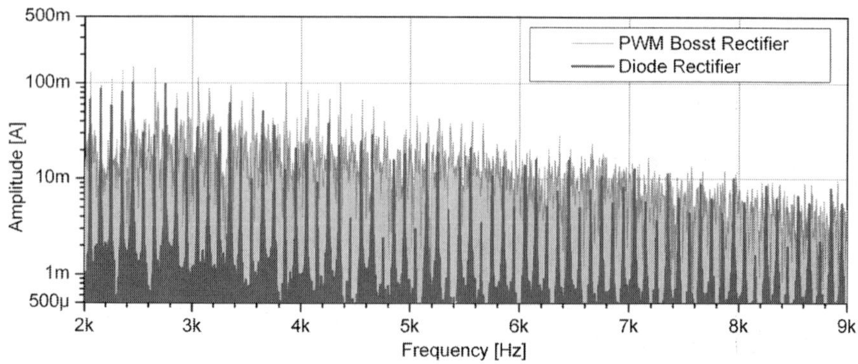

Figure 17. Comparison of harmonic emission of three phase six pulse diode rectifier and PWM boost converter in frequency range $2-9kHz$.

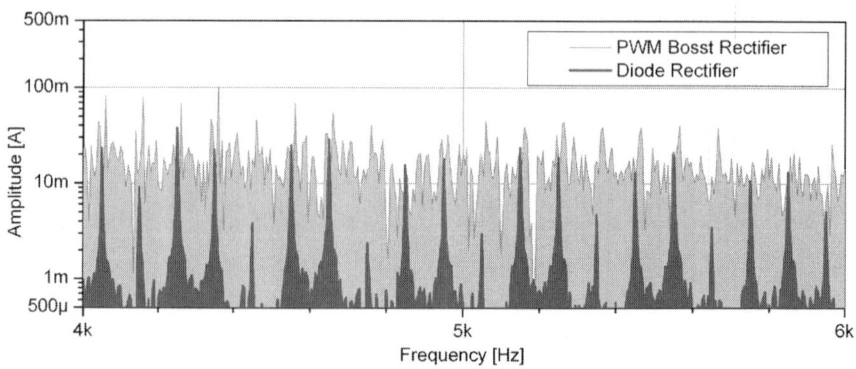

Figure 18. Comparison of harmonic emission of three phase six pulse diode rectifier and PWM boost converter - more detailed view at some exemplary frequency sub-range.

The obtained results show that the use of PWM boost converter do not change significantly current harmonic amplitudes for the frequencies close to integer multiples of the fundamental power frequency with relation to diode rectifier results, they are roughly at similar level. However, inter-harmonic emission for PWM boost converter is more or less at the same level as harmonics (Figure 18), whereas for diode rectifier inter-harmonic levels were in average at least more than ten times lower, which results with increase of power spectrum density in the whole frequency band. By employing the grouping method of harmonic content proposed in standard [8] the total power of harmonic emission within each of 200*Hz* wide frequency sub-range can be calculated using formula 4 . This standardized analysis shows a significant increase of total spectral power emission of PWM boost rectifier in relation to diode rectifier (Figure 19), whereas the maximum individual amplitudes of DFT product for both converters are at similar level (Figures 17 and 18).

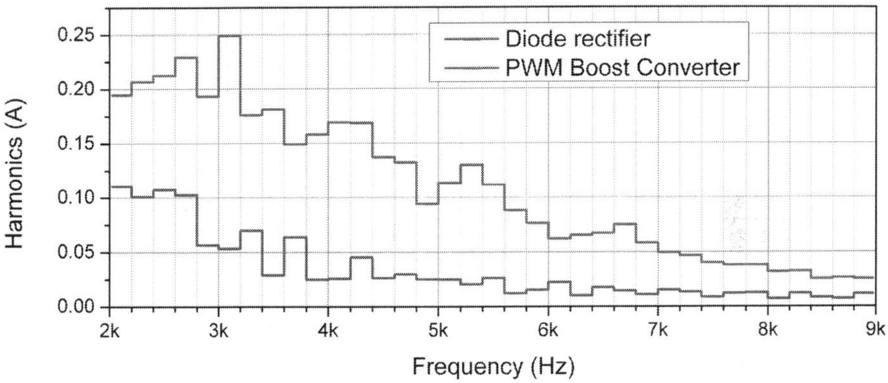

Figure 19. Comparison of harmonic emission of diode rectifier and PWM boost converter.

3.3.3. PWM carrier frequency range

Values of PWM carrier frequency used in typical applications are mainly correlated with converter's input voltage and its rated power. In contemporary applications of PWM boost rectifiers modulation frequency values are usually in a range of few kHz for high power converters (above hundreds of *kW*) up to tens of kHz low power converters (below *kW*). This frequency range is located just above power quality frequency range and includes significant part of CISPR A range (Figure 1). PWM carrier frequency and its integer multiples define frequency sub-rages, where increased current harmonic emissions usually appear. There are known different method of decreasing this emission, nevertheless it is difficult to eliminate them entirely by for example passive filtering.

In the evaluated converter modulation carrier frequency was set to 15*kHz* and more than ten times higher current harmonic amplitudes in analysed DFT product has been observed for this frequency, and about few times higher for frequencies close to PWM carrier frequency, between 13*kHz* and 15*kHz* (Figure 20). For analysing DFT product in accordance to CISPR 16 standard within CISPR A frequency band, 200*Hz* resolution band width should be used. Power spectral density (PSD) calculated according to this rule is presented in Figure 21. The obtained results show that harmonic emission in this frequency range is significantly higher in relation to diode rectifier converter.

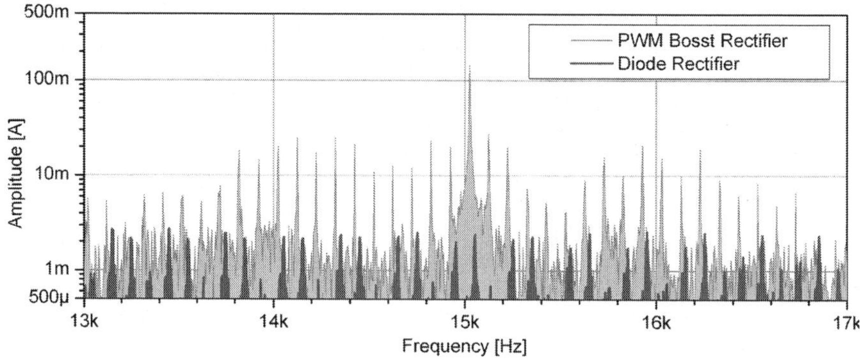

Figure 20. Current harmonic distortion of PWM boost converter in the frequency range close to PWM carrier frequency with comparison to classic diode rectifier distortions.

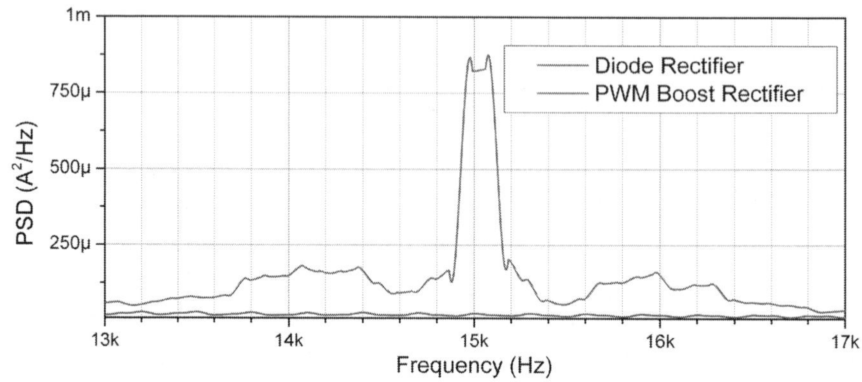

Figure 21. Power spectral density of current harmonic calculated for $200Hz$ resolution badnwidth.

4. Conclusions

PWM boost AC-DC converters are increasingly used in contemporary application because of its considerable advantages, like bidirectional power transfer with unity power factor operation and low level of low order harmonic distortions emission. Systematic significant increase of the overall power quota converted from AC to DC and from DC to AC in the power system make these advantages more meaningful from the power quality point of view. This development trends also introduce some unfavourable effects, like increased emission in higher frequency ranges, which are presented in this chapter.

Increased harmonic emission in the frequency range between $2kHz$ and $9kHz$ as an effect of pulse width modulation method used for line current control in PWM boost converters becomes a fundamental problem to solve in such converters connected directly to the power grid. In recent years increased number of investigations focused on arising compatibility challenges in frequency band $2 - 9kHz$ has been reported and some new standardization methods has been initially proposed.

Results of investigations presented in this chapter demonstrate that line current and voltage ripples, as an effect of PWM modulation carrier frequency in PWM boost converters, can induce compatibility problems in numerous applications which usually cannot be easily solved by using conventional passive harmonic-filters or radio frequency interference (RFI) filters. Harmonic emission filtering in this frequency band require new specific types of filters to be used.

Author details

Jaroslaw Luszcz

Faculty of Electrical and Control Engineering, Gdansk University of Technology, Gdansk, Poland

References

[1] Benysek, G. [2007]. *Improvement in the Quality of Delivery of Electrical Energy using Power Electronics Systems*, Power Systems, Springer-Verlag.

[2] Bollen, M. & Hassan, F. [2011]. *Integration of Distributed Generation in the Power System*, Wiley-IEEE Press.

[3] Bollen, M., Yang, Y. & Hassan, F. [2008]. Integration of distributed generation in the power system - a power quality approach, *Harmonics and Quality of Power, 2008. ICHQP 2008. 13th International Conference on*, pp. 1 –8.

[4] Cichowlas, M., Malinowski, M., Kazmierkowski, M. & Blaabjerg, F. [2003]. Direct power control for three-phase pwm rectifier with active filtering function, *Applied Power Electronics Conference and Exposition, 2003. APEC '03. Eighteenth Annual IEEE*, Vol. 2, pp. 913 –918 vol.2.

[5] Cichowlas, M., Malinowski, M., Kazmierkowski, M., Sobczuk, D., Rodriguez, P. & Pou, J. [2005]. Active filtering function of three-phase pwm boost rectifier under different line voltage conditions, *Industrial Electronics, IEEE Transactions on* **52**(2): 410 – 419.

[6] *IEC 61000-3-4 (1998): Electromagnetic compatibility (EMC) - Part 3-4: Limits - Limitation of emission of harmonic currents in low-voltage power supply systems for equipment with rated current greater than 16 A* [1998]. International Electrotechnical Commission Std.

[7] *IEC 61000-3-6 (2008): Electromagnetic compatibility (EMC) - Part 3-6: Limits - Assessment of emission limits for the connection of distorting installations to MV, HV and EHV power systems* [2008]. International Electrotechnical Commission Std.

[8] *IEC 61000-4-7 (2009): Electromagnetic compatibility (EMC) - Part 4-7: Testing and measurement techniques - General guide on harmonics and interharmonics measurements and instrumentation, for power supply systems and equipment connected thereto* [2009]. International Electrotechnical Commission Std.

[9] *IEEE Standard 519 (1992): Recommended Practices and Requirements for Harmonic Control in Electrical Power Systems* [1993]. The Institute of Electrical and Electronics Engineers.

[10] Liserre, M., Dell'Aquila, A. & Blaabjerg, F. [2003]. An overview of three-phase voltage source active rectifiers interfacing the utility, *Power Tech Conference Proceedings, 2003 IEEE Bologna*, Vol. 3, p. 8 pp. Vol.3.

[11] M. H. J. Bollen, J. Z. e. a. [2010]. Power quality aspects of smart grids, *International Conference on Renewable Energies and Power Quality (ICREPQ'10) Granada (Spain).*

[12] McGranaghan, M. & Beaulieu, G. [2006]. Update on iec 61000-3-6: Harmonic emission limits for customers connected to mv, hv, and ehv, *Transmission and Distribution Conference and Exhibition, 2005/2006 IEEE PES*, pp. 1158 –1161.

[13] Olofsson, M. [2009]. Power quality and emc in smart grid, *Electrical Power Quality and Utilisation, 2009. EPQU 2009. 10th International Conference on*, pp. 1 –6.

[14] Smolenski, R. [2009]. Selected conducted electromagnetic interference issues in distributed power systems, *Bulletin of the Polish Academy of Sciences: Technical Sciences* Vol. 57(no 4): pp.383–393.

[15] Smolenski, R. [2012]. *Conducted Electromagnetic Interference (EMI) in Smart Grids*, Power Systems, Springer.

[16] Strauss, P., Degner, T., Heckmann, W., Wasiak, I., Gburczyk, P., Hanzelka, Z., Hatziargyriou, N., Romanos, T., Zountouridou, E. & Dimeas, A. [2009]. International white book on the grid integration of static converters, *Electrical Power Quality and Utilisation, 2009. EPQU 2009. 10th International Conference on*, pp. 1–6.

[17] Strzelecki, R. & Benysek, G. [2008]. *Power electronics in smart electrical energy networks*, Power Systems, Springer-Verlag.

[18] Tlusty, J. & Vybiralik, F. [2005]. Management of the voltage quality in the distribution system within dispersed generation sources, *Electricity Distribution, 2005. CIRED 2005. 18th International Conference and Exhibition on*, pp. 1 –4.

[19] Vekhande, V. & Fernandes, B. [2011]. Bidirectional current-fed converter for integration of dc micro-grid with ac grid, *India Conference (INDICON), 2011 Annual IEEE*, pp. 1 –5.

[20] Wasiak, I. & Hanzelka, Z. [2009]. *Integration of distributed energy sources with electrical power grid*, Vol. 57, Bulletin of the Polish Academy of Sciences - Technical Sciences, pp. 297–309.
URL: *http://bulletin.pan.pl/(57-4)297.pdf*